STUDENT'S SOLUTIONS MANUAL

WILLIAM CRAINE III

STATS: DATA AND MODELS
FOURTH EDITION

Richard De Veaux
Williams College

Paul Velleman
Cornell University

David Bock
Cornell University

PEARSON

Boston Columbus Hoboken Indianapolis New York San Francisco
Amsterdam Cape Town Dubai London Madrid Milan Munich Paris Montreal Toronto
Delhi Mexico City São Paulo Sydney Hong Kong Seoul Singapore Taipei Tokyo

Reproduced by Pearson from electronic files supplied by the author.

ISBN-13: 978-0-321-98997-0
ISBN-10: 0-321-98997-X

1 2 3 4 5 6 RRD 19 18 17 16 15

www.pearsonhighered.com

Contents

Chapter 1 – Stats Starts Here

Section 1.1

1. **Grocery shopping.** Discount cards at grocery stores allow the stores to collect information about the products that the customer purchases, what other products are purchased at the same time, whether or not the customer uses coupons, and the date and time that the products are purchased. This information can be linked to demographic information about the customer that was volunteered when applying for the card, such as the customer's name, address, sex, age, income level, and other variables. The grocery store chain will use that information to better market their products. This includes everything from printing out coupons at the checkout that are targeted to specific customers to deciding what television, print, or Internet advertisements to use.

Section 1.2

3. **Super Bowl.** When collecting data about the Super Bowl, the games themselves are the *who*.

Section 1.3

5. **Grade level.**

    a) If we are, for example, comparing the percentage of first-graders who can tie their own shoes to the percentage of second-graders who can tie their own shoes, grade-level is treated as categorical. It is just a way to group the students. We would use the same methods if we were comparing boys to girls or brown-eyed kids to blue-eyed kids.

    b) If we were studying the relationship between grade-level and height, we would be treating grade level as quantitative.

7. **Voters.** The response is a categorical variable.

9. **Medicine.** The company is studying a quantitative variable.

Chapter Exercises

11. **The News.** Answers will vary.

13. **Gaydar.** *Who* – 40 undergraduate women. *What* – Whether or not the women could identify the sexual orientation of men based on a picture. *Population of interest* – All women.

15. Bicycle Safety. *Who* – 2,500 cars. *What* – Distance from the bicycle to the passing car (in inches). *Population of interest* – All cars passing bicyclists.

17. Honesty. *Who* – Workers who buy coffee in an office. *What* – amount of money contributed to the collection tray. *Population of interest* – All people in honor system payment situations.

19. Not-so-diet soda. *Who* – 474 participants. *What* – whether or not the participant drank two or more diet sodas per day, waist size at the beginning of the study, and waist size at the end of the study. *Population of interest* – All people.

21. Weighing bears. *Who* – 54 bears. *What* – Weight, neck size, length (no specified units), and sex. *When* – Not specified. *Where* – Not specified. *Why* - Since bears are difficult to weigh, the researchers hope to use the relationships between weight, neck size, length, and sex of bears to estimate the weight of bears, given the other, more observable features of the bear.
How – Researchers collected data on 54 bears they were able to catch. *Variables* – There are 4 variables; weight, neck size, and length are quantitative variables, and sex is a categorical variable. No units are specified for the quantitative variables. *Concerns* – The researchers are (obviously!) only able to collect data from bears they were able to catch. This method is a good one, as long as the researchers believe the bears caught are representative of all bears, in regard to the relationships between weight, neck size, length, and sex.

23. Arby's menu. *Who* – Arby's sandwiches. *What* – type of meat, number of calories (in calories), and serving size (in ounces). *When* – Not specified. *Where* – Arby's restaurants. *Why* – These data might be used to assess the nutritional value of the different sandwiches. *How* – Information was gathered from each of the sandwiches on the menu at Arby's, resulting in a census. *Variables* – There are three variables. Number of calories and serving size are quantitative variables, and type of meat is a categorical variable.

25. Babies. *Who* – 882 births. *What* – Mother's age (in years), length of pregnancy (in weeks), type of birth (caesarean, induced, or natural), level of prenatal care (none, minimal, or adequate), birth weight of baby (unit of measurement not specified, but probably pounds and ounces), gender of baby (male or female), and baby's health problems (none, minor, major).
When – 1998-2000. *Where* – Large city hospital. *Why* – Researchers were investigating the impact of prenatal care on newborn health. *How* – It appears that they kept track of all births in the form of hospital records, although it is not specifically stated. *Variables* – There are three quantitative variables: mother's age, length of pregnancy, and birth weight of baby. There are four categorical variables: type of birth, level of prenatal care, gender of baby, and baby's health problems.

27. **Herbal medicine.** *Who* – experiment volunteers. *What* – herbal cold remedy or sugar solution, and cold severity. *When* – Not specified. *Where* – Major pharmaceutical firm. *Why* – Scientists were testing the efficacy of an herbal compound on the severity of the common cold.
    How – The scientists set up a controlled experiment. *Variables* – There are two variables. Type of treatment (herbal or sugar solution) is categorical, and severity rating is quantitative. *Concerns* – The severity of a cold seems subjective and difficult to quantify. Also, the scientists may feel pressure to report negative findings about the herbal product.

29. **Streams.** *Who* – Streams. *What* – Name of stream, substrate of the stream (limestone, shale, or mixed), acidity of the water (measured in pH), temperature (in degrees Celsius), and BCI (unknown units). *When* – Not specified. *Where* – Upstate New York. *Why* – Research was conducted for an Ecology class. *How* – Not specified. *Variables* – There are five variables. Name and substrate of the stream are categorical variables, and acidity, temperature, and BCI are quantitative variables.

31. **Refrigerators.** *Who* – 353 refrigerators. *What* – Brand, cost (probably in dollars), size (in cu. ft.), type, estimated annual energy cost (probably in dollars), overall rating, and repair history (in percent requiring repair over the past five years). *When* – 2013. *Where* – United States.
    Why – The information was compiled to provide information to the readers of *Consumer Reports*. *How* – Not specified. *Variables* – There are 7 variables. Brand, type, and overall rating are categorical variables. Cost, size, estimated energy cost, and repair history are quantitative variables.

33. **Kentucky Derby 2014.** *Who* – Kentucky Derby races. *What* – Year, winner, jockey, trainer, owner, and time (in minutes, seconds, and hundredths of a second. *When* – 1875 – 2013. *Where* – Churchill Downs, Louisville, Kentucky. *Why* – It is interesting to examine the trends in the Kentucky Derby. *How* – Official statistics are kept for the race each year. *Variables* – There are 6 variables. Winner, jockey, trainer and owner are categorical variables. Date and duration are quantitative variables.

34. **Indianapolis 500 .** *Who* – Indy 500 races. *What* – Year, driver, time (in minutes, seconds, and hundredths of a second), and speed (in miles per hour). *When* – 1911 – 2013.
    Where – Indianapolis, Indiana. *Why* – It is interesting to examine the trends in Indy 500 races.
    How – Official statistics are kept for the race every year. *Variables* – There are 4 variables. Driver is a categorical variable. Year, time, and speed are quantitative variables.

Chapter 2 – Displaying and Describing Categorical Data

Section 2.1

1. **Automobile fatalities.**

Subcompact and Mini	0.1128
Compact	0.3163
Intermediate	0.3380
Full	0.2193
Unknown	0.0137

3. **Movie genres.**

   a) 2008   **b)** 1996   **c)** 2006   **d)** 2012

Section 2.2

5. **Movies again.**

   a) $170/348 \approx 48.9\%$ of these films were rated R.

   b) $41/348 \approx 11.8\%$ of these films were R-rated comedies.

   c) $41/170 \approx 24.1\%$ of the R-rated films were comedies.

   d) $41/90 \approx 45.6\%$ of the comedies were R-rated.

Chapter Exercises

7. **Graphs in the news.** Answers will vary.

9. **Tables in the news.** Answers will vary.

11. **Movie genres.**

   a) A pie chart seems appropriate from the movie genre data. Each movie has only one genre, and the 193 movies constitute a "whole".

   b) "Other" is the least common genre. It has the smallest region in the chart.

13. **Genres, again.**

   a) SciFi/Fantasy has a higher bar than Action/Adventure, so it is the more common genre.

   b) This is easier to see on the bar chart. The percentages are so close that the difference is nearly indistinguishable in the pie chart.

15. **Magnet Schools.**

   There were 1755 qualified applicants for the Houston Independent School District's magnet schools program. 53% were accepted, 17% were wait-listed, and the other 30% were turned away for lack of space.

17. Causes of death 2011.

a) Yes, it is reasonable to assume that heart and respiratory disease caused approximately 29.4% of U.S. deaths in 2007, since there is no possibility for overlap. Each person could only have one cause of death.

b) Since the percentages listed add up to 62.3%, other causes must account for 37.7% of US deaths.

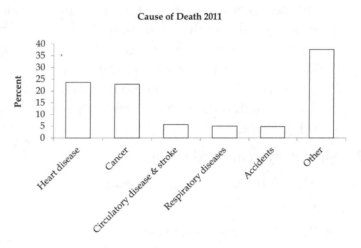

c) A bar chart is a good choice (with the inclusion of the "Other" category). Since causes of US deaths represent parts of a whole, a pie chart would also be a good display.

19. Oil spills as of 2013.

a) Grounding, accounting for approximately 150 spills, is the most frequent cause of oil spillage for these 459 spills. A substantial number of spills, approximately 140, were caused by Collision. Less prevalent causes of oil spillage in descending order of frequency were Hull or equipment failures, Fire & Explosions, and Other/Unknown causes.

b) A pie chart is an appropriate display of the data, since there is only a single cause attributed to each spill, and all spills are represented in some category.

c) There were more spills due to Grounding than Collisions. This is much easier to see on the bar chart.

21. Global warming.

Perhaps the most obvious error is that the percentages in the pie chart only add up to 93%, when they should, of course, add up to 100%. Furthermore, the three-dimensional perspective view distorts the regions in the graph, violating the area principle. The regions corresponding to No Solid Evidence and Due to Human Activity should be roughly the same size, at 32% and 34% of respondents, respectively. However, the angle for the 32% region looks much bigger. Always use simple, two-dimensional graphs. Additionally, the graph does not include a title.

23. Teen smokers.

According to the Monitoring the Future study, teen smoking brand preferences differ somewhat by region. Although Marlboro is the most popular brand in each region, with about 58% of teen smokers preferring this brand in each region, teen smokers from the South prefer Newports at a higher percentage than teen smokers from the West, 22.5% to approximately 10%, respectively. Camels are more popular in the West, with 9.5% of teen smokers preferring this brand, compared to only 3.3% in the South. Teen smokers in the West are also more likely to have no particular brand than teen smokers in the South. 12.9% of teen smokers in the West have no particular brand, compared to only 6.7% in the South. Both regions have about 9% of teen smokers that prefer one of over 20 other brands.

25. Movies by genre and rating.

a) The table uses column percents, since each column adds to 100%, while the rows do not.

b) 25.86% of these movies are comedies.

c) 28.57% of the PG-rated movies were comedies.

d) i) 27.36% of the PG-13 movies were comedies.

   ii) You cannot determine this from the table.

   iii) None (0%) of the dramas were G-rated.

   iv) You cannot determine this from the table.

27. Seniors.

a) A table with marginal totals is to the right. There are 268 White graduates and 325 total graduates.
$268/325 \approx 82.5\%$ of the graduates are white.

Plans	White	Minority	TOTAL
4-year college	198	44	242
2-year college	36	6	42
Military	4	1	5
Employment	14	3	17
Other	16	3	19
TOTAL	268	57	325

b) There are 42 graduates planning to attend 2-year colleges. $42/325 \approx 12.9\%$

c) 36 white graduates are planning to attend 2-year colleges. $36/325 \approx 11.1\%$

d) 36 white graduates are planning to attend 2-year colleges and there are 268 whites graduates. $36/268 \approx 13.4\%$

e) There are 42 graduates planning to attend 2-year colleges, and 36 of them are white.
$36/42 \approx 85.7\%$

29. More about seniors.

a) For white students, 73.9% plan to attend a 4-year college, 13.4% plan to attend a 2-year college, 1.5% plan on the military, 5.2% plan to be employed, and 6.0% have other plans.

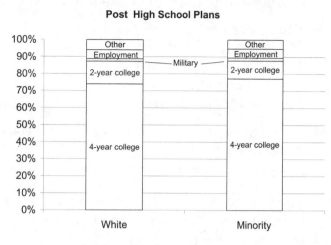

b) For minority students, 77.2% plan to attend a 4-year college, 10.5% plan to attend a 2-year college, 1.8% plan on the military, 5.3% plan to be employed, and 5.3% have other plans.

c) A segmented bar chart is a good display of these data.

d) The conditional distributions of plans for Whites and Minorities are similar: White – 74% 4-year college, 13% 2-year college, 2% military, 5% employment, 6% other.
Minority – 77% 4-year college, 11% 2-year college, 2% military, 5% employment, 5% other.
Caution should be used with the percentages for Minority graduates, because the total is so small. Each graduate is almost 2%. Still, the conditional distributions of plans are essentially the same for the two groups. There is little evidence of an association between race and plans for after graduation.

31. Magnet schools revisited.

a) There were 1755 qualified applicants to the Houston Independent School District's magnet schools program. Of those, 292, or about 16.6% were Asian.

b) There were 931 students accepted to the magnet schools program. Of those, 110, or about 11.8% were Asian.

c) There were 292 Asian applicants. Of those, 110, or about 37.7%, were accepted.

d) There were 1755 total applicants. Of those, 931, or about 53%, were accepted.

33. Back to school.

There were 1,755 qualified applicants for admission to the magnet schools program. 53% were accepted, 17% were wait-listed, and the other 30% were turned away. While the overall acceptance rate was 53%, 93.8% of Blacks and Hispanics were accepted, compared to only 37.7% of Asians, and 35.5% of whites. Overall, 29.5% of applicants were Black or Hispanics, but only 6% of those turned away were Black or Hispanic. Asians accounted for 16.6% of applicants, but 25.3% of those turned away. It appears that the admissions decisions were not independent of the applicant's ethnicity.

35. Weather forecasts.

a) The table shows the marginal totals. It rained on 34 of 365 days, or 9.3% of the days.

		Actual Weather		Total
		Rain	No Rain	
Forecast	Rain	27	63	90
	No Rain	7	268	275
	Total	34	331	365

b) Rain was predicted on 90 of 365 days. $90/365 \approx 24.7\%$ of the days.

c) The forecast of Rain was correct on 27 of the days it actually rained and the forecast of No Rain was correct on 268 of the days it didn't rain. So, the forecast was correct a total of 295 times. $295/365 \approx 80.8\%$ of the days.

d) On rainy days, rain had been predicted 27 out of 34 times (79.4%). On days when it did not rain, forecasters were correct in their predictions 268 out of 331 times (81.0%). These two percentages are very close. There is no evidence of an association between the type of weather and the ability of the forecasters to make an accurate prediction.

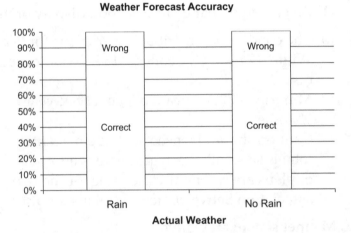

e) 52.9% of all twin births were preterm, while only 39.7% of births in which inadequate medical care was received were preterm. This is evidence of an association between level of prenatal care and twin birth outcome. If these variables were independent, we would expect the percentages to be roughly the same. Generally, those mothers who received adequate medical care were more likely to have preterm births than mothers who received intensive medical care, who were in turn more likely to have preterm births than mothers who received inadequate health care. This does *not* imply that mothers should receive inadequate health care do decrease their chances of having a preterm birth, since it is likely that women that have some complication *during* their pregnancy (that might lead to a preterm birth), would seek intensive or adequate prenatal care.

37. Blood pressure.

a) The marginal distribution of blood pressure for the employees of the company is the total column of the table, converted to percentages.

Blood pressure	under 30	30 - 49	over 50	Total
low	27	37	31	95
normal	48	91	93	232
high	23	51	73	147
Total	98	179	197	474

20% low, 49% normal and 31% high blood pressure.

b) The conditional distribution of blood pressure within each age category is:
Under 30 : 28% low, 49% normal, 23% high
30 – 49 : 21% low, 51% normal, 28% high
Over 50 : 16% low, 47% normal, 37% high

c) A segmented bar chart of the conditional distributions of blood pressure by age category is at the right.

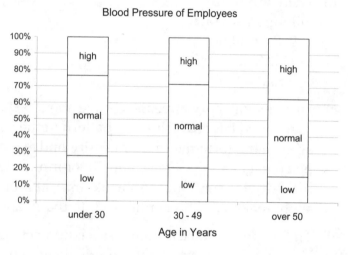

Blood Pressure of Employees

d) In this company, as age increases, the percentage of employees with low blood pressure decreases, and the percentage of employees with high blood pressure increases.

e) No, this does not prove that people's blood pressure increases as they age. Generally, an association between two variables does not imply a cause-and-effect relationship. Specifically, these data come from only one company and cannot be applied to all people. Furthermore, there may be some other variable that is linked to both age and blood pressure. Only a controlled experiment can isolate the relationship between age and blood pressure.

39. Anorexia.

These data provide no evidence that Prozac might be helpful in treating anorexia. About 71% of the patients who took Prozac were diagnosed as "Healthy", while about 73% of the patients who took a placebo were diagnosed as "Healthy". Even though the percentage was higher for the placebo patients, this does not mean that Prozac is hurting patients. The difference between 71% and 73% is not likely to be statistically significant.

41. Driver's licenses 2011.

Registered U.S. Drivers by Age and Gender

Age in Years

a) There are 10.0 million drivers under 20 and a total of 208.3 million drivers in the U.S. That's about 4.8% of U.S. drivers under 20.

b) There are 103.5 million males out of 208.4 million total U.S. drivers, or about 49.7%.

c) Each age category appears to have about 50% male and 50% female drivers. The segmented bar chart shows a pattern in the deviations from 50%. At younger ages, males form the slight majority of drivers. This percentage shrinks until the percentages are 50% male and 50% for middle aged drivers. The percentage of male drivers continues to shrink until, at around age 45, female drivers hold a slight majority. This continues into the 85 and over category.

d) There appears to be a slight association between age and gender of U.S. drivers. Younger drivers are slightly more likely to be male, and older drivers are slightly more likely to be female.

43. Hospitals.

a) The marginal totals have been added to the table:

		Discharge delayed		
		Large Hospital	**Small Hospital**	**Total**
Procedure	**Major surgery**	120 of 800	10 of 50	130 of 850
	Minor surgery	10 of 200	20 of 250	30 of 450
	Total	130 of 1000	30 of 300	160 of 1300

160 of 1300, or about 12.3% of the patients had a delayed discharge.

b) Yes. Major surgery patients were delayed 130 of 850 times, or about 15.3% of the time.
Minor Surgery patients were delayed 30 of 450 times, or about 6.7% of the time.

c) Large Hospital had a delay rate of 130 of 1000, or 13%.
Small Hospital had a delay rate of 30 of 300, or 10%.
The small hospital has the lower overall rate of delayed discharge.

d) Large Hospital: Major Surgery 15% delayed and Minor Surgery 5% delayed. Small Hospital: Major Surgery 20% delayed and Minor Surgery 8% delayed. Even though small hospital had the lower overall rate of delayed discharge, the large hospital had a lower rate of delayed discharge for each type of surgery.

e) No. While the overall rate of delayed discharge is lower for the small hospital, the large hospital did better with *both* major surgery and minor surgery.

f) The small hospital performs a higher percentage of minor surgeries than major surgeries. 250 of 300 surgeries at the small hospital were minor (83%). Only 200 of the large hospital's 1000 surgeries were minor (20%). Minor surgery had a lower delay rate than major surgery (6.7% to 15.3%), so the small hospital's overall rate was artificially inflated. Simply put, it is a mistake to look at the overall percentages. The real truth is found by looking at the rates after the information is broken down by type of surgery, since the delay rates for each type of surgery are so different. The larger hospital is the better hospital when comparing discharge delay rates.

45. Graduate admissions.

a) 1284 applicants were admitted out of a total of 3014 applicants. 1284/3014 = 42.6%

Program	Males Accepted (of applicants)	Females Accepted (of applicants)	Total
1	511 of 825	89 of 108	600 of 933
2	352 of 560	17 of 25	369 of 585
3	137 of 407	132 of 375	269 of 782
4	22 of 373	24 of 341	46 of 714
Total	**1022 of 2165**	**262 of 849**	**1284 of 3014**

b) 1022 of 2165 (47.2%) of males were admitted. 262 of 849 (30.9%) of females were admitted.

c) Since there are four comparisons to make, the table at the right organizes the percentages of males and females accepted in each program. Females are accepted at a higher rate in every program.

Program	Males	Females
1	61.9%	82.4%
2	62.9%	68.0%
3	33.7%	35.2%
4	5.9%	7%

d) The comparison of acceptance rate within each program is most valid. The overall percentage is an unfair average. It fails to take the different numbers of applicants and different acceptance rates of each program. Women tended to apply to the programs in which gaining acceptance was difficult for everyone. This is an example of Simpson's Paradox.

Chapter 3 – Displaying and Summarizing Quantitative Data

Section 3.1

1. Details.

Boxplots don't tell us much about the shape of a distribution, beyond a basic idea of symmetry or skewness. The given boxplot could be displaying a distribution that has multiple modes (the first histogram), is reasonably unimodal (the second histogram), has gaps and clusters (the third histogram), or has outliers (the fourth histogram). We simply can't determine shape from a boxplot.

Section 3.2

3. Outliers.

IQR = Q3 – Q1 = 116 – 98 = 18.
Using the Outlier Rule (1.5 IQRs beyond quartiles):

Upper Fence: $Q3+1.5(IQR)=116+1.5(18)$
$=116+27$
$=143$

Since the maximum, 160 minutes, is above the upper fence, there is at least one high outlier.

Lower Fence: $Q1-1.5(IQR)=98-1.5(18)$
$=98-27$
$=71$

Since the minimum, 43 minutes, is below the lower fence, there is at least one low outlier.

Section 3.3

5. Adoptions II.

The mean number of adoptions is expected to by higher than the median number of adoptions, since the distribution of the number of adoptions is skewed to the right.

Section 3.4

7. Test score spreads.

The IQR of the test scores will not be affected, but the standard deviation of the test scores will increase.

Chapter Exercises

9. Histogram. Answers will vary.

11. In the news. Answers will vary.

13. Thinking about shape.

a) The distribution of the number of speeding tickets each student in the senior class of a college has ever had is likely to be unimodal and skewed to the right. Most students will have very few speeding tickets (maybe 0 or 1), but a small percentage of students will likely have comparatively many (3 or more?) tickets.

b) The distribution of player's scores at the U.S. Open Golf Tournament would most likely be unimodal and slightly skewed to the right. The best golf players in the game will likely have around the same average score, but some golfers might be off their game and score 15 strokes above the mean. (Remember that high scores are undesirable in the game of golf!)

c) The weights of female babies in a particular hospital over the course of a year will likely have a distribution that is unimodal and symmetric. Most newborns have about the same weight, with some babies weighing more and less than this average. There may be slight skew to the left, since there seems to be a greater likelihood of premature birth (and low birth weight) than post-term birth (and high birth weight).

d) The distribution of the length of the average hair on the heads of students in a large class would likely be bimodal and skewed to the right. The average hair length of the males would be at one mode, and the average hair length of the females would be at the other mode, since women typically have longer hair than men. The distribution would be skewed to the right, since it is not possible to have hair length less than zero, but it is possible to have a variety of lengths of longer hair.

15. Sugar in cereals.

a) The distribution of the sugar content of breakfast cereals is bimodal, with a cluster of cereals with sugar content around 6% sugar and another cluster of cereals around 48% sugar. The lower cluster shows a bit of skew to the right. Most cereals in the lower cluster have between 0% and 10% sugar. The upper cluster is symmetric, with center around 45% sugar.

b) There are two different types of breakfast cereals, those for children and those for adults. The children's cereals are likely to have higher sugar contents, to make them taste better (to kids, anyway!). Adult cereals often advertise low sugar content.

17. Vineyards.

a) There is information displayed about 36 vineyards and it appears that 28 of the vineyards are smaller than 90 acres. That's around 78% of the vineyards. (75% would be a good estimate!)

b) The distribution of the size of 36 Finger Lakes vineyards is skewed to the right. Most vineyards are smaller than 90 acres, with a few larger ones, from 90 to 160 acres. One vineyard was larger than all the rest, over 240 acres. The mode of the distribution is between 0 and 30 acres.

19. Heart attack stays.

a) The distribution of length of stays is skewed to the right, so the mean is larger than the median.

b) The distribution of the length of hospital stays of female heart attack patients is bimodal and skewed to the right, with stays ranging from 1 day to 36 days. The distribution is centered around 8 days, with the majority of the hospital stays lasting between 1 and 15 days. There are a relatively few hospital stays longer than 27 days. Many patients have a stay of only one day, possibly because the patient died.

c) The median and IQR would be used to summarize the distribution of hospital stays, since the distribution is strongly skewed.

21. Super Bowl points 2013.

a) The median number of points scored in the first 48 Super Bowl games is 45 points.

b) The first quartile of the number of points scored in the first 48 Super Bowl games is 35 points. The third quartile is 54.5 (or 55) points.

c) In the first 48 Super Bowl games, the lowest number of points scored was 21, and the highest number of points scored was 75. The median number of points scored was 45, and the middle 50% of Super Bowls has between 35 and 55 points scored.

23. Summaries.

a) The mean cost of the compact refrigerators is $144.44.

b) The median cost of the compact refrigerators is $150. The first quartile is $130, and the third quartile is $150.

c) The range of the cost of the compact refrigerators is $180 – $120 = $60. The IQR is $150 – $130 = $20.

25. Mistake.

a) As long as the boss's true salary of $200,000 is still above the median, the median will be correct. The mean will be too large, since the total of all the salaries will decrease by $2,000,000 - $200,000 = $1,800,000, once the mistake is corrected.

b) The range will likely be too large. The boss's salary is probably the maximum, and a lower maximum would lead to a smaller range. The IQR will likely be unaffected, since the new maximum has no effect on the quartiles. The standard deviation will be too large, because the $2,000,000 salary will have a large squared deviation from the mean.

27. Standard deviation I.

a) Set 2 has the greater standard deviation. Both sets have the same mean (6) but set two has values that are generally farther away from the mean.
SD(Set 1) = 2.24   SD(Set 2) = 3.16

b) Set 2 has the greater standard deviation. Both sets have the same mean (15), maximum (20), and minimum (10), but 11 and 19 are farther from the mean than 14 and 16.
SD(Set 1) = 3.61   SD(Set 2) = 4.53

c) The standard deviations are the same. Set 2 is simply Set 1 + 80. Although the measures of center and position change, the spread is exactly the same.
SD(Set 1) = 4.24   SD(Set 2) = 4.24

29. Pizza prices.

The mean and standard deviation would be used to summarize the distribution of pizza prices, since the distribution is unimodal and symmetric.

31. Pizza prices again.

a) The mean pizza price is closest to $2.60. That's the balancing point of the histogram.

b) The standard deviation in pizza prices is closest to $0.15, since that is the typical distance to the mean. There are no pizza prices as far as $0.50 or $1.00.

33. Movie lengths 2010.

a) A typical movie would be a little over 100 minutes long. This is near the center of the unimodal and slightly skewed histogram, with the outlier set aside.

b) You would be surprised to find that your movie ran for 150 minutes. Only 3 movies ran that long.

c) It's difficult to say which would be higher. While the distribution of movie lengths is generally skewed to the right, which would raise the mean, there is a low outlier, which would lower the mean. (The actual mean of 107.07 minutes is a bit higher than the median of 104.50 minutes.)

35. Movie lengths II 2010.

a) i) The distribution of movie running times is fairly consistent, with the middle 50% of running times between 98 and 116 minutes. The interquartile range is 18 minutes.

ii) The standard deviation of the distribution of movie running times is 16.6 minutes, which indicates that movies typically have running times fairly close to the mean running time.

b) Since the distribution of movie running times is generally skewed to the right and contains an outlier, the standard deviation is a poor choice of numerical summary for the spread. The interquartile range is better, since it is resistant to outliers.

37. Movie earnings 2013.

The industry publication is using the median, while the watchdog group is using the mean. It is likely that the mean is pulled higher by a few high earning movies.

39. Payroll.

a) The mean salary is $\dfrac{1200 + 700 + 6(400) + 4(500)}{12} = \$525$.

The median salary is the middle of the ordered list:
400   400   400   400   400   400   500   500   500   500   700   1200
The median is $450.

b) Only two employees, the supervisor and the inventory manager, earn more than the mean wage.

c) The median better describes the wage of the typical worker. The mean is affected by the two higher salaries.

d) The IQR is the better measure of spread for the payroll distribution. The standard deviation and the range are both affected by the two higher salaries.

41. Gasoline 2014.

a)   Gasoline Prices

```
31 |1
31 |5
32 |1233
32 |6678
33 |
33 |9
34 |23
34 |556
```
Key : 32 | 1 = $3.21/gal

b) The distribution of gas prices is bimodal, with two clusters, one centered around $3.45 per gallon, and another centered around $3.25 per gallon. The lowest and highest prices were $3.11 and $3.46 per gallon.

c) There is a gap in the distribution of gasoline prices. There were no stations that charged between $3.28 and $3.39.

43. States.

a) The distribution of state populations is skewed heavily to the right. Therefore, the median and IQR are the appropriate measures of center and spread.

b) The mean population must be larger than the median population. The extreme values on the right affect the mean greatly and have no effect on the median.

c) There are 50 entries in the stemplot, so the median must be between the 25th and 26th population values. Counting in the ordered stemplot gives median = 4.5 million people. The middle of the lower 50% of the list (25 state populations) is the 13th population, or 2 million people. The middle of the upper half of the list (25 state populations) is the 13th population from the top, or 7 million people. The IQR = Q3 − Q1 = 7 − 2 = 5 million people.

d) The distribution of population for the 50 U.S. States is unimodal and skewed heavily to the right. The median population is 4.5 million people, with 50% of states having populations between 2 and 7 million people. There are two outliers, a state with 37 million people, and a state with 25 million people. The next highest population is only 19 million.

45. A-Rod 2013.

The distribution of the number of homeruns hit by Alex Rodriguez during the 1994 – 2013 seasons is reasonably symmetric, with the exception of a second mode around 10 homeruns. A typical number of homeruns per season was in the high 30s to low 40s. With the exception of 3 seasons in which A-Rod hit 0 , 5, and 7 homeruns, his total number of homeruns per season was between 16 and the maximum of 57.

47. Major Hurricanes 2013.

a) A dotplot of the number of hurricanes each year from 1944 through 2013 is displayed. Each dot represents a year in which there were that many hurricanes.

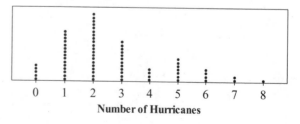

Major Hurricanes - 1944 - 2013

Number of Hurricanes

b) The distribution of the number of hurricanes per year is unimodal and skewed to the right, with center around 2 hurricanes per year. The number of hurricanes per year ranges from 0 to 8. There are no outliers. There may be a second mode at 5 hurricanes per year, but since there were only 6 years in which 5 hurricanes occurred, this may simply be natural variability.

49. A-Rod again 2013.

a) This is not a histogram. The horizontal axis should the number of home runs per year, split into bins of a convenient width. The vertical axis should show the frequency; that is, the number of years in which A-Rod hit a number of home runs within the interval of each bin. The display shown is a bar chart/time plot hybrid that simply displays the data table visually. It is of no use in describing the shape, center, spread, or unusual features of the distribution of home runs hit per year by A-Rod.

b) The histogram is at the right.

51. Acid rain.

The distribution of the pH readings of water samples in Allegheny County, Penn. is bimodal. A roughly uniform cluster is centered around a pH of 4.4. This cluster ranges from pH of 4.1 to 4.9. Another smaller, tightly packed cluster is centered around a pH of 5.6. Two readings in the middle seem to belong to neither cluster.

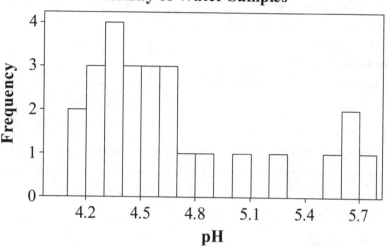

53. Final grades.

The width of the bars is much too wide to be of much use. The distribution of grades is skewed to the left, but not much more information can be gathered.

55. Zip codes.

Even though zip codes are numbers, they are not quantitative in nature. Zip codes are categories. A histogram is not an appropriate display for categorical data. The histogram the Holes R Us staff member displayed doesn't take into account that some 5-digit numbers do not correspond to zip codes or that zip codes falling into the same classes may not even represent similar cities or towns. The employee could design a better display by constructing a bar chart that groups together zip codes representing areas with similar demographics and geographic locations.

57. Math scores 2013.

a) Median: 285
    IQR: 9
    Mean: 284.36
    Standard deviation:
    6.84

b) Since the distribution
    of Math scores is
    skewed to the left, it
    is probably better to
    report the median
    and IQR.

c) The distribution of
    average math achievement scores for eighth graders in the United States is
    skewed slightly to the left, and roughly unimodal. The distribution is centered at
    285. Scores range from 269 to 301, with the middle 50% of the scores falling
    between 280 and 289.

59. Population growth 2010.

The distribution of population growth among the 50 United States and the
District of Columbia is unimodal and skewed to the right. Most states
experienced modest growth, as measured by percent change in population
between 2000 and 2010. Nearly every state experienced positive growth, with the
exception of Michigan. The median population growth was 7.8%, with the
middle 50% of states experiencing between 4.30% and 14.10% growth, for an IQR
of 9.80. The distribution contains one high outlier. Nevada experienced
population growth of 35.1%.

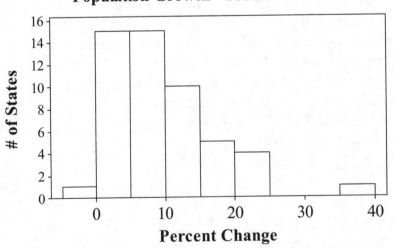

Chapter 4 – Understanding and Comparing Distributions

Section 4.1

1. Load factors, 2013.

The distribution of domestic load factors and the distribution of international load factors are both unimodal and skewed to the left. The distribution of international load factors may contain a low outlier. Because the distributions are skewed, the median and IQR are the appropriate measures of center and spread. The medians are very close, which tell us that typical international and domestic load factors are about the same. The IQRs show a bit more variability in the domestic load factors.

Section 4.2

3. Load factors 2013 by month.

Load factors are generally higher and less variable in the summer months (June – August). They are lower and more variable in the winter and spring.

Section 4.3

5. Extraordinary months.

Air travel immediately after the events of 9/11 was not typical of air travel in general. If we want to analyze monthly patterns, it might be best to set these months aside.

Section 4.4

7. Load factors 2013 over time.

a) After a period of little change in 2000-2001, load factors have been increasing steadily.

b) We would never assume that a pattern like this would continue. This case illustrates one of the reasons why we wouldn't assume this. Since load factors are percentages, they cannot exceed 100%. At the very least, the load factors would have to level out in the future.

Section 4.5

9. Exoplanets.

It is difficult to summarize data with a distribution this skewed. The extremely large values will dominate any summary or description.

Chapter Exercises

11. In the news. Answers will vary.

13. Time on the Internet. Answers will vary.

15. Pizza prices.

a) Pizza prices appear to be both higher on average, and more variable, in Baltimore than in the other three cities. Prices in Chicago may be slightly higher on average than in Dallas and Denver, but the difference is small.

b) There are low outliers in the distribution of pizza prices in Baltimore and Chicago. There is one high outlier in the distribution of pizza prices in Dallas. These outliers do not affect the overall conclusions reached in the previous part.

17. Rock concert accidents.

a) The histogram and boxplot of the distribution of "crowd crush" victims' ages both show that a typical crowd crush victim was approximately 18 - 20 years of age, that the range of ages is 36 years, that there are two outliers, one victim at age 36 - 38 and another victim at age 46 – 48.

b) This histogram shows that there may have been two modes in the distribution of ages of "crowd crush" victims, one at 18 – 20 years of age and another at 22 – 24 years of age. Boxplots, in general, can show symmetry and skewness, but not features of shape like bimodality or uniformity. Most victims were between 16 and 24 years of age.

c) The median is the better measure of center, since the distribution of ages has outliers. The median is more resistant to outliers than the mean.

d) The IQR is a better measure of spread, since the distribution of ages has outliers. The IQR is more resistant to outliers than the standard deviation.

19. Cereals.

a) The maximum sugar content is approximately 60% and the minimum sugar content is approximately 1%, so the range of sugar contents is about 60 – 1 = 59%.

b) The distribution of sugar content of cereals is bimodal, with modes centered around 5% and 45% sugar by weight.

c) Some cereals are healthy, low-sugar brands, and others are very sugary.

d) Yes. The minimum sugar content in the children's cereals is about 35% and the maximum sugar content of adult cereals is only 34%.

e) The range of sugar contents is about the same for the two types of cereals, approximately 28%, but the IQR is larger for the adult cereals. This is an indication of more variability in the sugar content of the middle 50% of adult cereals.

21. Population growth 2010 by region.

a) Comparative boxplots are at the right.

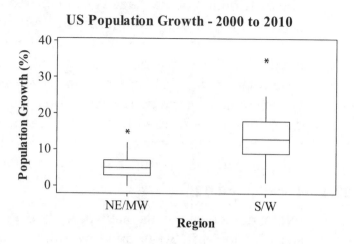

b) The distribution of population growth in NE/MW states is unimodal, symmetric and tightly clustered around 5% growth. The distribution of population growth in S/W states is much more spread out, with most states having population growth between 5% and 25%. A typical state had about 15% growth. There was an outliers, with 35% growth. Generally, the growth rates in the S/W states were higher and more variable than the rates in the NE/MW states.

23. Hospital stays.

a) The histograms of male and female hospital stay durations would be easier to compare if the were constructed with the same scale, perhaps from 0 to 20 days.

b) The distribution of hospital stays for men is skewed to the right, with many men having very short stays of about 1 or 2 days. The distribution tapers off to a maximum stay of approximately 25 days. The distribution of hospital stays for women is skewed to the right, with a mode at approximately 5 days, and tapering off to a maximum stay of approximately 22 days. Typically, hospital stays for women are longer than those for men.

c) The peak in the distribution of women's hospital stays can be explained by childbirth. This time in the hospital increases the length of a typical stay for women, and not for men.

25. Women's basketball.

a) Both girls have a median score of about 17 points per game, but Scyrine is much more consistent. Her IQR is about 2 points, while Alexandra's is over 10.

b) If the coach wants a consistent performer, she should take Scyrine. She'll almost certainly deliver somewhere between 15 and 20 points. But, if she wants to take a chance and needs a "big game", she should take Alexandra. Alex scores over 24 points about a quarter of the time. On the other hand, she scores under 11 points about as often.

27. Marriage age.

The distribution of marriage age of U.S. men is skewed right, with a typical man (as measured by the median) first marrying at around 24 years old. The middle 50% of male marriage ages is between about 23 and 26 years. For U.S. women, the distribution of marriage age is also skewed right, with median of around 21 years. The middle 50% of female marriage age is between about 20 and 23 years. When comparing the two distributions, the most striking feature is that the distributions are nearly identical in spread, but have different centers. Females typically seem to marry earlier than males. In fact, between 50% and 75% of the women marry at a younger age than *any* man.

29. Fuel economy 2012.

(Note: numerical details may vary.) In general, fuel economy is highest in mid-size cars, lower in SUVs, and lowest in pickup trucks. There are numerous outliers on both ends for cars, one high outlier for pickups and two high outliers for SUVs. The top 50% of cars get higher fuel economy than 75% of SUVs and all pickups. The distributions of fuel economy for cars and SUVs have approximately the same IQR, about 5 miles per gallon, but cars generally have higher combined fuel economy. Furthermore, there are several cars with much higher fuel economy. Pickups trucks have consistently low fuel economy.

31. Test scores.

Class A is Class 1. The median is 65, but has less spread than Class B, which is Class 2. Class C is Class 3, since it's median is higher, which corresponds to the skew to the left.

33. Graduation?

a) The distribution of the percent of incoming college freshman who graduate on time is roughly symmetric. The mean and the median are reasonably close to one another and the quartiles are approximately the same distance from the median.

b) Upper Fence: $Q3+1.5(IQR)=74.75+1.5(74.75-59.15)$

$$=74.75+23.4$$

$$=98.15$$

Lower Fence: $Q1-1.5(IQR)=59.15-1.5(74.75-59.15)$

$$=59.15-23.4$$

$$=35.75$$

Since the maximum value of the distribution of the percent of incoming freshmen who graduate on time is 87.4% and the upper fence is 98.15%, there are no high outliers. Likewise, since the minimum is 43.2% and the lower fence is 35.75%, there are no low outliers. Since the minimum and maximum percentages are within the fences, all percentages must be within the fences.

c) A boxplot of the distribution of the percent of incoming freshmen who graduate on time is at the right.

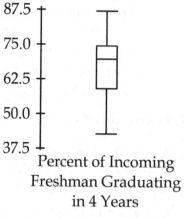

Percent of Incoming Freshman Graduating in 4 Years

d) The distribution of the percent of incoming freshmen who graduate on time is roughly symmetric, with mean of approximately 68% of freshmen graduating on time. Universities surveyed had between 43.2% and 87.4% of students graduating on time, with the middle 50% of universities reporting between 59.15% and 74.75% graduating on time.

35. Caffeine.

a) *Who* – 45 student volunteeers. *What* – Level of caffeine consumption and memory test score. *When* – Not specified. *Where* – Not specified. *Why* – The student researchers want to see the possible effects of caffeine on memory. *How* – It appears that the researchers imposed the treatment of level of caffeine consumption in an experiment. However, this point is not clear. Perhaps they allowed the subjects to choose their own level of caffeine.

b) *Variables* – Caffeine level is a categorical variable with three levels: no caffeine, low caffeine, and high caffeine. Test score is a quantitative variable, measured in number of items recalled correctly.

c)

   d) The groups consuming no caffeine and low caffeine had comparable memory test scores. A typical score from these groups was around 21. However, the scores of the group consuming no caffeine were more consistent, with a smaller range and smaller interquartile range than the scores of the group consuming low caffeine. The group consuming high caffeine had lower memory scores in general, with a median score of about 19. No one in the high caffeine group scored above 24, but 25% of each of the other groups scored above 24.

37. Derby speeds 2014.

   a) The median speed is the speed at which 50% of the winning horses ran slower. Find 50% on the left, move straight over to the graph and down to a speed of about 36.25 mph.

   b) Quartile 1 is at 25% on the left, and Quartile 3 is at 75% on the left. Matching these to the ogive, Q1 = 35 mph and Q3 = 36.8 mph, approximately.

   c) Range = Max – Min = 37.5 – 31.4 = 6.1 mph
      IQR = Q3 – Q1 = 36.8 – 35 = 1.8 mph

   d) An approximate boxplot of winning Kentucky Derby Speeds is at the right.

   e) The distribution of winning speeds in the Kentucky Derby is skewed to the left. The lowest winning speed is about 31.4 mph and the fastest speed is about 37.5 mph. The median speed is approximately 36.25 mph, and 75% of winning speeds are above 35 mph. Only a few percent of winners have had speeds below 33 mph. The middle 50% of winning speeds are between 35 and 36.8 mph.

39. Reading scores.

   a) The highest score for boys was 6, which is higher than the highest score for girls, 5.9.

   b) The range of scores for boys is greater than the range of scores for girls.
      Range = Max – Min   Range(Boys) = 4   Range(Girls) = 3.1

   c) The girls had the greater IQR.
      IQR = Q3 – Q1   IQR(Boys) = 4.9 – 3.9 = 1   IQR(Girls) = 5.2 – 3.8 = 1.4

   d) The distribution of boys' scores is more skewed. The quartiles are not the same distance from the median. In the distribution of girls' scores, Q1 is 0.7 units below the median, while Q3 is 0.7 units above the median.

e) Overall, the girls did better on the reading test. The median, 4.5, was higher than the median for the boys, 4.3. Additionally, the upper quartile score was higher for girls than boys, 5.2 compared to 4.9. The girls' lower quartile score was slightly lower than the boys' lower quartile score, 3.8 compared to 3.9.

f) The overall mean is calculated by weighting each mean by the number of students. $\dfrac{14(4.2) + 11(4.6)}{25} = 4.38$

41. Industrial experiment.

First of all, there is an extreme outlier in the distribution of distances for the slow speed drilling. One hole was drilled almost an inch away from the center of the target! If that distance is correct, the engineers at the computer production plant should investigate the slow speed drilling process closely. It may be plagued by extreme, intermittent inaccuracy. The outlier in the slow speed drilling process is so extreme that no graphical display can display the distribution in a meaningful way while including that outlier. That distance should be removed before looking at a plot of the drilling distances.

With the outlier set aside, we can determine that the slow drilling process is more accurate. The greatest distance from the target for the slow drilling process, 0.000098 inches, is still more accurate than the smallest distance for the fast drilling process, 0.000100 inches.

43. MPG.

a) Comparative boxplots are at the right. Comparative dotplots, histograms, or stemplots would also be acceptable.

b) In general, the Other cars got better gas mileage than the US Cars. The median of the distribution of mileage for US cars was approximately 22 miles per gallon, while the median for Other cars was approximately 24 miles per gallon. U.S. cars had a distribution that was less variable than the distribution of Other cars, with 75% of US cars having gas mileage that was lower than the median of cars from Other countries. Both groups had some high outliers, most likely hybrid models.

45. Fruit Flies.

a) The most fruit flies died around day 16. Over 60,000 fruit flies died that day.

b) The largest proportion of fruit flies died around day 65. About 20% of that previously surviving population died that day.

c) At around day 50, the number of fruit flies wasn't changing by very much day to day.

47. Assets.

a) The distribution of assets of 79 companies chosen from the *Forbes* list of the nation's top corporations is skewed so heavily to the right that the vast majority of companies have assets represented in the first bar of the histogram, 0 to 10 billion dollars. This makes meaningful discussion of center and spread impossible.

b) Re-expressing these data by, for example, logs or square roots might help make the distribution more nearly symmetric. Then a meaningful discussion of center might be possible.

49. Assets again.

  a) The distribution of logarithm of assets is preferable, because it is roughly unimodal and symmetric. The distribution of the square root of assets is still skewed right, with outliers.

  b) If $\sqrt{Assets} = 50$, then the companies assets are approximately $50^2 = 2500$ million dollars.

  c) If $\log(Assets) = 3$, then the companies assets are approximately $10^3 = 1000$ million dollars.

51. Stereograms.

  a) The two variables discussed in the description are fusion time and treatment group.

  b) Fusion time is a quantitative variable, measured in seconds. Treatment group is a categorical variable, with subjects either receiving verbal clues only, or visual and verbal clues.

  c) Both groups have distributions that are skewed to the right. Generally, the Visual/Verbal group had shorter fusion times than the No/Verbal group. The median for the Visual/Verbal group was approximately the same as the lower quartile for the No/Verbal group. The No/Verbal Group also had an extreme outlier, with at least one subject whose fusion time was approximately 50 seconds. There is evidence that visual information may reduce fusion time.

Chapter 5 – The Standard Deviation as a Ruler and the Normal Model

Section 5.1

1. Stats test.

Gregor scored 65 points on the test.

(Or, 75 – 2(5) = 65)

$$z = \frac{y - \mu}{\sigma}$$

$$-2 = \frac{y - 75}{5}$$

$$y = 65$$

3. Temperatures.

In January, with mean temperature 36° and standard deviation in temperature 10°, a high temperature of 55° is almost 2 standard deviations above the mean. In July, with mean temperature 74° and standard deviation 8°, a high temperature of 55° is more than two standard deviations below the mean. A high temperature of 55° is less likely to happen in July, when 55° is farther away from the mean.

Section 5.2

5. Shipments.

a) Adding 4 ounces will affect only the median. The new median will be 68 +4 = 72 ounces, and the IQR will remain at 40 ounces.

b) Changing the units will affect both the median and IQR. The median will be 72/16 = 4.5 pounds and the IQR will be 40/16 = 2.5 pounds.

Section 5.3

7. Guzzlers?

a) The Normal model for auto fuel economy is at the right.

b) Approximately 68% of the cars are expected to have highway fuel economy between 18.6 mpg and 31.0 mpg.

c) Approximately 16% of the cars are expected to have highway fuel economy above 31 mpg.

d) Approximately 13.5% of the cars are expected to have highway fuel economy between 31 mpg and 37 mpg.

e) The worst 2.5% of cars are expected to have fuel economy below approximately 12.4 mpg.

Section 5.4

9. Normal cattle.

a)

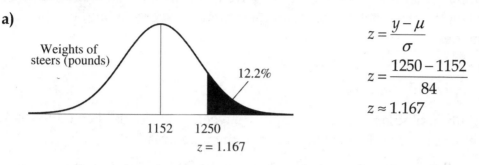

$$z = \frac{y - \mu}{\sigma}$$

$$z = \frac{1250 - 1152}{84}$$

$$z \approx 1.167$$

According to the Normal model, we expect 12.2% of steers to weigh over 1250 pounds.

b)

$$z = \frac{y - \mu}{\sigma}$$

$$z = \frac{1200 - 1152}{84}$$

$$z \approx 0.571$$

According to the Normal model, 71.6% of steers are expected to weigh under 1200 pounds.

c)

$$z = \frac{y - \mu}{\sigma}$$

$$z = \frac{1000 - 1152}{84}$$

$$z \approx -1.810$$

$$z = \frac{y - \mu}{\sigma}$$

$$z = \frac{1100 - 1152}{84}$$

$$z \approx -0.619$$

According to the Normal model, 23.3% of steers are expected to weigh between 1000 and 1100 pounds.

Section 5.5

11. Music library.

a) The Normal probability plot is not straight, so there is evidence that the distribution of the lengths of songs in Corey's music library is not Normal.

b) The distribution of the lengths of songs in Corey's music library appears to be skewed to the right. The Normal probability plot show that the longer songs in Corey's library are much longer than the lengths predicted by the Normal model. The song lengths are much longer than their quantile scores would predict for a Normal model.

Chapter Exercises.

13. Payroll.

a) The distribution of salaries in the company's weekly payroll is skewed to the right. The mean salary, $700, is higher than the median, $500.

b) The IQR, $600, measures the spread of the middle 50% of the distribution of salaries.

$$Q3 - Q1 = IQR$$
$$Q3 = Q1 + IQR$$
$$Q3 = \$350 + \$600$$
$$Q3 = \$950$$

50% of the salaries are found between $350 and $950.

c) If a $50 raise were given to each employee, all measures of center or position would increase by $50. The minimum would change to $350, the mean would change to $750, the median would change to $550, and the first quartile would change to $400. Measures of spread would not change. The entire distribution is simply shifted up $50. The range would remain at $1200, the IQR would remain at $600, and the standard deviation would remain at $400.

d) If a 10% raise were given to each employee, all measures of center, position, and spread would increase by 10%.

Minimum = $330   Mean = $770   Median = $550   Range = $1320
IQR = $660   First Quartile = $385   St. Dev. = $440

15. SAT or ACT?

Measures of center and position (lowest score, top 25% above, mean, and median) will be multiplied by 40 and increased by 150 in the conversion from ACT to SAT by the rule of thumb. Measures of spread (standard deviation and IQR) will only be affected by the multiplication.

Lowest score = 910   Mean = 1230   Standard deviation = 120
Top 25% above = 1350   Median = 1270   IQR = 240

17. Stats test, part II.

A *z*-score of 2.20 means that your score was 2.20 standard deviations above the mean.

19. Music library again.

On the Nickel, by Tom Waits has a z-score of 1.20

$$z = \frac{y - \mu}{\sigma}$$

$$z = \frac{380 - 242.4}{114.51}$$

$$z = 1.20$$

21. Combining test scores.

The z-scores, which account for the difference in the distributions of the two tests, are 1.5 and 0 for Derrick and 0.5 and 2 for Julie. Derrick's total is 1.5 which is less than Julie's 2.5.

23. Final Exams.

a) Anna's average is $\frac{83 + 83}{2} = 83$. Megan's average is $\frac{77 + 95}{2} = 86$.

Only Megan qualifies for language honors, with an average higher than 85.

b) On the French exam, the mean was 81 and the standard deviation was 5. Anna's score of 83 was 2 points, or 0.4 standard deviations, above the mean. Megan's score of 77 was 4 points, or 0.8 standard deviations below the mean.

On the Spanish exam, the mean was 74 and the standard deviation was 15. Anna's score of 83 was 9 points, or 0.6 standard deviations, above the mean. Megan's score of 95 was 21 points, or 1.4 standard deviations, above the mean.

Measuring their performance in standard deviations is the only fair way in which to compare the performance of the two women on the test.

Anna scored 0.4 standard deviations above the mean in French and 0.6 standard deviations above the mean in Spanish, for a total of 1.0 standard deviation above the mean.

Megan scored 0.8 standard deviations below the mean in French and 1.4 standard deviations above the mean in Spanish, for a total of only 0.6 standard deviations above the mean.

Anna did better overall, but Megan had the higher average. This is because Megan did very well on the test with the higher standard deviation, where it was comparatively easy to do well.

25. Cattle.

a) A steer weighing 1000 pounds would be about 1.81 standard deviations below the mean weight. $\quad z = \frac{y - \mu}{\sigma} = \frac{1000 - 1152}{84} \approx -1.81$

b) A steer weighing 1000 pounds is more unusual.
Its z-score of −1.81 is further from 0 than the 1250 pound steer's z-score of 1.17.

27. More cattle.

a) The new mean would be 1152 – 1000 = 152 pounds. The standard deviation would not be affected by subtracting 1000 pounds from each weight. It would still be 84 pounds.

b) The mean selling price of the cattle would be 0.40(1152) = $460.80. The standard deviation of the selling prices would be 0.40(84) = $33.60.

29. Cattle, part III.

Generally, the minimum and the median would be affected by the multiplication and subtraction. The standard deviation and the IQR would only be affected by the multiplication.

Minimum = 0.40(980) – 20 = $372.00   Median = 0.40(1140) – 20 = $436
Standard deviation = 0.40(84) = $33.60   IQR = 0.40(102) = $40.80

31. Professors.

The standard deviation of the distribution of years of teaching experience for college professors must be 6 years. College professors can have between 0 and 40 (or possibly 50) years of experience. A workable standard deviation would cover most of that range of values with ±3 standard deviations around the mean. If the standard deviation were 6 months ($\frac{1}{2}$ year), some professors would have years of experience 10 or 20 standard deviations away from the mean, whatever it is. That isn't possible. If the standard deviation were 16 years, ±2 standard deviations would be a range of 64 years. That's way too high. The only reasonable choice is a standard deviation of 6 years in the distribution of years of experience.

33. Small steer.

Any weight more than 2 standard deviations below the mean, or less than 1152 – 2(84) = 984 pounds might be considered unusually low. We would expect to see a steer below 1152 – 3(84) = 900 very rarely.

35. Trees.

a) The Normal model for the distribution of tree diameters is at the right.

b) Approximately 95% of the trees are expected to have diameters between 1.0 inch and 19.8 inches.

c) Approximately 2.5% of the trees are expected to have diameters less than an inch.

d) Approximately 34% of the trees are expected to have diameters between 5.7 inches and 10.4 inches.

e) Approximately 16% of the trees are expected to have diameters over 15 inches.

37. Trees, part II.

The use of the Normal model requires a distribution that is unimodal and symmetric. The distribution of tree diameters is neither unimodal nor symmetric, so use of the Normal model is not appropriate.

39. Winter Olympics 2010 downhill.

a) The 2010 Winter Olympics downhill times have mean of 117.34 seconds and standard deviation 2.456 seconds. 114.875 seconds is 1 standard deviation below the mean. If the Normal model is appropriate, 16% of the times should be below 99.7 seconds.

b) 8 out of 59 times (13.56%) are below 114.875 seconds.

c) The percentages in parts a and b do not agree because the Normal model is not appropriate in this situation.

d) The histogram of 2010 Winter Olympic Downhill times is skewed to the right. The Normal model is not appropriate for the distribution of times, because the distribution is not symmetric.

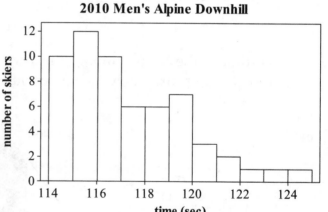

2010 Men's Alpine Downhill

41. Receivers 2013.

a) Approximately 2.5% of the receivers are expected to gain more yards than 2 standard deviations above the mean number of yards gained.

b) The distribution of the number of yards gained has mean 426.28 yards and standard deviation 408.34 yards. According to the Normal model, we expect 2.5% of the receivers, or 4.525 of them, to gain more than 2 standard deviations above the mean number of yards. This means more than 426.98 + 2(408.34) = 1244 yards. In 2013, 12 receivers ran for more than 1244 yards.

c) The distribution of the number of yards run by wide receivers is skewed heavily to the right. Use of the Normal model is not appropriate for this distribution, since it is not symmetric.

43. More cattle.

a)

$$z = \frac{y - \mu}{\sigma}$$

$$1.282 = \frac{y - 1152}{84}$$

$$y \approx 1259.7$$

According to the Normal model, the highest 10% of steer weights are expected to be above approximately 1259.7 pounds.

b)

$$z = \frac{y - \mu}{\sigma}$$

$$-0.842 = \frac{y - 1152}{84}$$

$$y \approx 1081.3$$

According to the Normal model, the lowest 20% of weights of steers are expected to be below approximately 1081.3 pounds.

c)

$$z = \frac{y - \mu}{\sigma}$$

$$-0.524 = \frac{y - 1152}{84}$$

$$y \approx 1108.0$$

$$z = \frac{y - \mu}{\sigma}$$

$$0.524 = \frac{y - 1152}{84}$$

$$y \approx 1196.0$$

According to the Normal model, the middle 40% of steer weights is expected to be between about 1108.0 pounds and 1196.0 pounds.

45. Cattle, finis.

a)

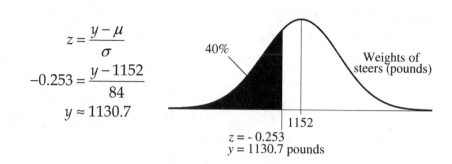

$$z = \frac{y - \mu}{\sigma}$$

$$-0.253 = \frac{y - 1152}{84}$$

$$y \approx 1130.7$$

According to the Normal model, the weight at the 40th percentile is 1130.7 pounds. This means that 40% of steers are expected to weigh less than 1130.7 pounds.

b)

$$z = \frac{y - \mu}{\sigma}$$

$$2.326 = \frac{y - 1152}{84}$$

$$y \approx 1347.4$$

According to the Normal model, the weight at the 99th percentile is 1347.4 pounds. This means that 99% of steers are expected to weigh less than 1347.4 pounds.

c)

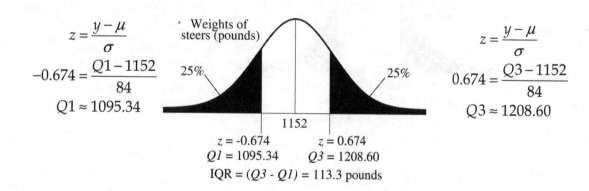

$$z = \frac{y - \mu}{\sigma}$$

$$-0.674 = \frac{Q1 - 1152}{84}$$

$$Q1 \approx 1095.34$$

$$z = \frac{y - \mu}{\sigma}$$

$$0.674 = \frac{Q3 - 1152}{84}$$

$$Q3 \approx 1208.60$$

According to the Normal model, the IQR of the distribution of weights of Angus steers is about 113.3 pounds.

47. Cholesterol.

a) The Normal model for cholesterol levels of adult American women is at the right.

b)

$$z = \frac{y - \mu}{\sigma}$$

$$z = \frac{200 - 188}{24}$$

$$z = 0.5$$

According to the Normal model, 30.85% of American women are expected to have cholesterol levels over 200.

c)

According to the Normal model, 17.00% of American women are expected to have cholesterol levels between 150 and 170.

d)

According to the Normal model, the interquartile range of the distribution of cholesterol levels of American women is approximately 32.38 points.

e)

$$z = \frac{y - \mu}{\sigma}$$

$$1.036 = \frac{y - 188}{24}$$

$$y = 212.87$$

According to the Normal model, the highest 15% of women's cholesterol levels are above approximately 212.87.

49. Kindergarten.

a)

$$z = \frac{y - \mu}{\sigma}$$

$$z = \frac{36 - 38.2}{1.8}$$

$$z = -1.222$$

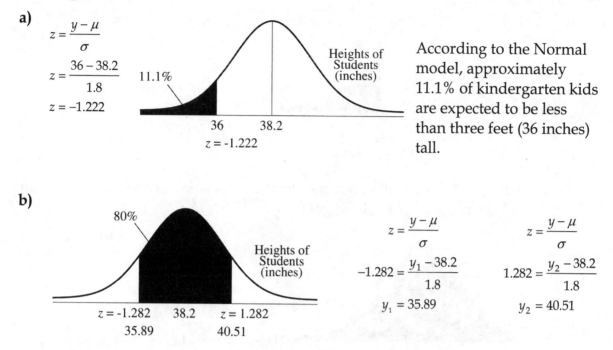

According to the Normal model, approximately 11.1% of kindergarten kids are expected to be less than three feet (36 inches) tall.

b)

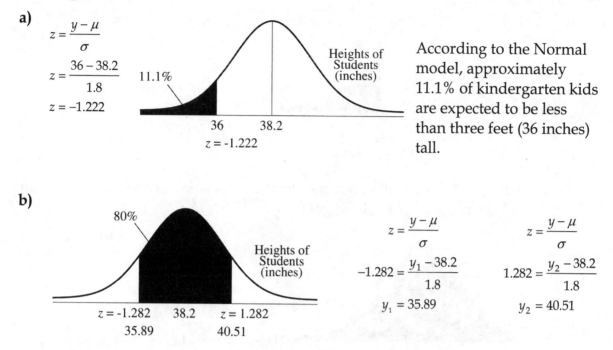

$$z = \frac{y - \mu}{\sigma}$$

$$-1.282 = \frac{y_1 - 38.2}{1.8}$$

$$y_1 = 35.89$$

$$z = \frac{y - \mu}{\sigma}$$

$$1.282 = \frac{y_2 - 38.2}{1.8}$$

$$y_2 = 40.51$$

According to the Normal model, the middle 80% of kindergarten kids are expected to be between 35.89 and 40.51 inches tall. (The appropriate values of $z = \pm1.282$ are found by using right and left tail percentages of 10% of the Normal model.)

c)

$$z = \frac{y - \mu}{\sigma}$$

$$1.282 = \frac{y - 38.2}{1.8}$$

$$y = 40.51$$

According to the Normal model, the tallest 10% of kindergarteners are expected to be at least 40.51 inches tall.

51. Eggs.

a)

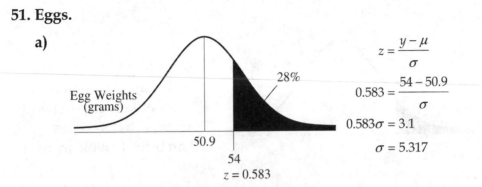

According to the Normal model, the standard deviation of the egg weights for young hens is expected to be 5.3 grams.

b)

According to the Normal model, the standard deviation of the egg weights for older hens is expected to be 6.4 grams.

c) The younger hens lay eggs that have more consistent weights than the eggs laid by the older hens. The standard deviation of the weights of eggs laid by the younger hens is lower than the standard deviation of the weights of eggs laid by the older hens.

Review of Part I – Exploring and Understanding Data

1. **Bananas.**

   a) A histogram of the prices of bananas from 15 markets, as reported by the USDA, appears at the right.

   Banana Prices

   b) The distribution of banana prices is skewed to the left, so median and IQR are appropriate measures of center and spread. Median = 49 cents per pound
   IQR = 6 cents per pound

   c) The distribution of the prices of bananas from 15 markets, as reported by the USDA, is unimodal and skewed to the left. The center of the distribution is approximately 50 cents, with the lowest price 42 cents per pound and the highest price 53 cents per pound.

3. **Singers by parts.**

   a) The two statistics could be the same if there were many sopranos of that height.

   b) The distribution of heights of each voice part is roughly symmetric. The basses and tenors are generally taller than the altos and sopranos, with the basses being slightly taller than the tenors. The sopranos and altos have about the same median height. Heights of basses and sopranos are more consistent than altos and tenors.

5. **Beanstalks.**

   a) The greater standard deviation for the distribution of women's heights means that their heights are more variable than the heights of men.

   b) The *z*-score for women to qualify is 2.4 compared with 1.75 for men, so it is harder for women to qualify.

7. **State University.**

   a) *Who* – Local residents near State University. *What* – Age, whether or not the respondent attended college, and whether or not the respondent had a favorable opinion of State University. *When* – Not specified. *Where* – Region around State University. *Why* – The information will be included in a report to the University's directors. *How* – 850 local residents were surveyed by phone.

b) There is one quantitative variable, age, probably measured in years. There are two categorical variables, college attendance (yes or no), and opinion of State University (favorable or unfavorable).

c) There are several problems with the design of the survey. No mention is made of a random selection of residents. Furthermore, there may be a non-response bias present. People with an unfavorable opinion of the university may hang up as soon as the staff member identifies himself or herself. Also, response bias may be introduced by the interviewer. The responses of the residents may be influenced, since employees of the university are asking the questions. A greater percentage of responses to the survey may be favorable than truly exist.

9. Fraud detection.

a) Even though they are numbers, the SIC code is a categorical variable. A histogram is a quantitative display, so it is not appropriate.

b) The Normal model will not work at all. The Normal model is for modeling distributions of unimodal and symmetric quantitative variables. SIC code is a categorical variable.

11. Cramming.

a) Comparitive boxplots of the distributions of Friday and Monday scores are at the right.

b) The distribution of scores on Friday was generally higher by about 5 points. Students fared worse on Monday after preparing for the test on Friday. The spreads are about the same, but the scores on Monday are slightly skewed to the right.

c) A histogram of the distribution of change in test score is at the right.

d) The distribution of changes in score is roughly unimodal and symmetric, and is centered near 4 points. Changes ranged from a student who scored 5 points higher on Monday, to two students who scored 13 and 14 points higher on Friday. Only three students did better on Monday.

13. Let's play cards.

a) Suit is a categorical variable.

b) In the game of Go Fish, the denomination is not ordered. Numbers are merely matched with one another. You may have seen children's Go Fish decks that have symbols or pictures on the cards instead of numbers. These work just fine.

c) In the game of Gin Rummy, the order of the cards is important. During the game, ordered "runs" of cards are assembled (with Jack = 11, Queen = 12, King = 13), and at the end of the hand, points are totaled from the denomination of the card (face cards = 10 points). However, even in Gin Rummy, the denomination of the card sometimes behaves like a categorical variable. When you are collecting 3s, for example, order doesn't matter.

15. Hard water.

a) The variables in this study are both quantitative. Annual mortality rate for males is measured in deaths per 100,000. Calcium concentration is measured in parts per million.

b) The distribution of calcium concentration is skewed right, with many towns having concentrations below 25 ppm. The rest of the towns have calcium concentrations which are distributed in a fairly uniform pattern from 25 ppm to 100 ppm, tapering off to a maximum concentration around 150 ppm. The distribution of mortality rates is unimodal and symmetric, with center approximately 1500 deaths per 100,000. The distribution has a range of 1000 deaths per 100,000, from 1000 to 2000 deaths per 100,000.

17. Seasons.

a) The two histograms have different horizontal and vertical scales. This makes a quick comparison impossible.

b) The center of the distribution of average temperatures in January in is the low 30s, compared to a center of the distribution of July temperatures in the low 70s. The January distribution is also much more spread out than the July distribution. The range is over 50 degrees in January, compared to a range of over 20 degrees in July. The distribution of average temperature in January is skewed slightly to the right, while the distribution of average temperature in July is roughly symmetric.

c) The distribution of difference in average temperature (July – January) for 60 large U.S. cities is slightly skewed to the left, with median at approximately 44 degrees. There are several low outliers, cities with very little difference between their average July and January temperatures. The single high outlier is a city with a large difference in average temperature between July and January. The middle 50% of differences are between approximately 38 and 46 degrees.

19. Old Faithful?

a) The distribution of duration of the 222 eruptions is bimodal, with modes at approximately 2 minutes and 4.5 minutes. The distribution is fairly symmetric around each mode.

b) The bimodal shape of the distribution of duration of the 222 eruptions suggests that there may be two distinct groups of eruption durations. Summary statistics would try to summarize these two groups as a single group, which wouldn't make sense.

c) The intervals between eruptions are generally longer for long eruptions than the intervals for short eruptions. Over 75% of the short eruptions had intervals of approximately 60 minutes or less, while almost all of the long eruptions had intervals of more than 60 minutes.

21. Liberty's nose.

a) The distribution of the ratio of arm length to nose length of 18 girls in a statistics class is unimodal and roughly symmetric, with center around 15. There is one low outlier, a ratio of 11.8. A boxplot is provided at the right. A histogram or stemplot is also an appropriate display.

b) In the presence of an outlier, the 5-number summary is the appropriate choice for summary statistics. The 5-number summary is 11.8, 14.4, 15.25, 15.7, 16.9. The IQR is 1.3.

c) The ratio of 9.3 for the Statue of Liberty is very low, well below the lowest ratio in the statistics class, 11.8, which is already a low outlier. Compared to the girls in the statistics class, the Statue of Liberty's nose is very long in relation to her arm.

23. Sample.

Overall, the follow-up group was insured only 11.1% of the time as compared to 16.6% for the not traced group. At first, it appears that group is associated with presence of health insurance. But for blacks, the follow-up group was quite close (actually slightly higher) in terms of being insured: 8.9% to 8.7%. The same is true for whites. The follow-up group was insured 83.3% of the time, compared to 82.5% of the not traced group. When broken down by race, we see that group is not associated with presence of health insurance for either race. This demonstrates Simpson's paradox, because the overall percentages lead us to believe that there is an association between health insurance and group, but we see the truth when we examine the situation more carefully.

25. Be quick!

a) The Normal model for the distribution of reaction times is at the right.

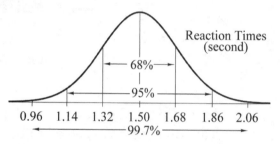

b) The distribution of reaction times is unimodal and symmetric, with mean 1.5 seconds, and standard deviation 0.18 seconds. According to the Normal model, 95% of drivers are expected to have reaction times between 1.14 seconds and 1.86 seconds.

c)

$$z = \frac{y - \mu}{\sigma}$$

$$z = \frac{1.25 - 1.50}{0.18}$$

$$z = -1.389$$

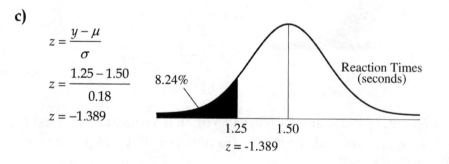

According to the Normal model, 8.24% of drivers are expected to have reaction times below 1.25 seconds.

d)

$$z = \frac{y - \mu}{\sigma}$$

$$z = \frac{1.6 - 1.5}{0.18}$$

$$z = 0.556$$

Reaction Times (seconds)

24.13%

1.50 1.60   1.80

$z = 0.556$ $z = 1.667$

$$z = \frac{y - \mu}{\sigma}$$

$$z = \frac{1.8 - 1.5}{0.18}$$

$$z = 1.667$$

According to the Normal model, 24.13% of drivers are expected to have reaction times between 1.6 seconds and 1.8 seconds.

e)

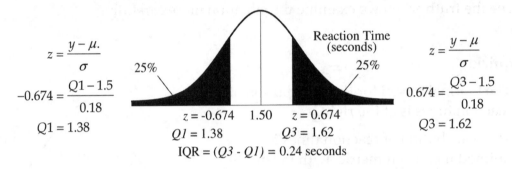

$$z = \frac{y - \mu.}{\sigma}$$

$$-0.674 = \frac{Q1 - 1.5}{0.18}$$

$$Q1 = 1.38$$

25%   Reaction Time (seconds)   25%

$z = -0.674$ 1.50 $z = 0.674$

$Q1 = 1.38$ $Q3 = 1.62$

IQR $= (Q3 - Q1) = 0.24$ seconds

$$z = \frac{y - \mu}{\sigma}$$

$$0.674 = \frac{Q3 - 1.5}{0.18}$$

$$Q3 = 1.62$$

According to the Normal model, the interquartile range of the distribution of reaction times is expected to be 0.24 seconds.

f)

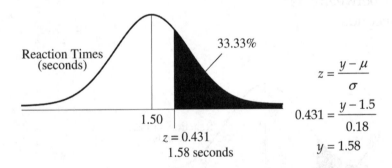

Reaction Times (seconds)

33.33%

1.50

$z = 0.431$

1.58 seconds

$$z = \frac{y - \mu}{\sigma}$$

$$0.431 = \frac{y - 1.5}{0.18}$$

$$y = 1.58$$

According to the Normal model, the slowest 1/3 of all drivers are expected to have reaction times of 1.58 seconds or more. (Remember that a high reaction time is a SLOW reaction time!)

27. Mail.

a) A histogram of the number of pieces of mail received at a school office is at the right.

Number of Pieces of Mail Received at a School Office

# of days

50.0   100.0   150.0

Number of pieces of mail

b) Since the distribution of number of pieces of mail is unimodal and symmetric, the mean and standard deviation are appropriate measures of center and spread. The mean number of pieces of mail is 100.25, and the standard deviation is 25.54 pieces.

c) The distribution of the number of pieces of mail received at the school office is unimodal and symmetric, with mean 100.25 and standard deviation 25.54. The lowest number of pieces received in a day was 52 and the highest was 151.

d) 23 of the 36 days (64%) had a number of pieces of mail received within one standard deviation of the mean, or within the interval 74.71 - 125.79. This is fairly close to the 68% predicted by the Normal model. The Normal model may be useful for modeling the number of pieces received by this school office.

29. Herbal medicine.

a) *Who* – 100 customers. *What* – Researchers asked whether or not the customer had taken the cold remedy and had customers rate the effectiveness of the remedy on a scale from 1 to 10. *When* – Not specified. *Where* – Store where natural health products are sold. *Why* – The researchers were from the Herbal Medicine Council, which sounds suspiciously like a group that might be promoting the use of herbal remedies. *How* – Researchers conducted personal interviews with 100 customers. No mention was made of any type of random selection.

b) "Have you taken the cold remedy?" is a categorical variable. Effectiveness on a scale of 1 to 10 is a categorical variable, as well, with respondents rating the remedy by placing it into one of 10 categories.

c) Very little confidence can be placed in the Council's conclusions. Respondents were people who already shopped in a store that sold natural remedies. They may be pre-disposed to thinking that the remedy was effective. Furthermore, no attempt was made to randomly select respondents in a representative manner. Finally, the Herbal Medicine Council has an interest in the success of the remedy.

31. Engines.

a) The count of cars is 38.

b) The mean displacement is higher than the median displacement, indicating a distribution of displacements that is skewed to the right. There are likely to be several very large engines in a group that consists of mainly smaller engines.

c) Since the distribution is skewed, the median and IQR are useful measures of center and spread. The median displacement is 148.5 cubic inches and the IQR is 126 cubic inches.

d) Your neighbor's car has an engine that is bigger than the median engine, but 227 cubic inches is smaller than the third quartile of 231, meaning that at least 25% of cars have a bigger engine than your neighbor's car. Don't be impressed!

e) Using the Outlier Rule (more than 1.5 IQRs beyond the quartiles) to find the fences:

Upper Fence: Q3 + 1.5(IQR) = 231 + 1.5(126) = 420 cubic inches.

Lower Fence: Q1 − 1.5(IQR) = 105 − 1.5(126) = − 84 cubic inches.

Since there are certainly no engines with negative displacements, there are no low outliers. Q1 + Range = 105 + 275 = 380 cubic inches. This means that the maximum must be less than 380 cubic inches. Therefore, there are no high outliers (engines over 420 cubic inches).

f) It is not reasonable to expect 68% of the car engines to measure within one standard deviation of the mean. The distribution engine displacements is skewed to the right, so the Normal model is not appropriate.

g) Multiplying each of the engine displacements by 16.4 to convert cubic inches to cubic centimeters would affect measures of position and spread. All of the summary statistics (except the count!) could be converted to cubic centimeters by multiplying each by 16.4.

33. Age and party 2011.

a) 3705 of 8414, or approximately 44.0%, of all voters surveyed were Republicans or leaned Republican.

b) This was a representative telephone survey conducted by Pew, a reputable polling firm. It is likely to be a reasonable estimate of the percentage of all voters who are Republicans.

c) 815 + 2416 = 3231 of 8414, or approximately 38.4%, of all voters surveyed were under 30 or over 65 years old.

d) 73 of 8414, or approximately 0.87%, of all voters surveyed were classified as "Neither" and under the age of 30.

e) 73 of the 733 people classified as "Neither", or 9.96%, were under the age of 30.

f) 73 of the 815 respondents under 30, or 8.96%, were classified as "Neither".

35. Age and party II.

a) The marginal distribution of party affiliation is:
Republican – 44.0% Democrat – 47.3%   Neither – 8.7%
(As counts: Republican – 3705   Democrat – 3976   Independent – 733)

b) Graphs are at the right.

c) It appears that older voters are more likely to lean Republican, and younger voters are more likely to lean Deomocrat.

d) No. There is an evidence of an association between party affiliation and age. Younger voters tend to be more Democratic and less Republican.

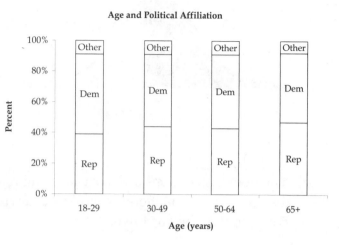

37. Some assembly required.

a)

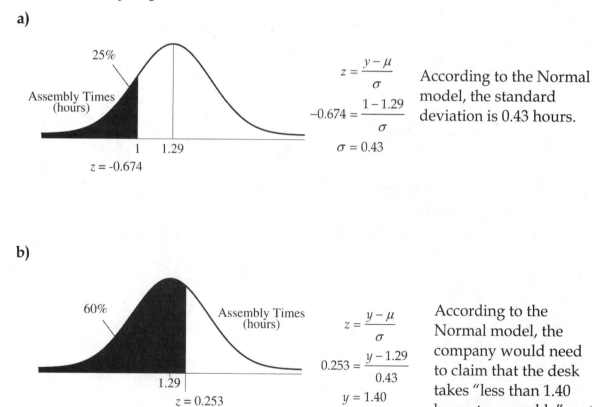

$$z = \frac{y - \mu}{\sigma}$$

$$-0.674 = \frac{1 - 1.29}{\sigma}$$

$$\sigma = 0.43$$

According to the Normal model, the standard deviation is 0.43 hours.

b)

$$z = \frac{y - \mu}{\sigma}$$

$$0.253 = \frac{y - 1.29}{0.43}$$

$$y = 1.40$$

According to the Normal model, the company would need to claim that the desk takes "less than 1.40 hours to assemble", not the catchiest of slogans!

c)

$$z = \frac{y - \mu}{\sigma}$$

$$0.253 = \frac{1 - \mu}{0.43}$$

$$\mu = 0.89$$

According to the Normal model, the company would have to lower the mean assembly time to 0.89 hour (53.4 minutes).

d) The new instructions and part-labeling may have helped lower the mean, but it also may have changed the standard deviation, making the assembly times more consistent as well as lower.

Chapter 6 – Scatterplots, Association, and Correlation

Section 6.1

1. **Association.**

   a) Either weight in grams or weight in ounces could be the explanatory or response variable. Greater weights in grams correspond with greater weights in ounces. The association between weight of apples in grams and weight of apples in ounces would be positive, straight, and perfect. Each apple's weight would simply be measured in two different scales. The points would line up perfectly.

   b) Circumference is the explanatory variable, and weight is the response variable, since one-dimensional circumference explains three-dimensional volume (and therefore weight). For apples of roughly the same size, the association would be positive, straight, and moderately strong. If the sample of apples contained very small and very large apples, the association's true curved form would become apparent.

   c) There would be no association between shoe size and GPA of college freshmen.

   d) Number of miles driven is the explanatory variable, and gallons remaining in the tank is the response variable. The greater the number of miles driven, the less gasoline there is in the tank. If a sample of different cars is used, the association is negative, straight, and moderate. If the data is gathered on different trips with the same car, the association would be strong.

3. **Bookstore sales.**

   a) The scatterplot is to the right.

   b) There is a positive association between bookstore sales and the number of sales people working.

   c) There is a linear association between bookstore sales and the number of sales people working.

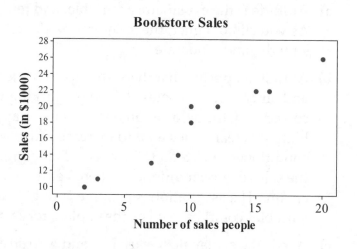

   d) There is a strong association between bookstore sales and the number of sales people working.

   e) The relationship between bookstore sales and the number of sales people working has no outliers.

Section 6.2

5. Correlation facts.

a) True.

b) False. The correlation will remain the same.

c) False. Correlation has no units.

Section 6.3

7. Bookstore sales again.

This conclusion is not justified. Correlation does not demonstrate causation. The analyst argues that the number of sales staff working causes sales to be higher. It is possible (perhaps more plausible) that the store hired more people as sales increased. The causation may run in the opposite direction of the analyst's argument.

Section 6.4

9. Salaries and logs.

Since $\log_{10} 10,000 = 4$, $\log_{10} 100,000 = 5$, and $\log_{10} 1,000,000 = 6$, the plotted points will be (1, 4), (15, 5), and (30, 6). The plot of these three points will lie very close to a straight line.

Chapter Exercises

11. Association III.

a) Altitude is the explanatory variable, and temperature is the response variable. As you climb higher, the temperature drops. The association is negative, straight, and moderate.

b) At first, it appears that there should be no association between ice cream sales and air conditioner sales. When the lurking variable of temperature is considered, the association becomes more apparent. When the temperature is high, ice cream sales tend to increase. Also, when the temperature is high, air conditioner sales tend to increase. Therefore, there is likely to be an increase in the sales of air conditioners whenever there is an increase in the sales of ice cream. The association is positive, straight, and moderate. Either one of the variables could be used as the explanatory variable.

c) Age is the explanatory variable, and grip strength is the response variable. The association is neither negative nor positive, but is curved, and moderate in strength, due to the variability in grip strength among people in general. The very young would have low grip strength, and grip strength would increase as age increased. After reaching a maximum (when physical prowess peaks), grip strength would decline again, with the elderly having low grip strengths.

d) Blood alcohol content is the explanatory variable, and reaction time is the response variable. As blood alcohol level increases, so does the time it takes to react to a stimulus. The association is positive, probably curved, and strong. The scatterplot would probably be almost linear for low concentrations of alcohol in the blood, and then begin to rise dramatically, with longer and longer reaction times for each incremental increase in blood alcohol content.

13. Scatterplots.

a) None of the scatterplots show little or no association, although # 4 is very weak.

b) #3 and #4 show negative association. Increases in one variable are generally related to decreases in the other variable.

c) #2, #3, and #4 each show a straight association.

d) #2 shows a moderately strong association.

e) #1 and #3 each show a very strong association. #1 shows a curved association and #3 shows a straight association.

15. Performance IQ scores vs. brain size.

The scatterplot of IQ scores *vs.* Brain Sizes is scattered, with no apparent pattern. There appears to be little or no association between the IQ scores and brain sizes displayed in this scatterplot.

17. Firing pottery.

a) A histogram of the number of broken pieces is at the right.

b) The distribution of the number broken pieces per batch of pottery is skewed right, centered around 1 broken piece per batch. Batches had from 0 and 6 broken pieces. The scatterplot does not show the center or skewness of the distribution.

c) The scatterplot shows that the number of broken pieces increases as the batch number increases. If the 8 daily batches are numbered sequentially, this indicates that batches fired later in the day generally have more broken pieces. This information is not visible in the histogram.

Number of Broken Pieces per Batch (24 batches)

19. Matching.

a) 0.006   **b)** 0.777   **c)** – 0.923   **d)** – 0.487

21. Politics.

The candidate might mean that there is an **association** between television watching and crime. The term correlation is reserved for describing linear associations between quantitative variables. We don't know what type of variables "television watching" and "crime" are, but they seem categorical. Even if the variables are quantitative (hours of TV watched per week, and number of crimes committed, for example), we aren't sure that the relationship is linear. The politician also seems to be implying a cause-and-effect relationship between television watching and crime. Association of any kind does not imply causation.

23. Roller coasters 2014.

a) It is appropriate to calculate correlation. Both height of the drop and speed are quantitative variables, the scatterplot shows an association that is straight enough, and there are not outliers.

b) There is a strong, positive, straight association between drop and speed; the greater the height of the initial drop, the higher the top speed.

25. Streams and hard water.

It is not appropriate to summarize the strength of the association between water hardness and pH with a correlation, since the association is curved, not Straight Enough.

27. Cold nights.

The correlation is between the number of days since January 1 and temperature is likely to be near zero. We expect the temperature to be low in January, increase through the spring and summer, then decrease again. The relationship is not Straight Enough, so correlation is not an appropriate measure of strength.

29. Prediction units.

The correlation between prediction error and year would not change, since the correlation is based on z-scores. The z-scores are the same whether the prediction errors are measured in nautical miles or miles.

31. Correlation errors.

a) If the association between GDP and infant mortality is linear, a correlation of -0.772 shows a moderate, negative association. Generally, as GDP increases, infant mortality rate decreases.

b) Continent is a categorical variable. Correlation measures the strength of linear associations between quantitative variables.

33. Height and reading.

a) Actually, this *does* mean that taller children in elementary school are better readers. However, this does *not* mean that height causes good reading ability.

b) Older children are generally both taller and are better readers. Age is the lurking variable.

35. Correlations conclusions I.

a) No. We don't know this from correlation alone. The relationship between age and income may be non-linear, or the relationship may contain outliers.

b) No. We can't tell the form of the relationship between age and income. We need to look at the scatterplot.

c) No. The correlation between age and income doesn't tell us anything about outliers.

d) Yes. Correlation is based on z-scores, and is unaffected by changes in units.

37. Baldness and heart disease.

Even though the variables baldness and heart disease were assigned numerical values, they are categorical. Correlation is only an appropriate measure of the strength of linear association between quantitative variables. Their conclusion is meaningless.

39. Income and housing.

a) There is a positive, moderately strong, linear relationship between *Housing Cost Index* and *Median Family Income*, with several states whose *Housing Cost Index* seems high for their *Median Family Income*, and one state whose *Housing Cost Index* seems low for their *Median Family Income*.

b) Correlation is based on z-scores. The correlation would still be 0.65.

c) Correlation is based on z-scores, and is unaffected by changes in units. The correlation would still be 0.65.

d) Washington, D.C. would be a moderately high outlier, with *Housing Cost Index* high for its *Median Family Income*. Since it doesn't fit the pattern, the correlation would decrease slightly if Washington, D.C. were included.

e) No. We can only say that higher *Housing Cost Index* scores are associated with higher *Median Family Income*, but we don't know why. There may be other variables at work.

f) No. We can say that there is a consistent monotone pattern, but correlation — even nonparametric correlation — does not demonstrate causation.

41. Fuel economy 2014.

a) A scatterplot of combined fuel economy vs. engine displacement is at the right.

Fuel Economy - 2014

b) There is a strong, negative, straight association between engine displacement and mileage of the selected vehicles. There is one high outlier. Cars with larger engines tend to have lower mileage.

c) Since the relationship is linear, with no outliers, correlation is an appropriate measure of strength. The correlation between engine displacement and mileage of the selected vehicles is $r = -0.848$.

d) There is a strong linear relationship in the negative direction between engine displacement and combined gas mileage. Lower fuel efficiency is associated with larger engine displacement.

43. Burgers.

a) There is no apparent association between the number of grams of fat and the number of milligrams of sodium in several brands of fast food burgers. The correlation is only $r = 0.199$, which is close to zero, an indication of no association. One burger had a much lower fat content than the other burgers, at 19 grams of fat, with 920 milligrams of sodium. Without this (comparatively) low fat burger, the correlation would have been $r = -0.325$.

b) Spearman's rho is slightly negative. Using ranks doesn't allow the outlier to have as strong an influence and the remaining points have little or no association.

45. Attendance 2013.

a) Number of runs scored and attendance are quantitative variables, the relationship between them appears to be straight (though weak), and there are no outliers, so calculating a correlation is appropriate.

b) The association between attendance and runs scored is positive, straight, and weak. Generally, as the number of runs scored increases, so does attendance.

c) There is evidence of an association between attendance and runs scored, but a cause-and-effect relationship between the two is not implied. There may be lurking variables that can account for the increases in each. For example, perhaps winning teams score more runs and also have higher attendance. We don't have any basis to make a claim of causation.

47. Thrills.

The scatterplot at the right shows that the association between duration and length is straight, positive, and moderate, with no outliers. Generally, rides on coasters with a greater length tend to last longer. The correlation between length and duration is 0.698, indicating a moderate association.

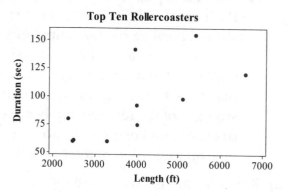

49. Thrills III.

a) We would expect that as one variable (say length of ride) increases, the rank will improve, which means it will decrease.

b) Drop has the strongest correlation ($r = -0.193$), but even that correlation is very weak. The scatterplot shows no apparent association. The number one ranked coaster, Bizarro, has a fairly typical drop. There appear to be other factors that influence the rank of coaster more than any of the ones measured in this data set.

c) There may be other variables that account for the ranking. For example, other quantitative variables such as number of loops and number of corkscrews, categorical variables such as whether the coaster is made of wood or steel and whether or not there are tunnels may all have an affect on the rank.

51. Planets (more or less).

a) The association between Position Number of each planet and its distance from the sun (in millions of miles) is very strong, positive and curved. The scatterplot is at the right.

b) The relationship between Position Number and distance from the sun is not linear. Correlation is a measure of the degree of *linear* association between two variables.

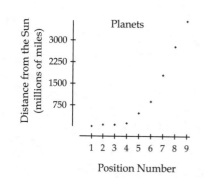

c) The scatterplot of the logarithm of distance versus Position Number (shown at the right) still shows a strong, positive relationship, but it is straighter than the previous scatterplot. It still shows a curve in the scatterplot, but it is straight enough that correlation may now be used as an appropriate measure of the strength of the relationship between logarithm of distance and Position Number, which will in turn give an indication of the strength of the association.

Planets

d) Kendall's tau is 1.0 because the points are perfectly monotonically increasing.

Chapter 7 – Linear Regression

Section 7.1

1. **True or false.**

   a) False. The line usually touches none of the points. The line minimizes the sum of least squares.

   b) True.

   c) False. Least squares means the sum of all the squared residuals is minimized.

Section 7.2

3. **Least square interpretations.**

   The weight of a newborn boy can be predicted as –5.94 kg plus 0.1875 kg per cm of length. This is a model fit to data. Parents should not be concerned if their newborn's length and weight don't fit this equation. No particular baby should be expected to fit this model exactly.

Section 7.3

5. **Bookstore sales revisited.**

   a) The slope of the line of best fit, b_1, is 0.914. (When using technology, the slope is 0.913. When calculated by hand, the standard deviations and correlation must be rounded, resulting in a slight inaccuracy.)

   $$b_1 = r\frac{s_y}{s_x}$$

   $$b_1 = (0.965)\frac{5.34}{5.64}$$

   b) The model predicts an increase in sales of 0.914($1000), or $914, for each additional sales person working.

   $$b_1 = 0.914$$

   c) The intercept, b_0, is 8.09. (When using technology, the intercept is 8.10. When calculated by hand, the means and slope must be rounded, resulting in a slight inaccuracy.)

   $$\hat{y} = b_0 + b_1 x$$
   $$\bar{y} = b_0 + b_1\bar{x}$$
   $$17.6 = b_0 + (0.914)(10.4)$$
   $$b_0 = 8.09$$

   d) The model predicts that average sales would be approximately $8.09($1000), or $809, when there were no sales people working. This doesn't make sense.

   e) $\widehat{Sales} = 8.09 + 0.914\, People$ (hand calculation)

   $\widehat{Sales} = 8.10 + 0.913\, People$ (technology)

   f) $\widehat{Sales} = 8.09 + 0.914\, People = 8.09 + 0.914(18) = 24.542$

   According to the model, we would expect sales to be approximately $24,540 when 18 sales people are working. ($24,530 using technology)

g) Residual $= sales - \widehat{sales} = \$25,000 - \$24,540 = \$460$ (Using technology, \$470)

h) Since the residual is positive, we have underestimated the sales.

Section 7.4

7. Sophomore slump?

The winners may be suffering from regression to the mean. Perhaps they weren't really better than other rookie executives, but just happened to have a lucky year the first year. When their second year performance landed them closer to the mean of the others, it looked like their performance had suffered.

Section 7.5

9. Bookstore sales once more.

a) The residuals are measured in the same units as the response variable, thousands of dollars.

b) The residual with the largest magnitude, 2.77, contributes most to the sum of the squared residuals.

c) The residual with the smalles magnitude, 0.07, contributes least to the sum of the squared residuals.

Section 7.6

11. Bookstore sales last time.

$R^2 = 93.12\%$ Approximately 93% of the variability in bookstore sales can be accounted for by the regression with the number of sales workers.

Section 7.7

13. Residual plots

a) The residual plot has a clear curved pattern. The linearity assumption is violated.

b) One point on the residual plot has a much larger residual than the others. The outlier condition is violated.

c) The residual plot shows a fanned shape. The equal spread condition is violated.

Chapter Exercises

15. Cereals.

$\widehat{Potassium} = 38 + 27\,Fiber = 38 + 27(9) = 281$ mg. According to the model, we expect cereal with 9 grams of fiber to have 281 milligrams of potassium.

17. More cereal.

A negative residual means that the potassium content is actually lower than the model predicts for a cereal with that much fiber.

19. Another bowl.

The model predicts that cereals will have approximately 27 more milligrams of potassium for each additional gram of fiber.

21. Cereal again.

$R^2 = (0.903)^2 \approx 0.815$ About 81.5% of the variability in potassium content is accounted for by the model.

23. Last bowl!

True potassium contents of cereals vary from the predicted values with a standard deviation of 30.77 milligrams.

25. Regression equations.

$\bar{x}$	s_x	$\bar{y}$	s_y	r	$\hat{y} = b_0 + b_1 x$
a) 10	2	20	3	0.5	$\hat{y} = 12.5 + 0.75x$
b) 2	0.06	7.2	1.2	–0.4	$\hat{y} = 23.2 - 8x$
c) 12	6	152	30	–0.8	$\hat{y} = 200 - 4x$
d) 2.5	1.2	25	100	0.6	$\hat{y} = -100 + 50x$

a)
$$b_1 = r\frac{s_y}{s_x}$$
$$b_1 = (0.5)\frac{3}{2}$$
$$b_1 = 0.75$$

$$\hat{y} = b_0 + b_1 x$$
$$\bar{y} = b_0 + b_1\bar{x}$$
$$20 = b_0 + 0.75(10)$$
$$b_0 = 12.5$$

b)
$$b_1 = r\frac{s_y}{s_x}$$
$$b_1 = (-0.4)\frac{1.2}{0.06}$$
$$b_1 = -8$$

$$\hat{y} = b_0 + b_1 x$$
$$\bar{y} = b_0 + b_1\bar{x}$$
$$7.2 = b_0 - 8(2)$$
$$b_0 = 23.2$$

c)
$$\hat{y} = b_0 + b_1 x$$
$$\bar{y} = b_0 + b_1\bar{x}$$
$$\bar{y} = 200 - 4(12)$$
$$\bar{y} = 152$$

$$b_1 = r\frac{s_y}{s_x}$$
$$-4 = (-0.8)\frac{s_y}{6}$$
$$s_y = 30$$

d)
$$\hat{y} = b_0 + b_1 x$$
$$\bar{y} = b_0 + b_1\bar{x}$$
$$\bar{y} = -100 + 50(2.5)$$
$$\bar{y} = 25$$

$$b_1 = r\frac{s_y}{s_x}$$
$$50 = r\frac{100}{1.2}$$
$$r = 0.6$$

27. Residuals.

a) The scattered residuals plot indicates an appropriate linear model.

b) The curved pattern in the residuals plot indicates that the linear model is not appropriate. The relationship is not linear.

c) The fanned pattern indicates that the linear model is not appropriate. The model's predicting power decreases as the values of the explanatory variable increase.

29. Real estate.

a) The explanatory variable (x) is size, measured in square feet, and the response variable (y) is price measured in thousands of dollars.

b) The units of the slope are thousands of dollars per square foot.

c) The slope of the regression line predicting price from size should be positive. Bigger homes are expected to cost more.

31. What slope?

The only slope that makes sense is 300 pounds per foot. 30 pounds per foot is too small. For example, a Honda Civic is about 14 feet long, and a Cadillac DeVille is about 17 feet long. If the slope of the regression line were 30 pounds per foot, the Cadillac would be predicted to outweigh the Civic by only 90 pounds! (The real difference is about 1500 pounds.) Similarly, 3 pounds per foot is too small. A slope of 3000 pounds per foot would predict a weight difference of 9000 pounds (4.5 tons) between Civic and DeVille. The only answer that is even reasonable is 300 pounds per foot, which predicts a difference of 900 pounds. This isn't very close to the actual difference of 1500 pounds, but at least it is in the right ballpark.

33. Real estate again.

71.4% of the variability in price can be accounted for by variability in size. (In other words, 71.4% of the variability in price can be accounted for by the linear model.)

35. Misinterpretations.

a) R^2 is an indication of the strength of the model, not the appropriateness of the model. A scattered residuals plot is the indicator of an appropriate model.

b) Regression models give predictions, not actual values. The student should have said, "The model predicts that a bird 10 inches tall is expected to have a wingspan of 17 inches."

37. Real estate redux.

a) The correlation between size and price is $r = \sqrt{R^2} = \sqrt{0.714} = 0.845$. The positive value of the square root is used, since the relationship is believed to be positive.

b) The price of a home that is one standard deviation above the mean size would be predicted to be 0.845 standard deviations (in other words r standard deviations) above the mean price.

c) The price of a home that is two standard deviations below the mean size would be predicted to be 1.69 (or 2×0.845) standard deviations below the mean price.

39. ESP.

a) First, since no one has ESP, you must have scored 2 standard deviations above the mean by chance. On your next attempt, you are unlikely to duplicate the extraordinary event of scoring 2 standard deviations above the mean. You will likely "regress" towards the mean on your second try, getting a lower score. If you want to impress your friend, don't take the test again. Let your friend think you can read his mind!

b) Your friend doesn't have ESP, either. No one does. Your friend will likely "regress" towards the mean score on his second attempt, meaning his score will probably go up. If the goal is to get a higher score, your friend should try again.

41. More real estate.

a) According to the linear model, the price of a home is expected to increase $61 (0.061 thousand dollars) for each additional square-foot in size.

b) $\widehat{Price} = 47.82 + 0.061\,Size$

$\widehat{Price} = 47.82 + 0.061(3000)$

$\widehat{Price} = 230.82$

According to the linear model, a 3000 square-foot home is expected to have a price of $230,820.

c) $\widehat{Price} = 47.82 + 0.061\,Size$

$\widehat{Price} = 47.82 + 0.061(1200)$

$\widehat{Price} = 121.02$

According to the linear model, a 1200 square-foot home is expected to have a price of $121,020. The asking price is $121,020 - $6000 = $115,020. $6000 is the (negative) residual.

43. Cigarettes.

a) A linear model is probably appropriate. The residuals plot shows some initially low points, but there is no clear curvature.

b) 92.4% of the variability in nicotine level is accounted for by variability in tar content. (In other words, 92.4% of the variability in nicotine level is accounted for by the linear model.)

45. Another cigarette.

a) The correlation between tar and nicotine is $r = \sqrt{R^2} = \sqrt{0.924} = 0.961$. The positive value of the square root is used, since the relationship is believed to be positive. Evidence of the positive relationship is the positive coefficient of tar in the regression output.

b) The average nicotine content of cigarettes that are two standard deviations below the mean in tar content would be expected to be about 1.922 (2×0.961) standard deviations below the mean nicotine content.

c) Cigarettes that are one standard deviation above average in nicotine content are expected to be about 0.961 standard deviations (in other words, r standard deviations) above the mean tar content.

47. Last cigarette.

a) $\widehat{Nicotine} = 0.15403 + 0.065052\,Tar$ is the equation of the regression line that predicts nicotine content from tar content of cigarettes.

b)

$\widehat{Nicotine} = 0.15403 + 0.065052\,Tar$

$\widehat{Nicotine} = 0.15403 + 0.065052(4)$

$\widehat{Nicotine} = 0.414$

The model predicts that a cigarette with 4 mg of tar will have about 0.414 mg of nicotine.

c) For each additional mg of tar, the model predicts an increase of 0.065 mg of nicotine.

d) The model predicts that a cigarette with no tar would have 0.154 mg of nicotine.

e)

$\widehat{Nicotine} = 0.15403 + 0.065052\,Tar$

$\widehat{Nicotine} = 0.15403 + 0.065052(7)$

$\widehat{Nicotine} = 0.6094$

The model predicts that a cigarette with 7 mg of tar will have 0.6094 mg of nicotine. If the residual is –0.5, the cigarette actually had 0.1094 mg of nicotine.

49. Income and housing revisited.

a) Yes. Both housing cost index and median family income are quantitative. The scatterplot is Straight Enough, although there may be a few outliers. The spread increases a bit for states with large median incomes, but we can still fit a regression line.

b) Using the summary statistics given in the problem, calculate the slope and intercept:

$b_1 = r\dfrac{s_{HCI}}{s_{MFI}}$

$b_1 = (0.65)\dfrac{116.55}{7072.47}$

$b_1 = 0.0107$

$\hat{y} = b_0 + b_1 x$

$\bar{y} = b_0 + b_1\bar{x}$

$338.2 = b_0 + 0.0107(46234)$

$b_0 = -156.50$

The regression equation that predicts HCI from MFI is

$\widehat{HCI} = -156.50 + 0.0107\,MFI$

c)

$$\widehat{HCI} = -156.50 + 0.0107\,MFI$$

$$\widehat{HCI} = -156.50 + 0.0107\,(44993)$$

$$\widehat{HCI} = 324.93$$

The model predicts that a state with median family income of $44993 have an average housing cost index of 324.93

d) The prediction is 223.09 too low. Washington has a positive residual.

e) The correlation is the slope of the regression line that relates z-scores, so the regression equation would be $\hat{z}_{HCI} = 0.65 z_{MFI}$.

f) The correlation is the slope of the regression line that relates z-scores, so the regression equation would be $\hat{z}_{MFI} = 0.65 z_{HCI}$.

51. Online clothes.

a) Using the summary statistics given in the problem, calculate the slope and intercept:

$$b_1 = r\frac{s_{Total}}{s_{Age}}$$

$$b_1 = (0.037)\frac{253.62}{8.51}$$

$$b_1 = 1.1027$$

$$\hat{y} = b_0 + b_1 x$$

$$\bar{y} = b_0 + b_1 \bar{x}$$

$$572.52 = b_0 + 1.1027\,(29.67)$$

$$b_0 = 539.803$$

The regression equation that predicts total online clothing purchase amount from age is

$$\widehat{Total} = 539.803 + 1.103\,Age$$

b) Yes. Both total purchases and age are quantitative variables, and the scatterplot is Straight Enough, even though it is quite flat. There are no outliers and the plot does not spread throughout the plot.

c)

$$\widehat{Total} = 539.803 + 1.103\,Age$$

$$\widehat{Total} = 539.803 + 1.103\,(18)$$

$$\widehat{Total} = 559.66$$

The model predicts that an 18-year-old will have $559.66 in total yearly online clothing purchases.

$$\widehat{Total} = 539.803 + 1.103\,Age$$

$$\widehat{Total} = 539.803 + 1.103\,(50)$$

$$\widehat{Total} = 594.95$$

The model predicts that a 50-year-old will have $594.95 in total yearly online clothing purchases.

d) $R^2 = (0.037)^2 \approx 0.0014 = 0.14\%..$

e) This model would not be useful to the company. The scatterplot is nearly flat. The model accounts for almost none of the variability in total yearly purchases.

53. SAT scores.

a) The association between SAT Math scores and SAT Verbal Scores was linear, moderate in strength, and positive. Students with high SAT Math scores typically had high SAT Verbal scores.

b) One student got a 500 Verbal and 800 Math. That set of scores doesn't seem to fit the pattern.

c) $r = 0.685$ indicates a moderate, positive association between SAT Math and SAT Verbal, but only because the scatterplot shows a linear relationship. Students who scored one standard deviation above the mean in SAT Math were expected to score 0.685 standard deviations above the mean in SAT Verbal. Additionally, $R^2 = (0.685)^2 = 0.469225$, so 46.9% of the variability in math score was accounted for by variability in verbal score.

d) The scatterplot of verbal and math scores shows a relationship that is straight enough, so a linear model is appropriate.

$$b_1 = r \frac{s_{Math}}{s_{Verbal}}$$

$$b_1 = (0.685)\frac{96.1}{99.5}$$

$$b_1 = 0.661593$$

$$\hat{y} = b_0 + b_1 x$$

$$\bar{y} = b_0 + b_1 \bar{x}$$

$$612.2 = b_0 + 0.661593(596.3)$$

$$b_0 = 217.692$$

The equation of the least squares regression line for predicting SAT Math score from SAT Verbal score is $\widehat{Math} = 217.692 + 0.662\,Verbal$.

e) For each additional point in verbal score, the model predicts an increase of 0.662 points in math score. A more meaningful interpretation might be scaled up. For each additional 10 points in verbal score, the model predicts an increase of 6.62 points in math score.

f)

$$\widehat{Math} = 217.692 + 0.662\,Verbal$$

$$\widehat{Math} = 217.692 + 0.662(500)$$

$$\widehat{Math} = 548.692$$

According to the model, a student with a verbal score of 500 was expected to have a math score of 548.692.

g)

$$\widehat{Math} = 217.692 + 0.662\,Verbal$$

$$\widehat{Math} = 217.692 + 0.662(800)$$

$$\widehat{Math} = 747.292$$

According to the model, a student with a verbal score of 800 was expected to have a math score of 747.292. She actually scored 800 on math, so her residual was 800 – 747.292 = 52.708 points

55. SAT, take 2.

a) $r = 0.685$. The correlation between SAT Math and SAT Verbal is a unitless measure of the degree of linear association between the two variables. It doesn't depend on the order in which you are making predictions.

b) The scatterplot of verbal and math scores shows a relationship that is straight enough, so a linear model is appropriate.

$$b_1 = r \frac{s_{Verbal}}{s_{Math}}$$

$$b_1 = (0.685)\frac{99.5}{96.1}$$

$$b_1 = 0.709235$$

$$\hat{y} = b_0 + b_1 x$$

$$\bar{y} = b_0 + b_1 \bar{x}$$

$$596.3 = b_0 + 0.709235(612.2)$$

$$b_0 = 162.106$$

The equation of the least squares regression line for predicting SAT Verbal score from SAT Math score is: $\widehat{Verbal} = 162.106 + 0.709\, Math$

c) A positive residual means that the student's actual verbal score was higher than the score the model predicted for someone with the same math score.

d)

$$\widehat{Verbal} = 162.106 + 0.709\, Math$$

$$\widehat{Verbal} = 162.106 + 0.709(500)$$

$$\widehat{Verbal} = 516.606$$

According to the model, a person with a math score of 500 was expected to have a verbal score of 516.606 points.

e)

$$\widehat{Math} = 217.692 + 0.662\, Verbal$$

$$\widehat{Math} = 217.692 + 0.662(516.606)$$

$$\widehat{Math} = 559.685$$

According to the model, a person with a verbal score of 516.606 was expected to have a math score of 559.685 points.

f) The prediction in part e) does not cycle back to 500 points because the regression equation used to predict math from verbal is a different equation than the regression equation used to predict verbal from math. One was generated by minimizing squared residuals in the verbal direction, the other was generated by minimizing squared residuals in the math direction. If a math score is one standard deviation above the mean, its predicted verbal score regresses toward the mean. The same is true for a verbal score used to predict a math score.

57. Wildfires 2012.

a) The scatterplot shows a roughly linear relationship between the year and the number of wildfires, so the linear model is appropriate. The relationship is very weak, however.

b) The model predicts an increase of an average of about 78.5 wildfires per year.

c) It seems reasonable to interpret the intercept. The model predicts about 75,609 wildfires in 1985, which is within the scope of the data, although it isn't very useful since we know the actual number of wildfires in 1985. There isn't much need for a prediction.

d) The standard deviation of the residuals is 11690 fires. That's a large residual, considering that these years show between 60,000 and 90,000 fires per year. The association just isn't very strong.

e) The model only accounts for about 0.032% of the variability in the number of fires each year. The rest of the variability is due to other factors that we don't have data about.

59. Used cars 2014.

a) We are attempting to predict the price in dollars of used Toyota Corollas from their age in years. A scatterplot of the relationship is at the right.

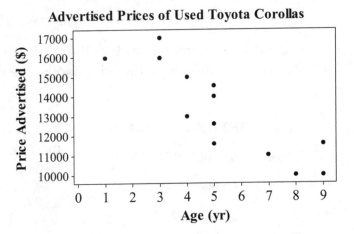

Advertised Prices of Used Toyota Corollas

b) There is a strong, negative, linear association between price and age of used Toyota Corollas.

c) The scatterplot provides evidence that the relationship is straight enough. A linear model will likely be an appropriate model.

d) Since $R^2 = 0.891$, simply take the square root to find r. $\sqrt{0.752} = 0.867$. Since association between age and price is negative, $r = -0.867$.

e) 75.2% of the variability in price of a used Toyota Corolla can be accounted for by variability in the age of the car.

f) The relationship is not perfect. Other factors, such as options, condition, and mileage explain the rest of the variability in price.

61. More used cars 2014.

a) The scatterplot from the previous exercise shows that the relationship is straight, so the linear model is appropriate.

```
Predictor        Coef   SE Coef      T      P
Constant     17674.0     836.2   21.14  0.000
Age           -844.5     146.1   -5.78  0.000


S = 1224.82    R-Sq = 75.2%    R-Sq(adj) = 73.0%
```

The regression equation to predict the price of a used Toyota Corolla from its age is $\widehat{Price} = 17674 - 844.5\,Years$.

Computer regression output used is at the right.

b) According to the model, for each additional year in age, the car is expected to drop $844.5 in price.

c) The model predicts that a new Toyota Corolla (0 years old) will cost $17,674.

d)

$\widehat{Price} = 17674 - 844.5\,Years$

$\widehat{Price} = 17674 - 844.5\,(7)$

$\widehat{Price} = 11762.5$

According to the model, an appropriate price for a 7-year old Toyota Corolla is $11,762.50.

e) Buy the car with the negative residual. Its actual price is lower than predicted.

f) $\widehat{Price} = 17674 - 844.5\,Years$

$\widehat{Price} = 17674 - 844.5\,(10)$

$\widehat{Price} = 9229$

According to the model, a 10-year-old Corolla is expected to cost $9229. The car has an actual price of $8500, so its residual is $8500 − $9229 = − $729

g) The model would not be useful for predicting the price of a 25-year-old Corolla. The oldest car in the list is 9 years old. Predicting a price after 25 years would be an extrapolation.

63. Burgers revisited.

a) The scatterplot of calories vs. fat content in fast food hamburgers is at the right. The relationship appears linear, so a linear model is appropriate.

Fat and Calories of Fast Food Burgers

Dependent variable is: Calories
No Selector
R squared = 92.3%   R squared (adjusted) = 90.7%
s = 27.33 with 7 - 2 = 5 degrees of freedom

Source	Sum of Squares	df	Mean Square	F-ratio
Regression	44664.3	1	44664.3	59.8
Residual	3735.73	5	747.146	

Variable	Coefficient	s.e. of Coeff	t-ratio	prob
Constant	210.954	50.10	4.21	0.0084
Fat	11.0555	1.430	7.73	0.0006

b) From the computer regression output, $R^2 = 92.3\%$. 92.3% of the variability in the number of calories can be explained by the variability in the number of grams of fat in a fast food burger.

c) From the computer regression output, the regression equation that predicts the number of calories in a fast food burger from its fat content is:

$$\widehat{Calories} = 210.954 + 11.0555\,Fat$$

d) The residuals plot at the right shows no pattern. The linear model appears to be appropriate.

e) The model predicts that a fat free burger would have 210.954 calories. Since there are no data values close to 0, this extrapolation isn't of much use.

f) For each additional gram of fat in a burger, the model predicts an increase of 11.056 calories.

g) $\widehat{Calories} = 210.954 + 11.0555\,Fat = 210.954 + 11.0555(28) = 520.508$

The model predicts a burger with 28 grams of fat will have 520.508 calories. If the residual is +33, the actual number of calories is 520.508 + 33 ≈ 553.5 calories.

65. A second helping of burgers.

a) The model from Exercise 63 was for predicting number of calories from number of grams of fat. In order to predict grams of fat from the number of calories, a new linear model needs to be generated.

b) The scatterplot at the right shows the relationship between number fat grams and number of calories in a set of fast food burgers. The association is strong, positive, and linear. Burgers with higher numbers of calories typically have higher fat contents. The relationship is straight enough to apply a linear model.

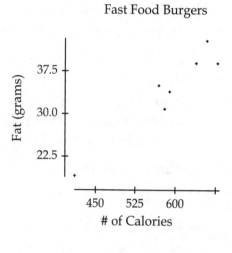

Calories and Fat in Fast Food Burgers

Dependent variable is: **Fat**
No Selector
R squared = 92.3%   R squared (adjusted) = 90.7%
s = 2.375 with 7 - 2 = 5 degrees of freedom

Source	Sum of Squares	df	Mean Square	F-ratio
Regression	337.223	1	337.223	59.8
Residual	28.2054	5	5.64109	

Variable	Coefficient	s.e. of Coeff	t-ratio	prob
Constant	-14.9622	6.433	-2.33	0.0675
Calories	0.083471	0.0108	7.73	0.0006

The linear model for predicting fat from calories is: $\widehat{Fat} = -14.9622 + 0.083471\,Cal$
The model predicts that for every additional 100 calories, the fat content is expected to increase by about 8.3 grams.

The residuals plot shows no pattern, so the model is appropriate. $R^2 = 92.3\%$, so 92.3% of the variability in fat content can be accounted for by the model.

$$\widehat{Fat} = -14.9622 + 0.083471\,Calories$$
$$\widehat{Fat} = -14.9622 + 0.083471(600)$$
$$\widehat{Fat} \approx 35.1$$

According to the model, a burger with 600 calories is expected to have 35.1 grams of fat.

67. New York bridges

a) Overall, the model predicts the condition score of new bridges to be 4.95, a little below the cutoff of 5. Of course, looking at the scatterplot, we can see that the cluster of new bridges all have scores above 5. Still, the model is not a very encouraging one in regards to the conditions of New York City bridges.

b) According to the model, the condition of the bridges in New York City is decreasing by an average of 0.0048 per year.

c) We shouldn't place too much faith in the model. First of all, the scatterplot shows some curvature, so the linear model may not be appropriate. Even if it is appropriate, R^2 of 2.6% is very low, and the standard deviation of the residuals, 0.6708, is quite high in relation to the scope of the data values themselves. This association is very weak.

69. Climate change 2013.

a) The correlation between CO_2 level and mean temperature is
$$r = \sqrt{R^2} = \sqrt{0.736} = 0.858.$$

b) 85.8% of the variability in mean temperature can be accounted for by variability in CO_2 level.

c) Since the scatterplot of CO_2 level and mean temperature shows a relationship that is straight enough, use of the linear model is appropriate. The linear regression model that predicts mean temperature from CO_2 level is:
$$\widehat{MeanTemp} = 10.64 + 0.0103\,CO_2$$

d) The model predicts that an increase in CO_2 level of 1 ppm is associated with an increase of 0.0103 °C in mean temperature.

e) According to the model, the mean temperature is predicted to be 10.64 °C when there is no CO_2 in the atmosphere. This is an extrapolation outside of the range of data, and isn't very meaningful in context, since there is always CO_2 in the atmosphere. We want to use this model to study the change in CO_2 level and how it relates to the change in temperature.

f) The residuals plot shows no apparent patterns. The linear model appears to be an appropriate one.

g) $\widehat{MeanTemp} = 10.64 + 0.0103\,CO_2 = 10.64 + 0.0103\,(400) = 14.76$

According to the model, the temperature is predicted to be 14.76 °C when the CO_2 level is 400 ppm.

h) No. The model describes the association between CO_2 level and temperature. It does not imply that there is a cause and effect relationship between them.

71. Body fat.

a) The scatterplot of % body fat and weight of 20 male subjects, at the right, shows a strong, positive, linear association. Generally, as a subject's weight increases, so does % body fat. The association is straight enough to justify the use of the linear model.

Weight and Body Fat

The linear model that predicts % body fat from weight is: $\widehat{\%Fat} = -27.3763 + 0.249874\,Weight$

b) The residuals plot, at the right, shows no apparent pattern. The linear model is appropriate.

c) According to the model, for each additional pound of weight, body fat is expected to increase by about 0.25%.

d) Only 48.5% of the variability in % body fat can be accounted for by the model. The model is not expected to make predictions that are accurate.

e)

$\widehat{\%Fat} = -27.3763 + 0.249874\,Weight$

$\widehat{\%Fat} = -27.3763 + 0.249874(190)$

$\widehat{\%Fat} = 20.09976$

According to the model, the predicted body fat for a 190-pound man is 20.09976%.
The residual is $21 - 20.09976 \approx 0.9\%$.

73. Heptathlon revisited.

a) Both high jump height and 800 meter time are quantitative variables, the association is straight enough to use linear regression.

Dependent variable is: **High Jump**
No Selector
R squared = 16.4%   R squared (adjusted) = 12.9%
s = 0.0617 with 26 - 2 = 24 degrees of freedom

Source	Sum of Squares	df	Mean Square	F-ratio
Regression	0.017918	1	0.017918	4.71
Residual	0.091328	24	0.003805	

Variable	Coefficient	s.e. of Coeff	t-ratio	prob
Constant	2.68094	0.4225	6.35	< 0.0001
800m	-6.71360e-3	0.0031	-2.17	0.0401

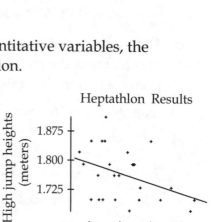

Heptathlon Results

The regression equation to predict high jump from 800m results is:

$$\widehat{Highjump} = 2.681 - 0.00671\,Time.$$

According to the model, the predicted high jump decreases by an average of 0.00671 meters for each additional second in 800 meter time.

b) $R^2 = 16.4\%$. This means that 16.4% of the variability in high jump height is accounted for by the variability in 800 meter time.

c) Yes, good high jumpers tend to be fast runners. The slope of the association is negative. Faster runners tend to jump higher, as well.

d) The residuals plot is fairly patternless. The scatterplot shows a slight tendency for less variation in high jump height among the slower runners than the faster runners. Overall, the linear model is appropriate.

Residuals plot

e) The linear model is not particularly useful for predicting high jump performance. First of all, 16.4% of the variability in high jump height is accounted for by the variability in 800 meter time, leaving 83.6% of the variability accounted for by other variables. Secondly, the residual standard deviation is 0.062 meters, which is not much smaller than the standard deviation of all high jumps, 0.066 meters. Predictions are not likely to be accurate.

75. Hard water.

a) There is a fairly strong, negative, linear relationship between calcium concentration (in ppm) in the water and mortality rate (in deaths per 100,000). Towns with higher calcium concentrations tended to have lower mortality rates.

b) The linear regression model that predicts mortality rate from calcium concentration is $\widehat{Mortality} = 1676 - 3.23\,Calcium$.

c) The model predicts a decrease of 3.23 deaths per 100,000 for each additional ppm of calcium in the water. For towns with no calcium in the water, the model predicts a mortality rate of 1676 deaths per 100,000 people.

d) Exeter had 348.6 fewer deaths per 100,000 people than the model predicts.

e)

$$\widehat{Mortality} = 1676 - 3.23\,Calcium$$
$$\widehat{Mortality} = 1676 - 3.23(100)$$
$$\widehat{Mortality} = 1353$$

The town of Derby is predicted to have a mortality rate of 1353 deaths per 100,000 people.

f) 43% of the variability in mortality rate can be explained by variability in calcium concentration.

77. Least squares.

If the 4 x-values are plugged into $\hat{y} = 7 + 1.1x$, the 4 predicted values are $\hat{y} = 18, 29, 51$ and 62, respectively. The 4 residuals are –8, 21, –31, and 18. The squared residuals are 64, 441, 961, and 324, respectively. The sum of the squared residuals is 1790. Least squares means that no other line has a sum lower than 1790. In other words, it's the best fit.

Chapter 8 – Regression Wisdom

Section 8.1

1. Credit card spending.

The different segments are not scattered at random throughout the residual plot. Each segment may have a different relationship, which would affect the accuracy of any predictions made with the model that ignores the differences between segments.

3. Market segments.

Yes, it is clear that the relationship between January and December spending is not the same for all five segments. Using one overall model to predict January spending would be very misleading.

Section 8.2

5. Cell phone costs.

Your friend is extrapolating. It is impossible to know if a trend like this will continue so far into the future.

7. Revenue and large venues.

a) According to the model, a venue that seats 10,000 would be expected to generate $354,472 in revenue, if it were to sell out.

$$\widehat{Revenue} = -14,228 + 36.87\,TicketSales$$
$$\widehat{Revenue} = -14,228 + 36.87(10,000)$$
$$\widehat{Revenue} = 354,472$$

b) An extrapolation this far from the data is unreliable. We only have data up to about 3000 seats. 10,000 seats is well above that.

Section 8.3

9. Abalone.

This observation was influential. After it was removed, the correlation and the slope of the regression line both changed by a large amount.

Section 8.4

11. Skinned knees.

No. There is a lurking variable, seasonal temperature. In warm weather, more children will go outside and play, and if there are more children playing, there will be more skinned knees.

Section 8.5

13. Grading.

Individual student scores will vary greatly. The class averages will have much less variability and may disguise important patterns.

Chapter Exercises.

15. Marriage age 2011.

a) The trend in age at first marriage for American women is very strong over the entire time period recorded on the graph, but the direction and form are different for different time periods. The trend appears to be somewhat linear, and consistent at around 22 years, up until about 1940, when the age seemed to drop dramatically, to under 21. From 1940 to about 1970, the trend appears non-linear and slightly positive. From 1975 to the present, the trend again appears linear and positive. The marriage age rose rapidly during this time period.

b) The association between age at first marriage for American women and year is strong over the entire time period recorded on the graph, but some time periods have stronger trends than others.

c) The correlation, or the measure of the degree of linear association is not high for this trend. The graph, as a whole, is non-linear. However, certain time periods, like 1975 to present, have a high correlation.

d) Overall, the linear model is not appropriate. The scatterplot is not Straight Enough to satisfy the condition. You could fit a linear model to the time period from 1975 to 1995, but this seems unnecessary. The ages for each year are reported, and, given the fluctuations in the past, extrapolation seems risky.

17. Human Development Index 2012.

a) Fitting a linear model to the association between HDI and GDPPC would be misleading, since the relationship is not straight.

b) If you fit a linear model to these data, the residuals plot will be curved downward.

19. Good model?

a) The student's reasoning is not correct. A scattered residuals plot, not high R^2, is the indicator of an appropriate model. Once the model is deemed appropriate, R^2 is used as a measure of the strength of the model.

b) The model may not allow the student to make accurate predictions. The data may be curved, in which case the linear model would not fit well.

21. Movie dramas.

a) The units for the slopes of these lines are millions of dollars per minutes of running time.

b) The slopes of the regression lines are about the same. Dramas and movies from other genres have costs for longer movies that increase at the same rate.

c) The regression line for dramas has a lower *y*-intercept. Regardless of running time, dramas cost about 20 million dollars less than other genres of movies of the same running time.

23. Oakland passengers 2013.

a) There are several features to comment on in this plot. There is a strong monthly pattern around the general trend. From 1997 to 2008, passengers increased fairly steadily with a notable exception of Sept. 2001, probably due to the attack on the twin towers. Then sometime in late 2008, departures dropped dramatically, possibly due to the economic crisis. Recently, they have been recovering, but not at the same rate as their previous increase.

b) The trend was fairly linear until late 2008, then passengers dropped suddenly.

c) The trend since 2009 has been linear (overall, ignoring monthly oscillations) If the increase continues to be linear, the predictions should be reasonable for the short term.

25. Unusual points.

a) 1) The point has high leverage and a small residual.
   2) The point is not influential. It has the *potential* to be influential, because its position far from the mean of the explanatory variable gives it high leverage. However, the point is not *exerting* much influence, because it reinforces the association.
   3) If the point were removed, the correlation would become weaker. The point heavily reinforces the positive association. Removing it would weaken the association.
   4) The slope would remain roughly the same, since the point is not influential.

b) 1) The point has high leverage and probably has a small residual.
   2) The point is influential. The point alone gives the scatterplot the appearance of an overall negative direction, when the points are actually fairly scattered.
   3) If the point were removed, the correlation would become weaker. Without the point, there would be very little evidence of linear association.
   4) The slope would increase, from a negative slope to a slope near 0. Without the point, the slope of the regression line would be nearly flat.

   c) **1)** The point has moderate leverage and a large residual.
     2) The point is somewhat influential. It is well away from the mean of the explanatory variable, and has enough leverage to change the slope of the regression line, but only slightly.
     3) If the point were removed, the correlation would become stronger. Without the point, the positive association would be reinforced.
     4) The slope would increase slightly, becoming steeper after the removal of the point. The regression line would follow the general cloud of points more closely.

   d) **1)** The point has little leverage and a large residual.
     2) The point is not influential. It is very close to the mean of the explanatory variable, and the regression line is anchored at the point $(\bar{x}, \bar{y})$, and would only pivot if it were possible to minimize the sum of the squared residuals. No amount of pivoting will reduce the residual for the stray point, so the slope would not change.
     3) If the point were removed, the correlation would become slightly stronger, decreasing to become more negative. The point detracts from the overall pattern, and its removal would reinforce the association.
     4) The slope would remain roughly the same. Since the point is not influential, its removal would not affect the slope.

27. The extra point.

   1) Point e is very influential. Its addition will give the appearance of a strong, negative correlation like $r = -0.90$.

   2) Point d is influential (but not as influential as point e). Its addition will give the appearance of a weaker, negative correlation like $r = -0.40$.

   3) Point c is directly below the middle of the group of points. Its position is directly below the mean of the explanatory variable. It has no influence. Its addition will leave the correlation the same, $r = 0.00$.

   4) Point b is almost in the center of the group of points, but not quite. Its addition will give the appearance of a very slight positive correlation like $r = 0.05$.

   5) Point a is very influential. Its addition will give the appearance of a strong, positive correlation like $r = 0.75$.

29. What's the cause?

   1) High blood pressure may cause high body fat.

   2) High body fat may cause high blood pressure.

   3) Both high blood pressure and high body fat may be caused by a lurking variable, such as a genetic or lifestyle trait.

31. Reading.

a) The principal's description of a strong, positive trend is misleading. First of all, "trend" implies a change over time. These data were gathered during one year, at different grade levels. To observe a trend, one class's reading scores would have to be followed through several years. Second, the strong, positive relationship only indicates the yearly improvement that would be expected, as children get older. For example, the 4th graders are reading at approximately a 4th grade level, on average. This means that the school's students are progressing adequately in their reading, not extraordinarily. Finally, the use of average reading scores instead of individual scores increases the strength of the association.

b) The plot appears very straight. The correlation between grade and reading level is very high, probably between 0.9 and 1.0.

c) If the principal had made a scatterplot of all students' scores, the correlation would have likely been lower. Averaging reduced the scatter, since each grade level has only one point instead of many, which inflates the correlation.

d) If a student is reading at grade level, then that student's reading score should equal his or her grade level. The slope of that relationship is 1. That would be "acceptable", according to the measurement scale of reading level. Any slope greater than 1 would indicate above grade level reading scores, which would certainly be acceptable as well. A slope less than 1 would indicate below grade level average scores, which would be unacceptable.

33. Heating.

a) The model predicts a decrease in $2.13 in heating cost for an increase in temperature of 1° Fahrenheit. Generally, warmer months are associated with lower heating costs.

b) When the temperature is 0° Fahrenheit, the model predicts a monthly heating cost of $133.

c) When the temperature is around 32° Fahrenheit, the predictions are generally too high. The residuals are negative, indicating that the actual values are lower than the predicted values.

d)

$\widehat{Cost} = 133 - 2.13(Temp)$

$\widehat{Cost} = 133 - 2.13(10)$

$\widehat{Cost} = \$111.70$

According to the model, the heating cost in a month with average daily temperature 10° Fahrenheit is expected to be $111.70.

e) The residual for a 10° day is approximately –$6, meaning that the actual cost was $6 less than predicted, or $111.70 – $6 = $105.70.

f) The model is not appropriate. The residuals plot shows a definite curved pattern. The association between monthly heating cost and average daily temperature is not linear.

g) A change of scale from Fahrenheit to Celsius would not affect the relationship. Associations between quantitative variables are the same, no matter what the units.

35. Interest rates 2014.

a) $r = \sqrt{R^2} = \sqrt{0.776} = 0.88$. The correlation between rate and year is +0.88, since the scatterplot shows a positive association.

b) According to the model, interest rates during this period increased at about 0.25% per year, starting from an interest rate of about 0.61% in 1950.

c) The linear regression equation predicting interest rate from year is

$\widehat{Rate} = 0.61149 + 0.24788(Year - 1950)$

$\widehat{Rate} = 0.61149 + 0.24788(50)$

$\widehat{Rate} = 13.00549$

According to the model, the interest rate is predicted to be about 13% in the year 2000.

d) This prediction is not likely to have been a good one. Extrapolating 20 years beyond the final year in the data would be risky, and unlikely to be accurate.

37. Interest rates 2014 revisited.

a) The values of R^2 are approximately the same, so the models fit comparably well, but they have very different slopes.

b) The model that predicts the interest rate on 3-month Treasury bills from the number of years since 1950 is $\widehat{Rate} = 18.87316 - 0.30317(Year - 1950)$. This model predicts the interest rate to be 3.71%, a rate much lower than the prediction from the other model.

c) We can trust the newer prediction, since it is in the middle of the data used to generate the model. Additionally, the model accounts for 77% of the variability in interest rate. We would like to see a residuals plot, though.

d) The first model predicts 17.96% and the second model predicts –2.35%.

e) Since 2020 is 7 years after the last year included in the newer model, it would be extremely risky to use this, or any, model to make a prediction that far into the future. The first model predicts historically high, and the second gives a negative prediction. Neither is a legitimate interest rate prediction.

39. Gestation.

a) The association would be stronger if humans were removed. The point on the scatterplot representing human gestation and life expectancy is an outlier from the overall pattern and detracts from the association. Humans also represent an influential point. Removing the humans would cause the slope of the linear regression model to increase, following the pattern of the non-human animals much more closely.

b) The study could be restricted to non-human animals. This appears justifiable, since one could point to a number of environmental factors that could influence human life expectancy and gestation period, making them incomparable to those of animals.

c) The correlation is moderately strong. The model explains 72.2% of the variability in gestation period of non-human animals.

d) For every year increase in life expectancy, the model predicts an increase of approximately 15.5 days in gestation period.

e)

$$\widehat{Gest} = -39.5172 + 15.4980\,LifEx$$

$$\widehat{Gest} = -39.5172 + 15.4980\,(20)$$

$$\widehat{Gest} \approx 270.4428$$

According to the linear model, monkeys with a life expectancy of 20 years are expected to have gestation periods of about 270.5 days. Care should be taken when assessing the accuracy of this prediction. First of all, the residuals plot has not been examined, so the appropriateness of the model is questionable. Second, it is unknown whether or not monkeys were included in the original 17 non-human species studied. Since monkeys and humans are both primates, the monkeys may depart from the overall pattern as well.

41. Elephants and hippos.

a) Hippos are more of a departure from the pattern. Removing that point would make the association appear to be stronger.

b) The slope of the regression line would increase, pivoting away from the hippos point.

c) Anytime data points are removed, there must be a justifiable reason for doing so, and saying, "I removed the point because the correlation was higher without it" is not a justifiable reason.

d) Elephants are an influential point. With the elephants included, the slope of the linear model is 15.4980 days gestation per year of life expectancy. When they are removed, the slope is 11.6 days per year. The decrease is significant.

43. Marriage age 2010, revisited.

a) Modeling decisions may vary, but the important idea is using a subset of the data that allows us to make an accurate prediction for the year in which we are interested. We might model a subset to predict the marriage age in 2015, and model another subset to predict the marriage age in 1911.

In order to predict the average marriage age of American women in 2015, use the data points from the most recent trend only. The data points from 1955 – 2010 look straight enough to apply the linear regression model. Even though there is still curvature, it is nowhere near as bad as the curvature when all the data points are used.

Regression output from a computer program is given below, as well as a residual plot.

The linear model used to predict average female marriage age from year is:

$\widehat{Age} = -210.827 + 0.117832\,Year$. The residuals plot shows a pattern, but the residuals are small, and the value of R^2 is high. 96.1% of the variability in average female age at first marriage is accounted for by variability in the year. The model predicts that each year that passes is associated with an increase of 0.118 years in the average female age at first marriage.

$\widehat{Age} = -210.827 + 0.117832\,Year$

$\widehat{Age} = -210.827 + 0.117832\,(2020)$

$\widehat{Age} = 27.19$

According to the model, the average age at first marriage for women in 2015 will be 27.19 years old. Care should be taken with this prediction, however. It represents an extrapolation of 10 years beyond the highest year, and the residuals plot shows a pattern.

b) This prediction is for a year that is 10 years higher than the highest year for which we have an average female marriage age. Don't place too much faith in this extrapolation.

c) An extrapolation of more than 50 years into the future would be absurd. There is no reason to believe the trend would continue. In fact, given the situation, it is very unlikely that the pattern would continue in this fashion. The model given in part a) predicts that the average marriage age will be 32.5 years in 2065. Realistically, that seems quite high.

45. Life expectancy 2013.

Life Expectancy vs Birth Rate - 2013

a) The scatterplot of birth rate and life expectancy is at the right. The association is moderate, linear, and negative.

b) There is one outlier, Haiti, with a birthrate of 23.35 births per 1000 people, and a life expectancy of 62.4 years.

c) Computer regression output is given below.

```
Predictor      Coef    SE Coef      T      P
Constant     79.504     2.418    32.87   0.000
Birth Rate  -0.2640     0.1390   -1.90   0.071


S = 2.79773    R-Sq = 14.1%    R-Sq(adj) = 10.2%
```

c) The linear regression equation that predicts life expectancy from birth rate is:

$$\widehat{LifeExp} = 79.504 - 0.2640\,Birthrate$$

d) The residuals plot, at the right, is reasonably scattered.

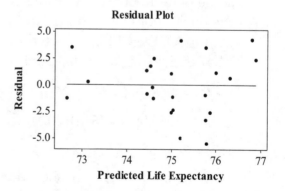

Residual Plot

e) The linear model is appropriate.

f) $R^2 = 14.1\%$, so 14.1% of the variability in life expectancy is explained by variability in the birthrate.

g) The government leaders should not suggest that women have fewer children in order to raise the life expectancy. Although there is evidence of an association between the birth rate and life expectancy, this does not mean that one causes the other. There may be lurking variables involved, such as economic conditions, social factors, or level of health care.

47. Inflation 2011.

a) The trend in Consumer Price Index is strong, non-linear, and positive. Generally, CPI has increased over the years, but the rate of increase has become much greater since approximately 1970. Other characteristics include fluctuations in CPI in the years prior to 1950.

b) In order to effectively predict the CPI in 2016, use only the most recent trend. The trend since 1970 is straight enough to apply the linear model. Prior to 1970, the trend is radically different from that of recent years, and is of no use in predicting CPI for the next decade.

Consumer Price Index

The linear model that predicts CPI from year is $\widehat{CPI} = -8966.98 + 4.56931\,Year$. $R^2 = 99.8\%$, meaning that the model predicts 99.8% of the variability in CPI. The residuals plot shows some pattern, but the residuals are small, so the linear model is appropriate. According to the model, the CPI is expected to increase by $4.57 each year, for 1970 – 2011.

$$\widehat{CPI} = -8966.98 + 4.56931\,Year$$

$$\widehat{CPI} = -8966.98 + 4.56931(2020)$$

$$\widehat{CPI} = 263.03$$

As with any model, care should be taken when extrapolating. If the pattern continues, the model predicts that the CPI in 2020 will be approximately $263.03.

Chapter 9 – Re-expressing Data: Get It Straight!

Section 9.1

1. Residuals.

a) The residuals plot shows no pattern. No re-expression is needed.

b) The residuals plot shows a curved pattern. Re-express to straighten the relationship.

c) The residuals plot shows a fan shape. Re-express to equalize spread.

3. BK protein.

The goal of this re-expression is to improve homoscedasticity. We desire more equal spread between groups.

Section 9.2

5. BK protein again.

The log re-expression is still preferable. The square root doesn't make the spreads as nearly equal. The reciprocal clearly goes too far on the Ladder of Powers.

Chapter Exercises.

7. Oakland passengers 2013 revisited.

a) The residuals cycle up and down because there is a yearly pattern in the number of passengers departing Oakland, California. There is also a sudden decrease in passenger traffic after 2008.

b) A re-expression should not be tried. A cyclic pattern such as this one cannot be helped by re-expression.

9. Models.

a) $\ln \hat{y} = 1.2 + 0.8x$

$\ln \hat{y} = 1.2 + 0.8(2)$

$\ln \hat{y} = 2.8$

$\hat{y} = e^{2.8} = 16.44$

b) $\sqrt{\hat{y}} = 1.2 + 0.8x$

$\sqrt{\hat{y}} = 1.2 + 0.8(2)$

$\sqrt{\hat{y}} = 2.8$

$\hat{y} = 2.8^2 = 7.84$

c) $\dfrac{1}{\hat{y}} = 1.2 + 0.8x$

$\dfrac{1}{\hat{y}} = 1.2 + 0.8(2)$

$\dfrac{1}{\hat{y}} = 2.8$

$\hat{y} = \dfrac{1}{2.8} = 0.36$

d) $\hat{y} = 1.2 + 0.8 \ln x$

$\quad \hat{y} = 1.2 + 0.8 \ln(2)$

$\quad \hat{y} = 1.75$

e) $\log \hat{y} = 1.2 + 0.8 \log x$

$\quad \log \hat{y} = 1.2 + 0.8 \log(2)$

$\quad \log \hat{y} = 1.440823997...$

$\quad\quad\quad \hat{y} = 10^{1.4408...}$

$\quad\quad\quad \hat{y} = 27.59$

11. Gas mileage.

a) The association between weight and gas mileage of cars is fairly linear, strong, and negative. Heavier cars tend to have lower gas mileage.

b) For each additional thousand pounds of weight, the linear model predicts a decrease of 7.652 miles per gallon in gas mileage.

c) The linear model is not appropriate. There is a curved pattern in the residuals plot. The model tends to underestimate gas mileage for cars with relatively low and high gas mileages, and overestimates the gas mileage of cars with average gas mileage.

13. Gas mileage revisited.

a) The residuals plot for the re-expressed relationship is much more scattered. This is an indication of an appropriate model.

b) The linear model that predicts the number of gallons per 100 miles in gas mileage from the weight of a car is: $\widehat{Gal/100} = 0.625 + 1.178(Weight)$.

c) For each additional 1000 pounds of weight, the model predicts that the car will require an additional 1.178 gallons to drive 100 miles.

d)

$\widehat{Gal/100} = 0.625 + 1.178(Weight)$

$\widehat{Gal/100} = 0.625 + 1.178(3.5)$

$\widehat{Gal/100} = 4.748$

According to the model, a car that weighs 3500 pounds (3.5 thousand pounds) is expected to require approximately 4.748 gallons to drive 100 miles, or 0.04748 gallons per mile.

This is $\dfrac{1}{0.04748} \approx 21.06$ miles per gallon.

15. GDP 2013.

a) Although more than 98% of the variation in GDP can be accounted for by the model, the residuals plot should be examined to determine whether or not the model is appropriate.

b) This is not a good model for these data. The residuals plot shows curvature.

17. Better GDP model?

There is still a pattern in the residuals. This much pattern still indicates an inappropriate model.

19. Brakes.

a) The association between speed and stopping distance is strong, positive, and appears straight. Higher speeds are generally associated with greater stopping distances. The linear regression model, with equation

$$\widehat{Distance} = -65.9 + 5.98(Speed),$$ has $R^2 = 96.9\%$, meaning that the model explains 96.9% of the variability in stopping distance. However, the residuals plot has a curved pattern. The linear model is not appropriate. A model using re-expressed variables should be used.

b) Stopping distances appear to be relatively higher for higher speeds. This increase in the rate of change might be able to be straightened by taking the square root of the response variable, stopping distance. The scatterplot of Speed versus $\sqrt{Distance}$ seems like it might be a bit straighter.

c) The model for the re-expressed data is $\sqrt{Distance} = 3.303 + 0.235(Speed)$. The residuals plot shows no pattern, and $R^2 = 98.4\%$, so 98.4% of the variability in the square root of the stopping distance can be explained by the model.

d)

$$\sqrt{Distance} = 3.303 + 0.235(Speed)$$

$$\sqrt{Distance} = 3.303 + 0.235(55)$$

$$\sqrt{Distance} = 16.228$$

$$Distance = 16.228^2 \approx 263.4$$

According to the model, a car traveling 55 mph is expected to require approximately 263.4 feet to come to a stop.

e)

$$\sqrt{Distance} = 3.303 + 0.235(Speed)$$

$$\sqrt{Distance} = 3.303 + 0.235(70)$$

$$\sqrt{Distance} = 19.753$$

$$Distance = 19.753^2 \approx 390.2$$

According to the model, a car traveling 70 mph is expected to require approximately 390.2 feet to come to a stop.

f) The level of confidence in the predictions should be quite high. R^2 is high, and the residuals plot is scattered. The prediction for 70 mph is a bit of an extrapolation, but should still be reasonably close.

21. Baseball salaries 2013.

a) The association between year and highest salary is strong and positive. Salaries were flat for many years, and began to increase in the 1970s, then increased more rapidly as in recent years.

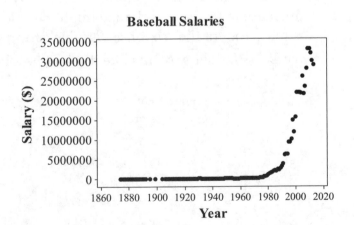

b) The association between year and adjusted salary is still very curved.

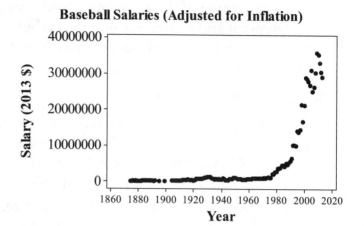

c) Re-expression using the logarithm of the adjusted salaries straightens the plot significantly.

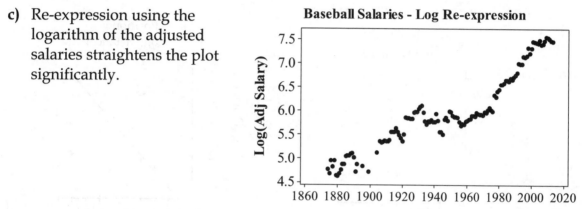

d) $\log(\widehat{AdjSalary}) = -30.022 + 0.0185(Year)$

e) $R^2 = 87.7\%$, so the model explains 87.7% of the variability in the logarithm of the inflation-adjusted salary.

f) The plot of year versus the residuals is at the right.

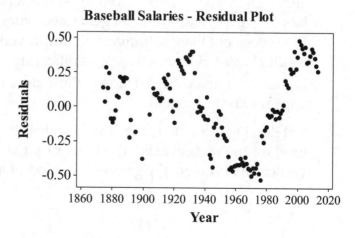

g) The residuals show a cyclic pattern. Salaries were lower than the model predicts at the beginning of the 20th century and again in the 1960s. They may recently hit a high point and started turning lower, but it is difficult to tell. The model based on the logarithmic re-expression may be the best model we can find, but it doesn't explain the pattern in highest baseball salary over time.

23. Planets, distances and order.

a) The association between planetary position and distance from the sun is strong, positive, and curved (below left). A good re-expression of the data is position versus Log(Distance). The scatterplot with regression line (below center) shows the straightened association. The equation of the model is

$\overline{\log(Distance)} = 1.245 + 0.271(Position)$. The residuals plot (below right) may have some pattern, but after trying several re-expressions, this is the best that can be done. $R^2 = 98.2\%$, so the model explains 98.2% of the variability in the log of the planet's distance from the sun.

Planetary Position and Distance

Position vs. Log(Distance)

b) At first glance, this model appears to provide little evidence to support the contention of the astronomers. Pluto appears to fit the pattern, although Pluto's distance from the sun is a bit less than expected. A model generated without Pluto does not have a dramatically improved residuals plot, does not have a significantly higher R^2, nor a different slope. Pluto does not appear to be influential.

Residuals Plot Log Re-expression

But don't forget that a logarithmic scale is being used for the vertical axis. The higher up the vertical axis you go, the greater the effect of a small change.

$$\widehat{\log(Distance)} = 1.24418 + 0.271229(Position)$$

$$\widehat{\log(Distance)} = 1.24418 + 0.271229(9)$$

$$\widehat{\log(Distance)} = 3.685241$$

$$\widehat{Distance} = 10^{3.685241} \approx 4844$$

According to the model, the 9th planet in the solar system is predicted to be approximately 4844 million miles away from the sun. Pluto is actually 3707 million miles away.

Pluto doesn't fit the pattern for position and distance in the solar system. In fact, the model made with Pluto included isn't a good one, because Pluto influences those predictions. The model without Pluto,

$\widehat{\log(Distance)} = 1.20267 + 0.283680(Position)$, works much better. It has a high R^2, and scattered residuals plot. This new model predicts that the 9th planet should be a whopping 5699 million miles away from the sun! There is evidence that the IAU is correct. Pluto doesn't behave like planet in its relation to position and distance.

25. Planets, and Eris.

A planet ninth from the sun was predicted, in a previous exercise, to be about 4844 million miles away from the sun. This distance is much shorter than the actual distance of Eris, about 6300 miles.

27. Logs (not logarithms).

a) The association between the diameter of a log and the number of board feet of lumber is strong, positive, and curved. As the diameter of the log increases, so does the number of board feet of lumber contained in the log.

Doyle Log Scale

The model used to generate the table used by the log buyers is based upon a square root re-expression. The values in the table correspond exactly to the model

$$\widehat{\sqrt{BoardFeet}} = -4 + Diameter.$$

Square root re-expression

b)

$$\widetilde{\sqrt{BoardFeet}} = -4 + Diameter$$

$$\widetilde{\sqrt{BoardFeet}} = -4 + (10)$$

$$\widetilde{\sqrt{BoardFeet}} = 6$$

$$\widetilde{BoardFeet} = 36$$

According to the model, a log 10″ in diameter is expected to contain 36 board feet of lumber.

c)

$$\widetilde{\sqrt{BoardFeet}} = -4 + Diameter$$

$$\widetilde{\sqrt{BoardFeet}} = -4 + (36)$$

$$\widetilde{\sqrt{BoardFeet}} = 32$$

$$\widetilde{BoardFeet} = 1024$$

According to the model, a log 36″ in diameter is expected to contain 1024 board feet of lumber.

Normally, we would be cautious of this prediction, because it is an extrapolation beyond the given data, but since this is a prediction made from an exact model based on the volume of the log, the prediction will be accurate.

29. Life expectancy history.

The association between year and life expectancy is strong, curved and positive. As the years passed, life expectancy for white males has increased.

Life Expectancy by Decade

The linear model,

$$\widehat{LifeExp} = 46.57 + 2.68(Dec),$$

explains 95.0% of the variability in the life expectancy of white males, but has a residuals plot that reveals a strong pattern.

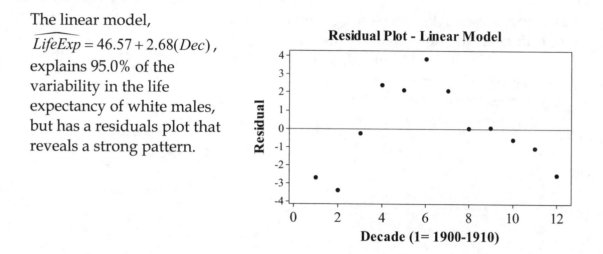

The association between Log(*Year*) and Log(*Life Expectancy*) is strong, positive, and reasonably straight. The model is $\widehat{\log_{10}(LifeExp)} = 1.646 + 0.215(\log_{10}(Year))$.

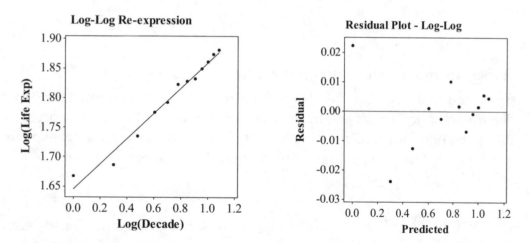

This model explains 97.4% of the variability in the logarithm of the Life Expectancy. The residuals plot shows some pattern, but seems more scattered than the residuals plot for the linear model.

31. Slower is cheaper?

The scatterplot shows the relationship between speed and mileage of the compact car. The association is extremely strong and curved, with mileage generally increasing as speed increases, until around 50 miles per hour, then mileage tends to decrease as speed increases. The linear model is a very poor fit, but the change in direction means that re-expression cannot be used to straighten this association.

33. Years to live, 2013.

a) The association between the age and estimated additional years of life for black males is strong, curved, and negative. Older men generally have fewer estimated years of life remaining.

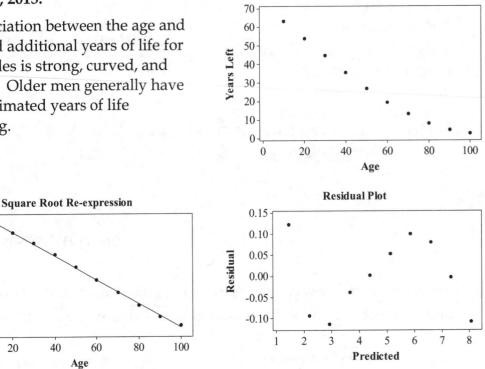

The square root re-expression of the data, $\widehat{\sqrt{YearsLeft}} = 8.79 - 0.0733(Age)$, straightens the data considerably, but has an extremely patterned residuals plot. The model is not a mathematically appropriate model, but fits so closely that it should be fine for predictions within the range of data. The model explains 99.8% of the variability in the estimated number of additional years.

b)

$$\widehat{\sqrt{YearsLeft}} = 8.79 - 0.0733(Age)$$

$$\widehat{\sqrt{YearsLeft}} = 8.79 - 0.0733(18)$$

$$\widehat{\sqrt{YearsLeft}} = 7.4706$$

$$\widehat{YearsLeft} = 7.4706^2 \approx 55.81$$

According to the model, an 18-year-old black male is expected to live and additional 55.81 years, for a total age of 73.81 years.

c) The residuals plot is extremely patterned, so the model is not appropriate. However, the residuals are very small, making for a tight fit. Since 18 years is within the range of the data, the prediction should be at least reasonable.

Review of Part II – Exploring Relationships Between Variables

1. College.

% over 50: $r = 0.69$	The only moderate, positive correlation in the list.
% under 20: $r = -0.71$	Moderate, negative correlation (-0.98 is too strong)
% Full-time Fac.: $r = 0.09$	No correlation.
% Gr. on time: $r = -0.51$	Moderate, negative correlation (not as strong as
%under 20)	%under 20)

3. Vineyards, more information.

a) There does not appear to be an association between ages of vineyards and the price of products. $r = \sqrt{R^2} = \sqrt{0.027} = 0.164$, indicating a very weak association, at best. The model only explains 2.7% of the variability in case price. Furthermore, the regression equation appears to be influenced by two outliers, products from vineyards over 30 years old, with relatively high case prices.

b) This analysis tells us nothing about vineyards worldwide. There is no reason to believe that the results for the Finger Lakes region are representative of the vineyards of the world.

c) The linear equation used to predict case price from age of the vineyard is:
$$\widehat{CasePrice} = 92.765 + 0.567284\,Years$$

d) This model is not useful because only 2.7% of the variability in case price is accounted for by the ages of the vineyards. Furthermore, the slope of the regression line seems influenced by the presence of two outliers, products from vineyards over 30 years old, with relatively high case prices.

5. More twins 2014?

a) The association between year and the twin birth rate is strong, positive, and appears non-linear. Generally, the rate of twin births has increased over the years. The linear model that predicts the rate of twin births from the year is:
$$\widehat{Twins} = 17.77 + 0.551\,(Years\,Since\,1980)$$

b) For each year that passes, the model predicts that the twin birth rate will increase by an average of approximately 0.55 twin births per 1000 live births.

c) $\widehat{Twins} = 17.77 + 0.551(Years\,Since\,1980)$

$\widehat{Twins} = 17.77 + 0.551(34)$

$\widehat{Twins} = 36.504$

According to the model, the twin birth rate is expected to be 36.50 twin births per 1000 live births in the US in 2014. However, the scatterplot appears non-linear, and there is no reason to believe the rate of twin births will keep increasing at the same rate for 4 years beyond the last recorded year. Faith in this prediction is very low.

d) The residuals plot shows a definite curved pattern. The association is not linear, so the linear model is not appropriate.

7. Acid rain.

a) $r = \sqrt{R^2} = \sqrt{0.27} = -0.5196$. The association between pH and BCI appears negative in the scatterplot, so use the negative value of the square root.

b) The association between pH and BCI is negative, moderate, and linear. Generally, higher pH is associated with lower BCI. Additionally, BCI appears more variable for higher values of pH.

c) In a stream with average pH, the BCI would be expected to be average, as well.

d) In a stream where the pH is 3 standard deviations above average, the BCI is expected to be 1.56 standard deviations below the mean level of BCI. $(r\,(3) = -0.5196(3) = -1.56)$

9. A manatee model 2013.

a) The association between the number of powerboat registrations and the number of manatees killed is straight enough to try a linear model.

$\widehat{Kills} = -51.875 + 0.142881\,ThousandBoats$ is the best fitting model. The residuals plot is scattered, so the linear model is appropriate.

b) For every additional 10,000 powerboats registered, the model predicts that an additional 1.429 manatees will be killed on average.

c) The model predicts that if no powerboats were registered, the number of manatee deaths would be approximately -51.875. This is an extrapolation beyond the scope of the data, and doesn't have much contextual meaning.

d)

$$\widehat{Kills} = -51.875 + 0.142881\, ThousandBoats$$

$$\widehat{Kills} = -51.875 + 0.142881(949)$$

$$\widehat{Kills} \approx 83.72$$

The model predicted 83.72 manatee kills in 2009, when the number of powerboat registrations was 949,000. The actual number of kills was 97. The model under-predicted the number of kills by 13.28.

e) Negative residuals are better for the manatees. A negative residual suggests that the actual number of kills was below the number of kills predicted by the model.

f) Over time, the number of powerboat registrations has increased and the number of manatees killed has increased. The trend may continue, resulting in a greater number of manatee deaths in the future. Extrapolation is risky, however. The very trend we are seeing may result in political or societal action to attempt to decrease the number of manatee deaths.

11. Traffic.

a)

$$b_1 = r\frac{s_y}{s_x}$$

$$-0.352 = r\frac{9.68}{27.07}$$

$$r = -0.984$$

The correlation between traffic density and speed is $r = -0.984$.

b) $R^2 = (-0.984)^2 = 0.969$.

The variation in the traffic density accounts for 96.9% of the variation in speed.

c)

$$\widehat{Speed} = 50.55 - 0.352\, Density$$

$$\widehat{Speed} = 50.55 - 0.352(50)$$

$$\widehat{Speed} = 32.95$$

According to the linear model, when traffic density is 50 cars per mile, the average speed of traffic on a moderately large city thoroughfare is expected to be 32.95 miles per hour.

d)

$$\widehat{Speed} = 50.55 - 0.352\, Density$$

$$\widehat{Speed} = 50.55 - 0.352(56)$$

$$\widehat{Speed} = 30.84$$

According to the linear model, when traffic density is 56 cars per mile, the average speed of traffic on a moderately large city thoroughfare is expected to be 30.84 miles per hour. If traffic is actually moving at 32.5 mph, the residual is $32.5 - 30.84 = 1.66$ miles per hour.

e)

$$\widehat{Speed} = 50.55 - 0.352\,Density$$

$$\widehat{Speed} = 50.55 - 0.352\,(125)$$

$$\widehat{Speed} = 6.55$$

According to the linear model, when traffic density is 125 cars per mile, the average speed of traffic on a moderately large city thoroughfare is expected to be 6.55 miles per hour. The point with traffic density 125 cars per minute and average speed 55 miles per hour is considerably higher than the model would predict. If this point were included in the analysis, the slope would increase.

f) The correlation between traffic density and average speed would become weaker. The influential point (125, 55) is a departure from the pattern established by the other data points.

g) The correlation would not change if kilometers were used instead of miles in the calculations. Correlation is a "unitless" measure of the degree of linear association based on z-scores, and is not affected by changes in scale. The correlation would remain the same, $r = -0.984$.

13. Car correlations.

a) Weight, with a correlation of –0.903, seems to be most strongly associated with fuel economy, since the correlation has the largest magnitude (distance from zero). However, without looking at a scatterplot, we can't be sure that the relationship is linear. Correlation might not be an appropriate measure of the strength of the association if the association is non-linear.

b) The negative correlation between weight and fuel economy indicates that, generally, cars with higher weights tend to have lower mileages than cars with lower weights. Once again, this is only correct if the association between weight and fuel economy is linear.

c) $R^2 = (-0.903)^2 = 0.815$. The variation in weight accounts for 81.5% of the variation in mileage. Once again, this is only correct if the association between weight and fuel economy is linear.

15. Cars, one more time!

a) The linear model that predicts the horsepower of an engine from the weight of the car is: $\overline{Horsepower} = 3.49834 + 34.3144\,Weight$.

b) The weight is measured in thousands of pounds. The slope of the model predicts an increase of about 34.3 horsepower for each additional unit of weight. 34.3 horsepower for each additional thousand pounds makes more sense than 34.3 horsepower for each additional pound.

c) Since the residuals plot shows no pattern, the linear model is appropriate for predicting horsepower from weight.

d)

$$\widehat{Horsepower} = 3.49843 + 34.3144\,Weight$$

$$\widehat{Horsepower} = 3.49843 + 34.3144\,(2.595)$$

$$\widehat{Horsepower} \approx 92.544$$

According to the model, a car weighing 2595 pounds is expected to have 92.543 horsepower. The actual horsepower of the car is: 92.544 + 22.5 ≈ 115.0 horsepower.

17. Old Faithful again.

a) The association between the duration of eruption and the interval between eruptions of Old Faithful is fairly strong, linear, and positive. Long eruptions are generally associated with long intervals between eruptions. There are also two distinct clusters of data, one with many short eruptions followed by short intervals, the other with many long eruptions followed by long intervals, with only a few medium eruptions and intervals in between.

b) The linear model used to predict the interval between eruptions is:
$$\widehat{Interval} = 33.9668 + 10.3582\,Duration.$$

c) As the duration of the previous eruption increases by one minute, the model predicts an increase of about 10.4 minutes in the interval between eruptions.

d) $R^2 = 77.0\%$, so the model accounts for 77% of the variability in the interval between eruptions. The predictions should be fairly accurate, but not precise. Also, the association appears linear, but we should look at the residuals plot to be sure that the model is appropriate before placing too much faith in any prediction.

e)

$$\widehat{Interval} = 33.9668 + 10.3582\,Duration$$

$$\widehat{Interval} = 33.9668 + 10.3582\,(4)$$

$$\widehat{Interval} \approx 75.4$$

According to the model, if an eruption lasts 4 minutes, the next eruption is expected to occur in approximately 75.4 minutes.

f) The actual eruption at 79 minutes is 3.6 minutes later than predicted by the model. The residual is 79 – 75.4 = 3.6 minutes. In other words, the model under-predicted the interval.

19. How old is that tree?

a) The correlation between tree diameter and tree age is $r = 0.888$. Although the correlation is moderately high, this does not suggest that the linear model is appropriate. We must look at a scatterplot in order to verify that the relationship is straight enough to try the linear model. After finding the linear model, the residuals plot must be checked. If the residuals plot shows no pattern, the linear model can be deemed appropriate.

b) The association between diameter and age of these trees is fairly strong, somewhat linear, and positive. Trees with larger diameters are generally older.

c) The linear model that predicts age from diameter of trees is:
$\widehat{Age} = -0.974424 + 2.20552\,Diameter$. This model explains 78.9% of the variability in age of the trees.

d) The residuals plot shows a curved pattern, so the linear model is not appropriate. Additionally, there are several trees with large residuals.

e) The largest trees are generally above the regression line, indicating a positive residual. The model is likely to underestimate these values.

21. Big screen.

TV screen sizes might vary from 19 to 70 inches. A TV with a screen that was 10 inches larger would be predicted to cost 10(0.03) = +0.3, 10(0.3) = +3, 10(3) = +30, or 10(30) = +300. Notice that the TV costs are measure in hundreds of dollars, so the potential price changes for getting a TV 10 inches larger are $30, $300, $3000, and $30,000. Only $300 is reasonable, so the slope must be 0.3.

23. No smoking?

a) The model from Exercise 22 is for predicting the percent of expectant mothers who smoked during their pregnancies from the year, not the year from the percent.

b) The model that predicts the year from the percent of expectant mothers who smoked during pregnancy is: $\widehat{Year} = 2027.13 - 1.98233(\%)$. This model predicts that 0% of mothers will smoke during pregnancy in $2027.13 - 1.98233(0) \approx 2027$.

c) The lowest data point corresponds to 9.0% of expectant mothers smoking during pregnancy in 2011. The prediction for 0% is an extrapolation outside the scope of the data. There is no reason to believe that the model will be accurate at that point.

25. U.S. Cities.

There is a strong, roughly linear, negative association between mean January temperature and latitude. U.S. cities with higher latitudes generally have lower mean January temperatures. There are two outliers, cities with higher mean January temperatures than the pattern would suggest.

27. Winter in the city.

a) $R^2 = (-0.848)^2 \approx 0.719$. The variation in latitude explains 71.9% of the variability in average January temperature.

b) The negative correlation indicates that the as latitude increases, the average January temperature generally decreases.

c) $b_1 = r\dfrac{s_y}{s_x}$

$b_1 = (-0.848)\dfrac{13.49}{5.42}$

$b_1 = -2.1106125$

$\hat{y} = b_0 + b_1 x$

$\bar{y} = b_0 + b_1\bar{x}$

$26.44 = b_0 - 2.1106125(39.02)$

$b_0 = 108.79610$

The equation of the linear model for predicting January temperature from latitude is: $\widehat{JanTemp} = 108.796 - 2.111\,Latitude$

d) For each additional degree of latitude, the model predicts a decrease of approximately 2.1°F in average January temperature.

e) The model predicts that the mean January temperature will be approximately 108.8°F when the latitude is 0°. This is an extrapolation, and may not be meaningful.

f)

$\widehat{JanTemp} = 108.796 - 2.111\,Latitude$

$\widehat{JanTemp} = 108.796 - 2.111(40)$

$\widehat{JanTemp} \approx 24.4$

According to the model, the mean January temperature in Denver is expected to be 24.4°F.

g) In this context, a positive residual means that the actual average temperature in the city was higher than the temperature predicted by the model. In other words, the model underestimated the average January temperature.

29. Jumps 2012.

a) The association between Olympic long jump distances and high jump heights is strong, linear, and positive. Years with longer long jumps tended to have higher high jumps. There is one departure from the pattern. The year in which the Olympic gold medal long jump was the longest had a shorter gold medal high jump than we might have predicted.

b) There is an association between long jump and high jump performance, but it is likely that training and technique have improved over time and affected both jump performances.

c) The correlation would be the same, 0.912. Correlation is a measure of the degree of linear association between two quantitative variables and is unaffected by changes in units.

d) In a year when the high jumper jumped one standard deviation better than the average jump, the long jumper would be predicted to jump $r = 0.912$ standard deviations above the average long jump.

31. French.

a) Most of the students would have similar weights. Regardless of their individual French vocabularies, the correlation would be near 0.

b) There are two possibilities. If the school offers French at all grade levels, then the correlation would be positive and strong. Older students, who typically weigh more, would have higher scores on the test, since they would have learned more French vocabulary. If French is not offered, the correlation between weight and test score would be near 0. Regardless of weight, most students would have horrible scores.

c) The correlation would be near 0. Most of the students would have similar weights and vocabulary test scores. Weight would not be a predictor of score.

d) The correlation would be positive and strong. Older students, who typically weigh more, would have higher test scores, since they would have learned more French vocabulary.

33. Lunchtime.

The association between time spent at the table and number of calories consumed by toddlers is moderate, roughly linear, and negative. Generally, toddlers who spent a longer time at the table consumed fewer calories than toddlers who left the table quickly. The scatterplot between time at the table and calories consumed is straight enough to justify the use of the linear model. The linear model that predicts the time number of calories consumed by a toddler from the time spent at the table is $\widehat{Calories} = 560.7 - 3.08\,Time$. For each additional minute spent at the table, the model predicts that the number of calories consumed will be approximately 3.08 fewer. Only 42.1% of the variability in the number of calories consumed can be accounted for by the variability in time spent at the table. The residuals plot shows no pattern, so the linear model is appropriate, if not terribly useful for prediction.

35. Tobacco and alcohol.

The first concern about these data is that they consist of averages for regions in Great Britain, not individual households. Any conclusions reached can only be about the regions, not the individual households living there. The second concern is the data point for Northern Ireland. This point has high leverage, since it has the highest household tobacco spending and the lowest household alcohol spending. With this point included, there appears to be only a weak positive association between tobacco and alcohol spending. Without the point, the association is much stronger. In Great Britain, with the exception of Northern Ireland, higher levels of household spending on tobacco are associated with higher levels of household spending on tobacco. It is not necessary to make the linear model, since we have the household averages for the regions in Great Britain, and the model wouldn't be useful for predicting in other countries or for individual households in Great Britain.

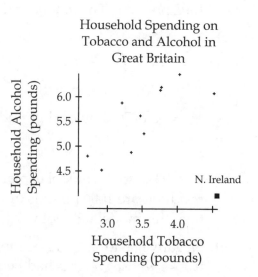

Household Spending on Tobacco and Alcohol in Great Britain

37. Models.

a) $\hat{y} = 2 + 0.8 \ln x$

$\hat{y} = 2 + 0.8 \ln(10)$

$\hat{y} \approx 3.842$

b) $\log \hat{y} = 5 - 0.23x$

$\log \hat{y} = 5 - 0.23(10)$

$\log \hat{y} = 2.7$

$\hat{y} = 10^{2.7} \approx 501.187$

c) $\dfrac{1}{\sqrt{\hat{y}}} = 17.1 - 1.66x$

$\dfrac{1}{\sqrt{\hat{y}}} = 17.1 - 1.66(10) = 0.5$

$\hat{y} = \dfrac{1}{0.5^2} = 4$

39. Vehicle weights.

a)

$\widehat{Wt} = 10.85 + 0.64\,scale$

$\widehat{Wt} = 10.85 + 0.64(31.2)$

$\widehat{Wt} = 30.818$

According to the model, a truck with a scale weight of 31,200 pounds is expected to weigh 30,818 pounds.

b) If the actual weight of the truck is 32,120 pounds, the residual is 32,120 – 30,818 = 1302 pounds. The model underestimated the weight.

c)

$\widehat{Wt} = 10.85 + 0.64\,scale$

$\widehat{Wt} = 10.85 + 0.64(35.590)$

$\widehat{Wt} = 33.6276$ thousand pounds

The predicted weight of the truck is 33,627.6 pounds. If the residual is –2440 pounds, the actual weight of the truck is 33,627.6 – 2440 = 31,187.6 pounds.

d) $R^2 = 93\%$, so the model explains 93% of the variability in weight, but some of the residuals are 1000 pounds or more. If we need to be more accurate than that, then this model will not work well.

e) Negative residuals will be more of a problem. Police would be issuing tickets to trucks whose weights had been overestimated by the model. The U.S. justice system is based upon the principle of innocence until guilt is proven. These truckers would be unfairly ticketed, and that is worse than allowing overweight trucks to pass.

41. Down the drain.

The association between diameter of the drain plug and drain time of this water tank is strong, curved, and negative. Tanks with larger drain plugs have lower drain times. The linear model is not appropriate for the curved association, so several re-expressions of the data were tried. The best one was the reciprocal square root re-expression, resulting in the equation

$$\frac{1}{\sqrt{DrainTime}} = 0.00243 + 0.219\,Diameter.$$

The re-expressed data is nearly linear, and although the residuals plot might still indicate some pattern and has one large residual, this is the best of the models examined. The model explains 99.7% of the variability in drain time.

Chapter 10 – Understanding Randomness

Section 10.1

1. Random outcomes.

  a) Yes, who takes out the trash cannot be predicted before the flip of a coin.

  b) No, it is not random, since you will probably name a favorite team.

  c) Yes, your new roommate cannot be predicted before names are drawn.

Section 10.2

3. Components.

  Rolling the pair of dice is the component.

5. Response variable.

  To simulate, you could roll dice and note whether or not "doubles" came up. A trial would be completed once "doubles" came up. You would count up the number of rolls until "doubles" for the response variable. Alternatively, you could use the digits 1, 2, 3, 4, 5, 6 on a random digits table and disregard digits 7, 8, 9, and 0. Using the table, note the first legal digit for the first die, and then note the next legal digit for the second die. A double would indicate rolling doubles.

Chapter Exercises.

7. The lottery.

  In state lotteries, a machine pops up numbered balls. If the lottery were truly random, the outcome could not be predicted and the outcomes would be equally likely. It is random only if the balls generate numbers in equal frequencies.

9. Birth defects.

  Answers may vary. Generate two-digit random numbers, 00-99. Let 00-02 represent a defect. Let 03-99 represent no defect.

11. Geography.

  a) Looking at pairs of digits, the first state number is 45, Vermont. The next set is ignored since there is no 92nd state. The next state number is 10, Georgia.

  b) Continuing along, the next state number is 17, Kentucky. The next state number, 10, is ignored, since Georgia was already assigned. The final state number is 22, Michigan.

13. Play the lottery.

If the lottery is random, it doesn't matter if you play the same favorite "lucky" numbers or if you play different numbers each time. All numbers are equally likely (or, rather, UNLIKELY) to win.

15. Bad simulations.

a) The outcomes are not equally likely. For example, the probability of getting 5 heads in 9 tosses is not the same as the probability of getting 0 heads, but the simulation assumes they are equally likely.

b) The even-odd assignment assumes that the player is equally likely to score or miss the shot. In reality, the likelihood of making the shot depends on the player's skill.

c) Suppose a hand has four aces. This might be represented by 1,1,1,1, and any other number. The likelihood of the first ace in the hand is not the same as for the second or third or fourth. But with this simulation, the likelihood is the same for each.

17. Wrong conclusion.

The conclusion should indicate that the simulation **suggests** that the average length of the line would be 3.2 people. Future results might not match the simulated results exactly.

19. Election.

a) Answers will vary. A component is one voter voting. An outcome is a vote for our candidate. Using two random digits, 00-99, let 01-55 represent a vote for your candidate, and let 55-99 and 00 represent a vote for the underdog.

b) A trial is 100 votes. Examine 100 two-digit random numbers and count how many simulated votes are cast for each candidate. Whoever gets the majority of the votes wins the trial.

c) The response variable is whether the underdog wins or not. To calculate the experimental probability, divide the number of trials in which the simulated underdog wins by the total number of trials.

21. Cereal.

Answers will vary. A component is the simulation of the picture in one box of cereal. One possible way to model this component is to generate random digits 0-9. Let 0 and 1 represent Hope Solo, 2-4 represent Danica Patrick, and 5-9 represent Blake Griffin. Each trial will consist of 5 random digits, and the response variable will be whether or not a complete set of pictures is simulated. Trials in which at least one of each picture is simulated will be a success. The total number of successes divided by the total number of trials will be the simulated probability of ending up with a complete set of pictures. According to the simulation, the probability of getting a complete set of pictures is expected to be about 51.5%.

23. Multiple choice

Answers will vary. A component is one multiple-choice question. One possible way to model this component is to generate random digits 0-9. Let digits 0-7 represent a correct answer, and let digits 8 and 9 represent an incorrect answer. Each trial will consist of 6 random digits. The response variable is whether or not all 6 simulated questions are answered correctly (all 6 digits are 0-7). The total number of successes divided by the total number of trials will be the simulated probability of getting all 6 questions right. According to the simulation, the probability of getting all 6 multiple-choice questions correct is expected to be about 26%.

25. Beat the lottery.

a) Answers based on your simulation will vary, but you should win about 10% of the time.

b) You should win at the same rate with any number.

27. It evens out in the end.

Answers based on your simulation will vary, but you should win about 10% of the time. Playing lottery numbers that have turned up the least in recent lottery drawers offers no advantage. Each new drawing is independent of recent drawings.

29. Driving test.

Answers will vary. A component is one drivers test, but this component will be modeled differently, depending on whether or not it is the first test taken. One possible way to model this component is to generate pairs of random digits 00-99. Let 01-34 represent passing the first test and let 35-99 and 00 represent failing the first test. Let 01-72 represent passing a retest, and let 73-99 and 00 represent failing a retest. To simulate one trial, generate pairs of random numbers until a pair is generated that represents passing a test. Begin each trial using the "first test" representation, and switch to the "retest" representation if failure is indicated on the first simulated test. The response variable is the number of simulated tests required to achieve the first passing test. The total number of simulated tests taken divided by the total number of trials is the simulated average number of tests required to pass. According to the simulation, the number of driving tests required to pass is expected to be about 1.9.

31. Basketball strategy.

Answers will vary. A component is one foul shot. One way to model this component would be to generate pairs of random digits 00-99. Let 01-72 represent a made shot, and let 73-99 and 00 represent a missed shot. The response variable is the number of shots made in a "one and one" situation. If the first shot simulated represents a made shot, simulate a second shot. If the first shot simulated represents a miss, the trial is over. The simulated average number of points is the total number of simulated points divided by the number of trials. According to the simulation, the player is expected to score about 1.24 points.

33. Free groceries.

Answers will vary. A component is the selection of one card with the prize indicated. One possible way to model the prize is to generate pairs of random digits 00-99. Let 01-10 represent $200, let 11-20 represent $100, let 21-40 represent $50, and let 41-99 and 00 represent $20. Repeated pairs of digits must be ignored. (For this reason, a simulation in which random digits 0-9 are generated with 0 representing $200, 1 representing $100, etc., is NOT acceptable. Each card must be individually represented.) A trial continues until the total simulated prize is greater than $500. The response variable is the number of simulated customers until the payoff is greater than $500. The simulated average number of customers is the total number of simulated customers divided by the number of trials. According to the simulation, about 10.2 winners are expected each week.

35. The family.

Answers will vary. Each child is a component. One way to model the component is to generate random digits 0-9. Let 0-4 represent a boy and let 5-9 represent a girl. A trial consists of generating random digits until a child of each gender is simulated. The response variable is the number of children simulated until this happens. The simulated average family size is the number of digits generated in each trial divided by the total number of trials. According to the simulation, the expected number of children in the family is about 3.

37. Dice game.

Answers will very. Each roll of the die is a component. One way of modeling this component is to generate random digits 0-9. The digits 1-6 correspond to the numbers on the faces of the die, and digits 7-9 and 0 are ignored. A trial consists of generating random numbers until the sum of the numbers is exactly 10. If the sum exceeds 10, the last roll must be ignored and simulated again, but still counted as a roll. The response variable is the number of rolls until the sum is exactly 10. The simulated average number of rolls until this happens is the total number of rolls simulated divided by the number of trials. According to the simulation, expect to roll the die about 7.5 times.

39. The hot hand.

Answers may vary. Each shot is a component. One way of modeling this component is to generate pairs of random digits 00-99. Let 01-65 represent a made shot, and let 66-99 and 00 represent a missed shot. A trial consists of 20 simulated shots. The response variable is whether or not the 20 simulated shots contained a run of 6 or more made shots. To find the simulated percentage of games in which the player is expected to have a run of 6 or more made shots, divide the total number of successes by the total number of trials. According to the simulation, the player is expected to make 6 or more shots in a row in about 40% of games. This isn't unusual. The announcer was wrong to characterize her performance as extraordinary.

41. Teammates.

Answers will vary. Each player chosen is a component. One way to model this component is to generate random numbers 0-9. Let 1 and 2 represent the first couple, 3 and 4 the second couple, 5 and 6 the third couple, and 7 and 8 the fourth couple. Ignore 9 and 0. A trial consists of generating random digits, ignoring repeats, and organizing them into pairs, until pairs representing the first three teams are generated. (The final team is assigned by default.) The response variable is whether or not each of the simulated teams is a pairing other than 1-2, 3-4, 5-6, or 7-8. The simulated percentage of the time this is expected to happen is the total number of successes (times that the pairings are *different* than the couples) divided by the total number of trials. According to the simulation, all players are expected to be paired with someone other than the person with whom he or she came to the party about 37.5% of the time.

43. Job discrimination?

Answers may vary. Each person hired is a component. One way of modeling this component is to generate pairs of random digits 00-99. Let 01-10 represent each of the 10 women, and let 11-22 represent each of the 12 men. Ignore 23-99 and 00. A trial consists of 3 usable pairs of numbers. Ignore repeated pairs of digits, since the same man or woman cannot be hired more than once. The response variable is whether or not all 3 simulated hires are women. The simulated percentage of the time that 3 women are expected to be hired is the number of successes divided by the number of trials. According to the simulation, the 3 people hired will all be women about 7.8% of time. This seems a bit strange, but not quite strange enough to be evidence of job discrimination.

Chapter 11 – Sample Surveys

Section 11.1

1. Texas A&M.

The A&M administrators should take a survey. They should sample a part of the student body, selecting respondents with a randomization method. They should be sure to draw a sufficiently large sample.

Section 11.2

3. A&M again.

The proportion in the sample is a statistic. The proportion of all students is the parameter of interest. The statistic estimates that parameter, but is not likely to be exactly the same.

Section 11.3

5. Sampling students.

This is not an SRS. Although each student may have an equal chance to be in the survey, groups of friends who choose to sit together will either all be in or out of the sample, so the selection is not independent.

Section 11.4

7. Sampling A&M students.

a) This is a cluster sample, with each selected dormitory as a cluster.

b) This is a stratified sample, stratified by class year.

c) This systematic sample, with a randomized starting point.

Section 11.6

9. Survey students.

Several terms are poorly defined. The survey needs to specify the meaning of "family" for this purpose and the meaning of "higher education." The term "seek" may also be poorly defined (for example, would applying to college but not being admitted qualify for seeking more education?)

Section 11.7

11. Student samples.

a) This would suffer from voluntary response bias. Only those students who saw the advertisement could even be part of the sample, and only those who choose to (and are able to) go to the website would actually be in the sample.

b) This would be a convenience sample, as well as suffer from voluntary response bias.

Chapter Exercises.

13. Roper.

  a) Roper is not using a simple random sample. The samples are designed to get 500 males and 500 females. This would be very unlikely to happen in a simple random sample.

  b) They are using stratified sample, with two strata, males and females.

15. Drug tests.

  a) This is a cluster sample, with teams being the clusters.

  b) Cluster sampling is a reasonable solution to the problem of randomly sampling players because an entire team can be sampled easily. It would be much more difficult to randomly sample players from many different teams on the same day.

17. Medical treatments.

  a) **Population** – Unclear, but possibly all U.S. adults.

  b) **Parameter** – Proportion who have used and have benefitted from alternative medical treatments.

  c) **Sampling Frame** – All Consumers Union subscribers.

  d) **Sample** – Those subscribers who responded.

  e) **Method** – Not specified, but probably a questionnaire mailed to all subscribers.

  f) **Left Out** – Those who are not Consumers Union subscribers.

  g) **Bias** – Voluntary response bias. Those who respond may have strong feelings one way or another.

19. Mayoral race.

  a) **Population** – City voters.

  b) **Parameter** – Not clear. They might be interested in the percentage of voters favoring various issues.

  c) **Sampling Frame** – All city residents

  d) **Sample** – As many residents as they can find in one block from each district. No randomization is specified, but hopefully a block is selected at random within each district.

  e) **Method** – Multistage sampling, stratified by district and clustered by block.

  f) **Left Out** – People not home during the time of the survey.

g) **Bias** – Convenience sampling. Once the block is randomly chosen as the cluster, every resident living in that block should be surveyed, not just those that were conveniently available. A random sample of each block could be also be taken, but we wouldn't refer to that as "cluster" sampling, but rather multi-stage, with stratification by district, a simple random sample of one block within each district, and another simple random sample of residents within the block.

21. **Roadblock.**

   a) **Population** – All cars.

   b) **Parameter** – Proportion of cars with up-to-date (or out-of-date) registrations, insurance, or safety inspections.

   c) **Sampling Frame** – Cars on that road.

   d) **Sample** – Cars stopped by the roadblock.

   e) **Method** – Cluster sample of an area, stopping all cars within the cluster.

   f) **Left Out** – Drivers that did not take that road, or traveled that road at a different time.

   g) **Bias** – Undercoverage. The cars stopped might not be representative of all cars because of time of day and location. The locations are probably not chosen randomly, so might represent areas in which it is easy to set up a roadblock, resulting in a convenience sample.

23. **Milk samples.**

   a) **Population** – Dairy farms.

   b) **Parameter** – Whether or not the milk contains dirt, antibiotics, or other foreign matter.

   c) **Sampling Frame** – All dairy farms

   d) **Sample** – Not specified, but probably a random sample of farms.

   e) **Method** – not specified

   f) **Left Out** – Nothing.

   g) **Bias** – Unbiased, as long as the day of inspection is randomly chosen. This might not be the case, however, since the farms might be spread out over a wide geographic area. Inspectors might tend to visit farms that are near one another on the same day, resulting in a convenience sample.

25. Another mistaken poll.

The newspaper's faulty prediction was more likely to be due to sampling error. The description of the sampling method suggests that samples should be representative of the voting population. Random chance in selecting the individuals who were polled means that sample statistics will vary from the population parameter, perhaps by quite a bit.

27. Parent opinion, part 2.

a) This sampling method suffers from voluntary response bias. Only those who see the show and feel strongly will call.

b) Although this method may result in a more representative sample than the method in part a), this is still a voluntary response sample. Only strongly motivated parents attend PTA meetings.

c) This is multistage sampling, stratified by elementary school and then clustered by grade. This is a good design, as long as the parents in the class respond. There should be follow-up to get the opinions of parents who do not respond.

d) This is systematic sampling. As long as a starting point is randomized, this method should produce reliable data.

29. Playground.

The managers will only get responses from people who come to the park to use the playground. Parents who are dissatisfied with the playground may not come.

31. Playground, act two.

The first sentence points our problems that the respondent may not have noticed, and might lead them to feel they should agree. The last phrase mentions higher fees, which could make people reject improvements to the playground.

33. Banning ephedra.

a) This is a voluntary response survey. The large sample will still be affected by any biases in the group of people that choose to respond.

b) The wording seems fair enough. It states the facts, and gives voice to both sides of the issue.

c) The sampling frame is, at best, those who visit this particular site, and even then depends of their volunteering to respond to the question.

d) This statement is true.

35. More survey questions.

a) The question seems unbiased.

b) The question is biased toward "yes" because of the phrase "great tradition". A better question: "Do you favor continued funding for the space program?"

37. Cell phone survey.

Cell phones are more likely to be used by younger individuals. This will result in an undercoverage bias. As cell phone use grows, this will be less of a problem. Also, many cell phone plans require the users to pay airtime for incoming calls. That seems like a sure way to irritate the respondent, and result in response bias toward negative responses.

39. Fuel economy.

a) The statistic calculated is the mean mileage for the last six fill-ups.

b) The parameter of interest is the mean mileage for the vehicle.

c) The driving conditions for the last six fill-ups might not be typical of the overall driving conditions. For instance, the last six fill-ups might all be in winter, when mileage might be lower than expected.

d) The EPA is trying to estimate the mean gas mileage for all cars of this make, model, and year.

41. Happy workers?

a) A small sample will probably consist mostly laborers, with few supervisors, and maybe no project managers. Also, there is a potential for response bias based on the interviewer if a member of management asks directly about discontent. Workers who want to keep their jobs will likely tell the management that everything is fine!

b) Assign a number from 001 to 439 to each employee. Use a random number table or software to select the sample.

c) The simple random sample might not give a good cross section of the different types of employees. There are relatively few supervisors and project managers, and we want to make sure their opinions are noted, as well as the opinions of the laborers.

d) A better strategy would be to stratify the sample by job type. Sample a certain percentage of each job type.

e) Answers will vary. Assign each person a number from 01-14, and generate 2 usable random numbers from a random number table or software.

43. A fish story.

> What conclusions they may be able to make will depend on whether fish with discolored scales are equally likely to be caught as those without. It also depends on the level of compliance by fisherman. If fish are not equally likely to be caught, or fishermen more disposed to bring discolored fish, the results will be biased.

45. More sampling methods.

a) A petition may pressure people into support. Additionally, some people may not be home on a Saturday, especially those who have taken their kids out to play in a distant park! We are undercovering a group made up of people who probably have a specific opinion.

b) If the food at the largest cafeteria is representative, this should be OK. However, those who really don't like the food won't be eating there. That group is undercovered.

Chapter 12 – Experiments and Observational Studies

Section 12.1

1. Steroids.

This is an observational study because the sports writer is not randomly assigning players to take steroids or not take steroids; the writer is merely observing home run totals between two eras. It would be unwise to conclude steroids caused any increases in home runs because we need to consider other factors besides steroids—factors possibly leading to more home runs include better equipment, players training more in the offseason, smaller ballparks, better scouting techniques, etc.

Section 12.2

3. Tips.

Each of the 40 deliveries is an experimental unit. He has randomized the experiment by flipping a coin to decide whether or not to phone.

5. Tips II.

The factor is calling, and the levels are whether or not he calls the customer. The response variable is the tip percentage for each delivery.

Section 12.3

7. Tips again.

By calling some customers but not others during the same run, the driver has controlled many variables, such as day of the week, season, and weather. The experiment was randomized because he flipped a coin to determine whether or not to phone and it was replicated because he did this for 40 deliveries.

Section 12.4

9. More tips.

Because customers don't know about the experiment, those that are called don't know that others are not, and vice versa. Thus, the customers are blind. That would make this a single-blind study. It can't be double-blind because the delivery driver must know whether or not he phones.

Section 12.5

11. Block that tip.

Yes. Driver is now a block. The experiment is randomized within each block. This is a good idea because some drivers might generally get higher tips than others, but the goal of the experiment is to study the effect of phone calls. Blocking on driver eliminates the variability in tips inherent to the driver.

Section 12.6

13. Confounded tips.

Answers may vary. The cost or size of a delivery may confound his results. Larger orders may generally tip a higher or lower percentage of the bill.

Chapter Exercises

15. Standardized test scores.

a) No, this is not an experiment. There are no imposed treatments. This is a retrospective observational study.

b) We cannot conclude that the differences in score are caused by differences in parental income. There may be lurking variables that are associated with both SAT score and parental income.

17. MS and vitamin D.

a) This is a retrospective observational study.

b) This is an appropriate choice, since MS is a relatively rare disease.

c) The subjects were U.S. military personnel, some of whom had developed MS.

d) The variables were the vitamin D blood levels and whether or not the subject developed MS.

19. Menopause.

a) This was a randomized, comparative, placebo-controlled experiment.

b) Yes, such an experiment is the right way to determine whether black cohosh is an effective treatment for hot flashes.

c) The subjects were 351 women, aged 45 to 55 who reported at least two hot flashes a day.

d) The treatments were black cohosh, a multi-herb supplement, plus advice to consume more soy foods, estrogen, and a placebo. The response was the women's self-reported symptoms, presumably the frequency of hot flashes.

21. a) This is an experiment, since treatments were imposed.

b) The subjects studied were 30 patients with bipolar disorder.

c) The experiment has 1 factor (omega-3 fats from fish oil), at 2 levels (high dose of omega-3 fats from fish oil and no omega-3 fats from fish oil).

d) 1 factor, at 2 levels gives a total of 2 treatments.

e) The response variable is "improvement", but there is no indication of how the response variable was measured.

f) There is no information about the design of the experiment.

g) The experiment is blinded, since the use of a placebo keeps the patients from knowing whether or not they received the omega-3 fats from fish oils. It is not stated whether or not the evaluators of the "improvement" were blind to the treatment, which would make the experiment double-blind.

h) Although it needs to be replicated, the experiment can determine whether or not omega-3 fats from fish oils cause improvements in patients with bipolar disorder, at least over the short term. The experiment design would be stronger is it were double-blind.

23. a) This is an observational study. The researchers are simply studying traits that already exist in the subjects, not imposing new treatments.

b) This is a prospective study. The subjects were identified first, then traits were observed.

c) The subjects are roughly 200 men and women with moderately high blood pressure and normal blood pressure. There is no information about the selection method.

d) The parameters of interest are difference in memory and reaction time scores between those with normal blood pressure and moderately high blood pressure.

e) An observational study has no random assignment, so there is no way to know that high blood pressure caused subjects to do worse on memory and reaction time tests. A lurking variable, such as age or overall health, might have been the cause. The most we can say is that there was an association between blood pressure and scores on memory and reaction time tests in this group, and recommend a controlled experiment to attempt to determine whether or not there is a cause-and-effect relationship.

25. a) This is an experiment, since treatments were imposed on randomly assigned groups.

b) 24 post-menopausal women were the subjects in this experiment.

c) There is 1 factor (type of drink), at 2 levels (alcoholic and non-alcoholic). (Supplemental estrogen is not a factor in the experiment, but rather a blocking variable. The subjects were not given estrogen supplements as part of the experiment.)

d) 1 factor, with 2 levels, is 2 treatments.

e) The response variable is an increase in estrogen level.

f) This experiment utilizes a blocked design. The subjects were blocked by whether or not they used supplemental estrogen. This design reduces variability in the response variable of estrogen level that may be associated with the use of supplemental estrogen.

g) This experiment does not use blinding.

h) This experiment indicates that drinking alcohol leads to increased estrogen level among those taking estrogen supplements.

27. a) This is an observational study.
   b) The study is retrospective. Results were obtained from pre-existing church records.
   c) The subjects of the study are women in Finland. The data were collected from church records dating 1640 to 1870, but the selection process is unknown.
   d) The parameter of interest is difference in average lifespan between mothers of sons and daughters.
   e) For this group, having sons was associated with a decrease in lifespan of an average of 34 weeks per son, while having daughters was associated with an unspecified increase in lifespan. As there is no random assignment, there is no way to know that having sons caused a decrease in lifespan.

29. a) This is an observational study. (Although some might say that the sad movie was "imposed" on the subjects, this was merely a stimulus used to trigger a reaction, not a treatment designed to attempt to influence some response variable. Researchers merely wanted to observe the behavior of two different groups when each was presented with the stimulus.)
   b) The study is prospective. Researchers identified subjects, and then observed them after the sad movie.
   c) The subjects in this study were people with and without depression. The selection process is not stated.
   d) The parameter of interest is the difference in crying response between depressed and nondepressed people exposed to sad situations.
   e) There is no apparent difference in crying response to sad movies for the depressed and nondepressed groups.

31. a) This is an experiment. Subjects were randomly assigned to treatments.
   b) The subjects were people experiencing migraines.
   c) There are 2 factors (pain reliever and water temperature). The pain reliever factor has 2 levels (pain reliever or placebo), and the water temperature factor has 2 levels (ice water and regular water).
   d) 2 factors, at 2 levels each, results in 4 treatments.
   e) The response variable is the level of pain relief.
   f) The experiment is completely randomized.
   g) The subjects are blinded to the pain reliever factor through the use of a placebo. The subjects are not blinded to the water factor. They will know whether they are drinking ice water or regular water.
   h) The experiment may indicate whether pain reliever alone or in combination with ice water give pain relief, but patients are not blinded to ice water, so the placebo effect may also be the cause of any relief seen due to ice water.

33. a) This is an experiment. Athletes were randomly assigned to one of two exercise programs.

   b) The subjects are athletes suffering hamstring injuries.

   c) There is one factor (type of exercise), at 2 levels (static stretching, and agility and trunk stabilization).

   d) 1 factor, at 2 levels, results in 2 treatments.

   e) The response variable is the time before the athletes were able to return to sports.

   f) The experiment is completely randomized.

   g) The experiment employs no blinding. The subjects know what kind of exercise they do.

   h) Assuming that the athletes actually followed the exercise program, this experiment can help determine which of the two exercise programs is more effective at rehabilitating hamstring injuries.

35. Omega-3.

The experimenters need to compare omega-3 results to something. Perhaps bipolarity is seasonal and would have improved during the experiment anyway.

37. Omega-3 revisited.

   a) Subjects' responses might be related to other factors, like diet, exercise, or genetics. Randomization should equalize the two groups with respect to unknown factors.

   b) More subjects would minimize the impact of individual variability in the responses, but the experiment would become more costly and time-consuming.

39. Omega-3 finis.

The researchers believe that people who engage in regular exercise might respond differently to the omega-3. This additional variability could obscure the effectiveness of the treatment.

41. Injuries.

Answers may vary. Use a random-number generator to randomly select 24 numbers from 01 to 24 without replication. Assign the first 8 numbers to the first group, the second 8 numbers to the second group, and the third 8 numbers to the third group. If an athlete states that he would prefer the other program, he should not be allowed to switch, since a successful experiment requires randomization. In fact, he should probably be eliminated from the study, since there is a risk that he will not follow the directions for his assigned program. He may instead opt to follow the protocols for the other program anyway, since he has an obvious preference. This would invalidate the results of the experiment.

43. Shoes.

a) First, the manufacturers are using athletes who have a vested interest in the success of the shoe by virtue of their sponsorship. They should try to find some volunteers that aren't employed by the company! Second, they should randomize the order of the runs, not run all the races with the new shoes second. They should blind the athletes by disguising the shoes, if possible, so they don't know which is which. The experiment could be double blinded, as well, by making sure that the timers don't know which shoes are being tested at any given time. Finally, they should replicate several times since times will vary under both shoe conditions.

b) First of all, the problems identified in part a would have to be remedied before *any* conclusions can be reached. Even if this is the case, the results cannot be generalized to all runners. This experiment compares effects of the shoes on speed for Olympic class runners, not runners in general.

45. Hamstrings.

a) Allowing the athletes to choose their own treatments could confound the results. Other issues such as severity of injury, diet, age, etc., could also affect time to heal, and randomization should equalize the two treatment groups with respect to any such variables.

b) A control group could have revealed whether either exercise program was better (or worse) than just letting the injury heal without exercise.

c) Although the athletes cannot be blinded, the doctors who approve their return to sports should not know which treatment the subject had engaged in.

d) It's difficult to say with any certainty, since we aren't sure if the distributions of return times are unimodal and roughly symmetric, and contain no outliers. Otherwise, the use of mean and standard deviation as measures of center and spread is questionable. Assuming mean and standard deviation are appropriate measures, the subjects who exercised with agility and trunk stabilization had a mean return time of 22.2 days compared to the static stretching group, with a mean return time of 37.4 days. The agility and trunk stabilization group also had a much more consistent distribution of return times, with a standard deviation of 8.3 days, compared to the standard deviation of 27.6 days for the static stretching group. This appears to be a statistically significant difference.

47. Mozart.

a) The differences in spatial reasoning scores between the students listening to Mozart and the students sitting quietly were more than would have been expected from ordinary sampling variation.

b)

Pre-tested volunteers — R a n d o m — Group 1 — Glass / Group 2 — Mozart / Group 3 — Silence → Test again and compare difference in score (Test 1 - Test 2)

c) The Mozart group seems to have the smallest median difference in spatial reasoning test score and thus the *least* improvement, but there does not appear to be a significant difference.

d) No, the results do not prove that listening to Mozart is beneficial. If anything, there was generally less improvement. The difference does not seem significant compared with the usual variation one would expect between the three groups. Even if type of music has no effect on test score, we would expect some variation between the groups.

49. Wine.

a) This is a prospective observational study. The researchers followed a group of children born at a Copenhagen hospital between 1959 and 1961.

b) The results of the Danish study report a link between high socioeconomic status, education, and wine drinking. Since people with high levels of education and higher socioeconomic status are also more likely to be healthy, the relation between health and wine consumption might be explained by the confounding variables of socioeconomic status and education.

c) Studies such as these prove none of these. While the variables have a relation, there is no indication of a cause-and-effect relationship. The only way to determine causation is through a controlled, randomized, and replicated experiment.

51. Dowsing.

a) Arrange the 20 containers in 20 separate locations. Number the containers 01 – 20, and use a random number generator to identify the 10 containers that should be filled with water.

b) We would expect the dowser to be correct about 50% of the time, just by guessing. A record of 60% (12 out of 20) does not appear to be significantly different than the 10 out of 20 expected.

c) Answers may vary. A high level of success would need to be observed. 90% to 100% success (18 to 20 correct identifications) would be convincing.

53. Reading.

Answers may vary. This experiment has 1 factor (reading program), at 2 levels (phonics and whole language), resulting in 2 treatments. The response variable is reading score on an appropriate reading test after a year in the program. After randomly assigning students to teachers, randomly assign half the reading teachers in the district to use each method. There may be variation in reading score based on school within the district, as well as by grade. Blocking by both school and grade will reduce this variation.

55. Weekend deaths.

a) The difference between death rate on the weekend and death rate during the week is greater than would be expected due to natural sampling variation.

b) This was a prospective observational study. The researchers identified hospitals in Ontario, Canada, and tracked admissions to the emergency rooms. This certainly cannot be an experiment. People can't be assigned to become injured on a specific day of the week!

c) Waiting until Monday, if you were ill on Saturday, would be foolish. There are likely to be confounding variables that account for the higher death rate on the weekends. For example, people might be more likely to engage in risky behavior on the weekend.

d) Alcohol use might have something to do with the higher death rate on the weekends. Perhaps more people drink alcohol on weekends, which may lead to more traffic accidents, and higher rates of violence during these days of the week.

57. Beetles.

Answers may vary. This experiment has 1 factor (pesticide), at 3 levels (pesticide A, pesticide B, no pesticide), resulting in 3 treatments. The response variable is the number of beetle larvae found on each plant. Randomly select a third of the plots to be sprayed with pesticide A, a third with pesticide B, and a third to be sprayed with no pesticide (since the researcher also wants to know whether the pesticides even work at all). To control the experiment, the plots of land should be as similar as possible, with regard to amount of sunlight, water, proximity to other plants, etc. If not, plots with similar characteristics should be blocked together. If possible, use some inert substance as a placebo pesticide on the control group, and do not tell the counters of the beetle larvae which plants have been treated with pesticides. After a given period of time, count the number of beetle larvae on each plant and compare the results.

Plots of corn —Random—⟨ Group 1 — pesticide A / Group 2 — pesticide B → Count the number of beetle larvae on each plant and compare / Group 3 — no pesticide

59. Safety switch.

Answers may vary. This experiment has 1 factor (hand), at 2 levels (right, left), resulting in 2 treatments. The response variable is the difference in deactivation time between left and right hand. Find a group of volunteers. Using a matched design, we will require each volunteer to deactivate the machine with his or her left hand, as well as with his or her right hand. Randomly assign the left or right hand to be used first. Hopefully, this will equalize any variability in time that may result from experience gained after deactivating the machine the first time. Complete the first attempt for the whole group. Now repeat the experiment with the alternate hand. Check the differences in time for the left and right hands. Since the response variable is difference in times for each hand, workers should be blocked into groups based on their dominant hand. Another way to account for this difference would be to use the absolute value of the difference as the response variable. We are interested in whether or not the difference is significantly different from the zero difference we would expect if the machine were just as easy to operate with either hand.

61. Skydiving, anyone?

a) There is 1 factor, jumping, with 2 levels, with and without a working parachute.

b) You would need some (dim-witted) volunteers skydivers as the subjects.

c) A parachute that looked real, but didn't open, would serve as the placebo.

d) 1 factor at 2 levels is 2 treatments, a good parachute and a placebo parachute.

e) The response variable is whether the skydiver survives the jump (or the extent of injuries).

f) All skydivers should jump from the same altitude, in similar weather conditions, and land on similar surfaces.

g) Make sure that you randomly assign the skydivers to the parachutes.

h) The skydivers (and the distributers of the parachutes) shouldn't know who got a working chute. Additionally, the people evaluating the subjects after the jumps should not be told who had a real chute, either.

Review of Part III – Gathering Data

1. The researchers performed a prospective observational study, since the children were identified at birth and examined at ages 8 and 20. There were indications of behavioral differences between the group of "preemies", and the group of full-term babies. The "preemies" were less likely to engage in risky behaviors, like use of drugs and alcohol, teen pregnancy, and conviction of crimes. This may point to a link between premature birth and behavior, but there may be lurking variables involved. Without a controlled, randomized, and replicated experiment, a cause-and-effect relationship cannot be determined.

3. The researchers at the Purina Pet Institute performed an experiment, matched by gender and weight. The experiment had one factor (diet), at two levels (allowing the dogs to eat as much as they want, or restricted diet), resulting in two treatments. One of each pair of similar puppies was randomly assigned to each treatment. The response variable was length of life. The researchers were able to conclude that, on average, dogs with a lower-calorie diet live longer.

5. This is a completely randomized experiment, with the treatment being receiving folic acid or not (one factor, two levels). Treatments were assigned randomly and the response variable is the number of precancerous growths, or simply the occurrence of additional precancerous growths. Neither blocking nor matching is mentioned, but in a study such as this one, it is likely that researchers and patients are blinded. Since treatments were randomized, it seems reasonable to generalize results to all people with precancerous polyps, though caution is warranted since these results contradict a previous study.

7. The fireworks manufacturers are sampling. No information is given about the sampling procedure, so hopefully the tested fireworks are selected randomly. It would probably be a good idea to test a few of each type of firework, so stratification by type seems likely. The population is all fireworks produced each day, and the parameter of interest is the proportion of duds. With a random sample, the manufacturers can make inferences about the proportion of duds in the entire day's production, and use this information to decide whether or not the day's production is suitable for sale.

9. This is an observational retrospective study. Researcher can conclude that for anyone's lunch, even when packed with ice, food temperatures are rising to unsafe levels.

11. This is an experiment, with a control group being the genetically engineered mice who received no antidepressant and the treatment group being the mice who received the drug. The response variable is the amount of plaque in their brains after one dose and after four months. There is no mention of blinding or matching. Conclusions can be drawn to the general population of mice and we should assume treatments were randomized. To conclude the same for humans would be risky, but researchers might propose an experiment on humans based on this study.

13. The researchers performed an experiment. There is one factor (gene therapy), at two levels (gene therapy and no gene therapy), resulting in two treatments. The experiment is completely randomized. The response variable is heart muscle condition. The researchers can conclude that gene therapy is responsible for stabilizing heart muscle in laboratory rats.

15. The orange juice plant depends on sampling to ensure the oranges are suitable for juice. The population is all of the oranges on the truck, and the parameter of interest is the proportion of unsuitable oranges. The procedure used is a random sample, stratified by location in the truck. Using this well-chosen sample, the workers at the plant can estimate the proportion of unsuitable oranges on the truck, and decide whether or not to accept the load.

17. Observational retrospective study, performed as a telephone-based randomized survey. Based on the excerpt, it seems reasonable to conclude that more education is associated with a higher Emotional Health Index score, but to insist on causality would be faulty reasoning.

19. Point spread.

Answers may vary. Perform a simulation to determine the gambler's expected winnings. A component is one game. To model that component, generate random digits 0 to 9. Since the outcome after the point spread is a tossup, let digits 0-4 represent a loss, and let digits 5-9 represent a win. A run consists of 5 games, so generate 5 random digits at a time. The response variable is the profit the gambler makes, after accounting for the $10 bet. If the outcome of the run is 0, 1, or 2 simulated wins, the profit is −$10. If the outcome is 3, 4, or 5 simulated wins, the profit is $0, $10, or $40, respectively. The total profit divided by the number of runs is the average weekly profit. According to the simulation (80 runs were performed), the gambler is expected to break even. His simulated losses equaled his simulated winnings. (In theory, the gambler is expected to lose about $2.19 per game.)

21. Everyday randomness.

Answers will vary. Most of the time, events described as "random" are anything but truly random.

23. Tips.

a) The waiters performed an experiment, since treatments were imposed on randomly assigned groups. This experiment has one factor (candy), at two levels (candy or no candy), resulting in two treatments. The response variable is the percentage of the bill given as a tip.

b) If the decision whether to give candy or not was made before the people were served, the server may have subconsciously introduced bias by treating the customers better. If the decision was made just before the check was delivered, then it is reasonable to conclude that the candy was the cause of the increase in the percentage of the bill given as a tip.

c) "Statistically significant" means that the difference in the percentage of tips between the candy and no candy groups was more than expected due to sampling variability.

25. Timing.

There will be voluntary response bias, and results will mimic those only of the visitors to Sodahead.com and not the general U.S. population. The question is leading responders to answer "yes" though many might understand that the president's timing for his vacation had nothing to do with the events of the week.

27. When to stop?

a) Answers may vary. A component in this simulation is rolling 1 die. To simulate this component, generate a random digit 1 to 6. To simulate a run, simulate 4 rolls, stopping if a 6 is rolled. The response variable is the sum of the 4 rolls, or 0 if a 6 is rolled. The average number of points scored is the sum of all rolls divided by the total number of runs. According to the simulation, the average number of points scored will be about 5.8.

b) Answers may vary. A component in this simulation is rolling 1 die. To simulate this component, generate a random digit 1 to 6. To simulate a run, generate random digits until the sum of the digits is at least 12, or until a 6 is rolled. The response variable is the sum of the digits, or 0 if a 6 is rolled. The average number of points scored is the sum of all rolls divided by the total number of runs. According to the simulation, the average number of points scored will be about 5.8, similar to the outcome of the method described in part a).

c) Answers may vary. Be careful when making your decision about the effectiveness of your strategy. If you develop a strategy with a higher simulated average number of points than the other two methods, this is only an indication that you may win in the long run. If the game is played round by round, with the winner of a particular round being declared as the player with the highest roll made during that round, the game is much more variable. For example, if Player B rolls a 12 in a particular game, Player A will always lose that game, provided he or she sticks to the strategy. A better way to get a feel for your chances of winning this type of game might be to simulate several rounds, recording whether each player won or lost the round. Then estimate the percentage of the time that each player is expected to win, according to the simulation.

29. Homecoming.

a) Since telephone numbers were generated randomly, every number that could possibly occur in that community had an equal chance of being selected. This method is "better" than using the phone book, because unlisted numbers are also possible. Those community members who deliberately do not list their phone numbers might not consider this method "better"!

b) Although this method results in a simple random sample of phone numbers, it does not result in a simple random sample of residences. Residences without a phone are excluded, and residences with more than one phone have a greater chance of being included.

c) No, this is not a SRS of local voters. People who respond to the survey may be of the desired age, but not registered to vote. Additionally, some voters who are contacted may choose not to participate.

d) This method does not guarantee an unbiased sample of households. Households in which someone answered the phone may be more likely to have someone at home when the phone call was generated. The attitude about homecoming of these households might not be the same as the attitudes of the community at large.

31. Smoking and Alzheimer's.

a) The studies do not prove that smoking offers any protection from Alzheimer's. The studies merely indicate an association. There may be other variables that can account for this association.

b) Alzheimer's usually shows up late in life. Since smoking is known to be harmful, perhaps smokers have died of other causes before Alzheimer's can be seen.

c) The only way to establish a cause-and-effect relationship between smoking and Alzheimer's is to perform a controlled, randomized, and replicated experiment. This is unlikely to ever happen, since the factor being studied, smoking, has already been proven harmful. It would be unethical to impose this treatment on people for the purposes of this experiment. A prospective observational study could be designed in which groups of smokers and nonsmokers are followed for many years and the incidence of Alzheimer's disease is tracked.

33. Sex and violence.

This experiment has one factor (program content), at three levels (violent, sexual, and neutral), resulting in three treatments. The response variable is the number of brand names recalled after watching the program. Numerous subjects will be randomly assigned to see shows with violent, sexual, or neutral content. They will see the same commercials. After the show, they will be interviewed for their recall of brand names in the commercials.

35. Age and party 2008.

a) The number of respondents is roughly the same for each age category. This may indicate a sample stratified by age category, although it may be a simple random sample.

b) 1530 Democrats were surveyed. $\frac{1530}{4002} \approx 38.2\%$ of the people surveyed were Democrats.

c) We don't know. If data were collected from voting precincts that are primarily Democratic or primarily Republican, that would bias the results. Because the survey was commissioned by NBC News, we can assume the data collected are probably reliable.

d) The pollsters were probably attempting to determine whether or not political party is associated with age.

37. Save the grapes.

This experiment has one factor (bird control device), at three levels (scarecrow, netting, and no device), resulting in three treatments. Randomly assign different plots in the vineyard to the different treatments, making sure to ensure adequate separation of plots, so that the possible effect of the scarecrow will not be confounded with the other treatments. The response variable to be measured at the end of the season is the proportion of bird-damaged grapes in each plot.

39. Acupuncture.

a) The "fake" acupuncture was the control group. In an experiment, all subjects must be treated as alike as possible. If there were no "fake" acupuncture, subjects would know that they had not received acupuncture, and might react differently. Of course, all volunteers for the experiment must be aware of the possibility of being randomly assigned to the control group.

b) Experiments always use volunteers. This is not a problem, since experiments are testing response to a treatment, not attempting to determine an unknown population parameter. The randomization in an experiment is random assignment to treatment groups, not random selection from a population. Voluntary response is a problem when sampling, but is not an issue in experimentation. In this case, it is probably reasonable to assume that the volunteers have similar characteristics to others in the population of people with chronic lower back pain.

c) There were differences in the amount of pain relief experienced by the two groups, and these differences were large enough that they could not be explained by natural variation alone. Researchers concluded that both proper and "fake" acupuncture reduced back pain.

41. Security.

a) To ensure that passengers from first-class, as well as coach, get searched, select 2 passengers from first-class and 12 from coach. Using this stratified random sample, 10% of the first-class passengers are searched, as are 10% of the coach passengers.

b) Answers will vary. Number the passengers alphabetically, with 2-digit numbers. Bergman = 01, Bowman = 02, and so on, ending with Testut = 20. Read the random digits in pairs, ignoring pairs 21 to 99 and 00, and ignoring repeated pairs.

```
65|43|67|11|          27|04|        The passengers selected for search
XX XX XX Fontana      XX Castillo   from first-class are Fontana and
                                    Castillo.
```

c) Number the passengers alphabetically, with 3 digit numbers, 001 to 120. Use the random number table to generate 3-digit numbers, ignoring numbers 121 to 999 and 000, and ignoring repeated numbers. Search the passengers corresponding to the first 12 valid numbers generated.

43. Par 4.

Answers may vary. A component in this simulation is a shot. Use pairs of random digits 00 to 99 to represent a shot. The way in which this component is simulated depends on the type of shot.

For the first shot, let pairs of digits 01 to 70 represent hitting the fairway, and let pairs of digits 71 to 99, and 00, represent not hitting the fairway.

If the first simulated shot hits the fairway, let 01 to 80 represent landing on the green on the second shot, and let 81 to 99, and 00, represent not landing on the green on the second shot. If the first simulated shot does not hit the fairway, let 01 to 40 represent landing on the green on the second shot, and let 41 to 99, and 00, represent not landing on the green on the second shot.

If the second simulated shot does not land on the green, let 01 to 90 represent landing on the green, and 91 to 99, and 00, represent not landing on the green. Keep simulating shots until the shot lands on the green.

Once on the green, let 01 to 20 represent sinking the putt on the first putt, and let 21 to 99, and 00, represent not sinking the putt on the first putt. If second putts are required, continue simulating putts until a putt goes in, with 01 to 90 representing making the putt, and 91 to 99, and 00, representing not making it.

A run consists of following the guidelines above until the final putt is made. The response variable is the number of shots required until the final putt is made.

The simulated average score on the hole is the total number of shots required divided by the total number of runs. According to 40 runs of this simulation, a pretty good golfer can be expected to average about 4.2 strokes per hole. Your simulation results may vary.

Chapter 13 – From Randomness to Probability

Section 13.1

1. **Flipping a coin.**

   In the long run, a fair coin will generate 50% heads and 50% tails, approximately. But for each flip we cannot predict the outcome.

3. **Flipping a coin II.**

   There is no law of averages for the short run. The first five flips do not affect the sixth flip.

Section 13.2

5. **Wardrobe.**

   a) There are a total of 10 shirts, and 3 of them are red. The probability of randomly selecting a red shirt is 3/10 = 0.30.

   b) There are a total of 10 shirts, and 8 of them are not black. The probability of randomly selecting a shirt that is not black is 8/10 = 0.80.

Section 13.3

7. **Cell phones and surveys.**

   a) If 25% of homes don't have a landline, then 75% of them do have a landline. The probability that all 5 houses have a landline is $(0.75)^5 \approx 0.237$.

   b) $P(\text{at least one without landline}) = 1 - P(\text{all landlines}) = 1 - (0.75)^5 \approx 0.763$.

   c) $P(\text{at least one with landline}) = 1 - P(\text{no landlines}) = 1 - (0.25)^5 \approx 0.999$

Chapter Exercises.

9. **Sample spaces.**

   a) S = { HH, HT, TH, TT} All of the outcomes are equally likely to occur.

   b) S = { 0, 1, 2, 3} All outcomes are not equally likely. A family of 3 is more likely to have, for example, 2 boys than 3 boys. There are three equally likely outcomes that result in 2 boys (BBG, BGB, and GBB), and only one that results in 3 boys (BBB).

   c) S = { H, TH, TTH, TTT} All outcomes are not equally likely. For example the probability of getting heads on the first try is $\frac{1}{2}$. The probability of getting three tails is $\left(\frac{1}{2}\right)^3 = \frac{1}{8}$.

   d) S = {1, 2, 3, 4, 5, 6} All outcomes are not equally likely. Since you are recording only the larger number (or the number if there is a tie) of two dice, 6 will be the larger when the other die reads 1, 2, 3, 4, or 5. The outcome 2 will only occur when the other die shows 1 or 2.

11. Roulette.

If a roulette wheel is to be considered truly random, then each outcome is equally likely to occur, and knowing one outcome will not affect the probability of the next. Additionally, there is an implication that the outcome is not determined through the use of an electronic random number generator.

13. Winter.

Although acknowledging that there is no law of averages, Knox attempts to use the law of averages to predict the severity of the winter. Some winters are harsh and some are mild over the long run, and knowledge of this can help us to develop a long-term probability of having a harsh winter. However, probability does not compensate for odd occurrences in the short term. Suppose that the probability of having a harsh winter is 30%. Even if there are several mild winters in a row, the probability of having a harsh winter is still 30%.

15. Cold streak.

There is no such thing as being "due to make a shot". This statement is based on the so-called law of averages, which is a mistaken belief that probability will compensate in the short term for odd occurrences in the past. The player's chance of making a shot does not change based on recent successes or failures.

17. Auto insurance.

   a) It would be foolish to insure your neighbor against automobile accidents for $1500. Although you might simply collect $1500, there is a good chance you could end up paying much more than $1500. That risk is not worth the $1500.

   b) The insurance company insures many people. The overwhelming majority of customers pay the insurance and never have a claim, or have claims that are lower than the cost of their payments. The few customers who do have a claim are offset by the many who simply send their premiums without a claim. The relative risk to the insurance company is low.

19. Spinner.

   a) This is a legitimate probability assignment. Each outcome has probability between 0 and 1, inclusive, and the sum of the probabilities is 1.

   b) This is a legitimate probability assignment. Each outcome has probability between 0 and 1, inclusive, and the sum of the probabilities is 1.

   c) This is not a legitimate probability assignment. Each outcome has probability between 0 and 1, inclusive, but the sum of the probabilities is greater than 1.

d) This is a legitimate probability assignment. Each outcome has probability between 0 and 1, inclusive, and the sum of the probabilities is 1. However, this game is not very exciting!

e) This probability assignment is not legitimate. The sum of the probabilities is 0, and there is one probability, –1.5 , that is not between 0 and 1, inclusive.

21. Electronics.

A family may have both a computer and an HDTV. The events are not disjoint, so the Addition Rule does not apply.

23. Speeders.

When cars are traveling close together, their speeds are not independent. For example, a car following directly behind another can't be going faster than the car ahead. Since the speeds are not independent, the Multiplication Rule does not apply.

25. College admissions.

a) Jorge had multiplied the probabilities.

b) Jorge assumes that being accepted to the colleges are independent events.

c) No. Colleges use similar criteria for acceptance, so the decisions are not independent. Students that meet these criteria are more likely to be accepted at all of the colleges. Since the decisions are not independent, the probabilities cannot be multiplied together.

27. Car repairs.

Since all of the events listed are disjoint, the addition rule can be used.

a) $P(\text{no repairs}) = 1 - P(\text{some repairs}) = 1 - (0.17 + 0.07 + 0.04) = 1 - (0.28) = 0.72$

b) $P(\text{no more than one repair}) = P(\text{no repairs or one repair}) = 0.72 + 0.17 = 0.89$

c) $P(\text{some repairs}) = P(\text{one or two or three or more repairs})$
$$= 0.17 + 0.07 + 0.04 = 0.28$$

29. More repairs.

Assuming that repairs on the two cars are independent from one another, the multiplication rule can be used. Use the probabilities of events from Exercise 27 in the calculations.

a) $P(\text{neither will need repair}) = (0.72)(0.72) = 0.5184$

b) $P(\text{both will need repair}) = (0.28)(0.28) = 0.0784$

c) $P(\text{at least one will need repair}) = 1 - P \text{ (neither will need repair)}$
$$= 1 - (0.72)(0.72) = 0.4816$$

31. Repairs, again.

a) The repair needs for the two cars must be independent of one another.

b) This may not be reasonable. An owner may treat the two cars similarly, taking good (or poor) care of both. This may decrease (or increase) the likelihood that each needs to be repaired.

33. Energy 2013.

a) P(response is "Increase oil, gas, and coal") $= 164/529 \approx 0.310$

b) P("Equally important" or "No opinion") $= 37/529 + 16/529 = 53/529 \approx 0.100$

35. More energy.

a) P(all three respond "Develop wind and solar") $= \left(\frac{312}{529}\right)\left(\frac{312}{529}\right)\left(\frac{312}{529}\right) \approx 0.205$

b) P(none respond "Equally important") $= \left(\frac{492}{529}\right)\left(\frac{492}{529}\right)\left(\frac{492}{529}\right) \approx 0.805$

c) In order to compute the probabilities, we must assume that responses are independent.

d) It is reasonable to assume that responses are independent, since the three people were chosen at random.

37. Polling.

a) P(household is contacted and household refuses to cooperate)
$= P$(household is contacted)P(household refuses | contacted)
$= (0.62)(1-0.14) = 0.5332$

b) P(fail to contact household or contacting and not getting interview)
$= P$(fail to contact) $+ P$(contact household)P(not getting interview | contacted)
$= (1-0.62)+(0.62)(1-0.14) = 0.9132$

c) The question in part b covers all possible occurrences *except* contacting the house and getting the interview.

P(failing to contact household or contacting and not getting the interview)
$= 1 - P$(contacting the household and getting the interview)
$= 1 - (0.62)(0.14) = 0.9132$

39. M&M's.

a) Since all of the events are disjoint (an M&M can't be two colors at once!), use the addition rule where applicable.

1. P(brown) $= 1 - P$(not brown) $= 1 - P$(yellow or red or orange or blue or green)
$= 1 - (0.20 + 0.20 + 0.10 + 0.10 + 0.10) = 0.30$

2. P(yellow or orange) $= 0.20 + 0.10 = 0.30$

   3. $P(\text{not green}) = 1 - P(\text{green}) = 1 - 0.10 = 0.90$

   4. $P(\text{striped}) = 0$

b) Since the events are independent (picking out one M&M doesn't affect the outcome of the next pick), the multiplication rule may be used.

   1. $P(\text{all three are brown}) = (0.30)(0.30)(0.30) = 0.027$

   2. $P(\text{the third one is the first one that is red}) = P(\text{not red and not red and red})$
   $$= (0.80)(0.80)(0.20) = 0.128$$

   3. $P(\text{no yellow}) = P(\text{not yellow and not yellow and not yellow})$
   $$= (0.80)(0.80)(0.80) = 0.512$$

   4. $P(\text{at least one is green}) = 1 - P(\text{none are green}) = 1 - (0.90)(0.90)(0.90) = 0.271$

41. Disjoint or independent?

a) For one draw, the events of getting a red M&M and getting an orange M&M are disjoint events. Your single draw cannot be both red and orange.

b) For two draws, the events of getting a red M&M on the first draw and a red M&M on the second draw are independent events. Knowing that the first draw is red does not influence the probability of getting a red M&M on the second draw.

c) Disjoint events can never be independent. Once you know that one of a pair of disjoint events has occurred, the other one cannot occur, so its probability has become zero. For example, consider drawing one M&M. If it is red, it cannot possible be orange. Knowing that the M&M is red influences the probability that the M&M is orange. It's zero. The events are not independent.

43. Dice.

a) $P(6) = \dfrac{1}{6}$, so $P(\text{all 6's}) = \left(\dfrac{1}{6}\right)\left(\dfrac{1}{6}\right)\left(\dfrac{1}{6}\right) \approx 0.005$

b) $P(\text{odd}) = P(1 \text{ or } 3 \text{ or } 5) = \dfrac{3}{6}$, so $P(\text{all odd}) = \left(\dfrac{3}{6}\right)\left(\dfrac{3}{6}\right)\left(\dfrac{3}{6}\right) = 0.125$

c) $P(\text{not divisible by 3}) = P(1 \text{ or } 2 \text{ or } 4 \text{ or } 5) = \dfrac{4}{6}$

   $P(\text{none divisible by 3}) = \left(\dfrac{4}{6}\right)\left(\dfrac{4}{6}\right)\left(\dfrac{4}{6}\right) \approx 0.296$

d) $P(\text{at least one 5}) = 1 - P(\text{no 5's}) \doteq 1 - \left(\dfrac{5}{6}\right)\left(\dfrac{5}{6}\right)\left(\dfrac{5}{6}\right) \approx 0.421$

e) $P(\text{not all 5's}) = 1 - P(\text{all 5's}) = 1 - \left(\dfrac{1}{6}\right)\left(\dfrac{1}{6}\right)\left(\dfrac{1}{6}\right) \approx 0.995$

45. Champion bowler.

Assuming each frame is independent of others, the multiplication rule may be used.

a) P(no strikes in 3 frames) = $(0.30)(0.30)(0.30) = 0.027$

b) P(makes first strike in the third frame) = $(0.30)(0.30)(0.70) = 0.063$

c) P(at least one strike in first three frames) = $1 - P$(no strikes) = $1 - (0.30)^3 = 0.973$

d) P(perfect game) = $(0.70)^{12} \approx 0.014$

47. Voters.

Since you are calling at random, one person's political affiliation is independent of another's. The multiplication rule may be used.

a) P(all Republicans) = $(0.29)(0.29)(0.29) \approx 0.024$

b) P(no Democrats) = $(1 - 0.37)(1 - 0.37)(1 - 0.37) \approx 0.25$

c) P(at least one Ind.) = $1 - P$(no Independents) = $1 - (0.77)(0.77)(0.77) \approx 0.543$

49. Lights.

Assume that the defective light bulbs are distributed randomly to all stores so that the events can be considered independent. The multiplication rule may be used.

P(at least one of five bulbs is defective) = $1 - P$(none are defective)

$$= 1 - (0.94)(0.94)(0.94)(0.94)(0.94) \approx 0.266$$

51. 9/11?

a) For any date with a valid three-digit date, the chance is 0.001, or 1 in 1000. For many dates in October through December, the probability is 0. For example, there is no way three digits will make 1015, to match October 15.

b) There are 65 days when the chance to match is 0. (October 10 through October 31, November 10 through November 30, and December 10 through December 31.) That leaves 300 days in a year (that is not a leap year) in which a match might occur.
P(no matches in 300 days) = $(0.999)^{300} \approx 0.741$.

c) P(at least one match in a year) = $1 - P$(no matches in a year) = $1 - 0.741 \approx 0.259$

d) P(at least one match on 9/11 in one of the 50 states)
= $1 - P$(no matches in 50 states) = $1 - (0.999)^{50} \approx 0.049$

Chapter 14 – Probability Rules!

Section 14.1

1. Pet ownership.

$$P(\text{dog or cat}) = P(\text{dog}) + P(\text{cat}) - P(\text{dog and cat})$$
$$= 0.25 + 0.29 - 0.12 = 0.42$$

Section 14.2

3. Sports.

	Football	No Football	Total
Basketball	27	13	40
No Basketball	38	22	60
Total	65	35	100

$$P(\text{football} \mid \text{basketball}) = \frac{P(\text{football and basketball})}{P(\text{basketball})} = \frac{\frac{27}{100}}{\frac{40}{100}} = 0.675$$

(Or, use the table. Of the 40 people who like to watch basketball, 27 people also like to watch football. 27/40 = 0.675)

5. Late to the train.

$$P(\text{let out late and missing train}) = P(\text{let out late}) \times P(\text{missing train} \mid \text{let out late})$$
$$= (0.30)(0.45) = 0.135$$

Section 14.3

7. Titanic.

The overall survival rate, $P(S)$, was 0.323, yet the survival rate for first class passengers, $P(S \mid FC)$, was 0.625. Since, $P(S) \neq P(S \mid FC)$, survival and ticket class are not independent. Rather, survival rate depended on class.

Section 14.4

9. **Facebook.**

	US	Not US	Total
Log on Every Day	0.20	0.30	0.50
Do Not Log on Every Day	0.10	0.40	0.50
Total	0.30	0.70	1.00

We have joint probabilities and marginal probabilities, not conditional probabilities, so a table is the better choice.

11. **Facebook again.**

A tree is better because we have conditional and marginal probabilities. The joint probabilities are found at the end of the branches.

Section 14.5

13. **Facebook final.**

$$P(\text{US} \mid \text{Log on every day}) = \frac{P(\text{US and Log on every day})}{P(\text{Log on every day})} = \frac{0.20}{0.20 + 0.30} = 0.40$$

Knowing that a person logs on every day increases probability that the person is from the United States.

Chapter Exercises.

15. Phones.

a) P(home phone or cell phone)
 = P(home) + P(cell) – P(home and cell)
 = 0.73 + 0.83 – 0.58 = 0.98

Or, from the Venn: 0.15 + 0.58 + 0.25 = 0.98

b) P(neither) = 1 – P(home or cell) = 1 – 0.98 = 0.02

Or, from the Venn: 0.02 (the region outside the circles)

c) P(cell but no home) = P(cell) – P(home and cell) = 0.83 – 0.58 = 0.25

Or, from the Venn: 0.25 (the region inside cell circle, yet outside home circle)

17. Amenities.

a) P(TV and no refrigerator)
 = P(TV) – P(TV and refrigerator)
 = 0.52 – 0.21 = 0.31

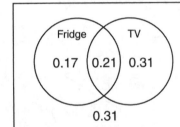

Or, from the Venn: 0.31
(inside the TV circle, yet outside the Fridge circle)

b) P(refrigerator or TV, but not both) =
 = [P(refrigerator) – P(refrigerator and TV)] +[P(TV) – P(refrigerator and TV)]
 = [0.38 – 0.21] + [0.52 – 0.21] = 0.48

Using the Venn diagram, simply add the probabilities in the two regions for Fridge only and TV only. P(refrigerator or TV, but not both) = 0.17 + 0.31 = 0.48

c) P(neither TV nor refrigerator) = 1 – P(either TV or refrigerator)
 = 1 – [P(TV) + P(ref.) – P(TV and ref.)]
 = 1 – [0.52 + 0.38 – 0.21]
 = 0.31

Or, from the Venn: 0.31 (the region outside the circles)

19. Global survey.

a) $P(USA) = \dfrac{1557}{7690} \approx 0.2025$

b) P(some high school or primary or less) $= \dfrac{4195}{7690} + \dfrac{1161}{7690} \approx 0.6965$

c)

$$P(\text{France or post-graduate}) = P(\text{France}) + P(\text{post-graduate}) - P(\text{both})$$

$$= \frac{1539}{7690} + \frac{379}{7690} - \frac{69}{7690} \approx 0.2404$$

d) $P(\text{France and primary school or less}) = \dfrac{309}{7690} \approx 0.0402$

21. Cards.

a) $P(\text{heart} \mid \text{red}) = \dfrac{P(\text{heart and red})}{P(\text{red})} = \dfrac{13/52}{26/52} = \dfrac{1}{2}$ A more intuitive approach is to think about only the red cards. Half of them are hearts.

b) $P(\text{red} \mid \text{heart}) = \dfrac{P(\text{red and heart})}{P(\text{heart})} = \dfrac{13/52}{13/52} = 1$ Think about only the hearts. They are all red!

c) $P(\text{ace} \mid \text{red}) = \dfrac{P(\text{ace and red})}{P(\text{red})} = \dfrac{2/52}{26/52} = \dfrac{2}{26} \approx 0.077$ Consider only the red cards. Of those 26 cards, 2 of them are aces.

d) $P(\text{queen} \mid \text{face}) = \dfrac{P(\text{queen and face})}{P(\text{face})} = \dfrac{4/52}{12/52} \approx 0.333$ There are 12 faces cards: 4 jacks, 4 queens, and 4 kings. Four of the 12 face cards are queens.

23. Health.

Construct a two-way table of the conditional probabilities, including the marginal probabilities.

a) $P(\text{both conditions}) = 0.11$

b) $P(\text{high BP}) = 0.11 + 0.16 = 0.27$

	Blood Pressure		
Cholesterol	High	OK	Total
High	0.11	0.21	0.32
OK	0.16	0.52	0.68
Total	0.27	0.73	1.00

c) $P(\text{high chol.} \mid \text{high BP}) = \dfrac{P(\text{high chol. and high BP})}{P(\text{high BP})} = \dfrac{0.11}{0.27} \approx 0.407$

Consider only the High Blood Pressure column. Within this column, the probability of having high cholesterol is 0.11 out of a total of 0.27.

d) $P(\text{high BP} \mid \text{high chol.}) = \dfrac{P(\text{high BP and high chol.})}{P(\text{high chol.})} = \dfrac{0.11}{0.32} \approx 0.344$

This time, consider only the high cholesterol row. Within this row, the probability of having high blood pressure is 0.11, out of a total of 0.32.

25. Global survey, take 2.

a) $P(\text{USA and postgraduate work}) = \dfrac{84}{7690} \approx 0.011$

b) $P(\text{USA} \mid \text{post-graduate}) = \dfrac{84}{379} \approx 0.222$

c) $P(\text{post-graduate} \mid \text{USA}) = \dfrac{84}{1557} \approx 0.054$

d) $P(\text{primary} \mid \text{China}) = \dfrac{506}{1502} \approx 0.337$

e) $P(\text{China} \mid \text{primary}) = \dfrac{506}{1161} \approx 0.436$

27. Sick kids.

Having a fever and having a sore throat are not independent events, so:

$P(\text{fever and sore throat}) = P(\text{Fever})\, P(\text{Sore Throat} \mid \text{Fever}) = (0.70)(0.30) = 0.21$

The probability that a kid with a fever has a sore throat is 0.21.

29. Cards.

a)

$P(\text{first heart drawn is on the third card}) = P(\text{no heart})P(\text{no heart})P(\text{heart})$

$$= \left(\frac{39}{52}\right)\left(\frac{38}{51}\right)\left(\frac{13}{50}\right) \approx 0.145$$

b)

$P(\text{all three cards drawn are red}) = P(\text{red})P(\text{red})P(\text{red})$

$$= \left(\frac{26}{52}\right)\left(\frac{25}{51}\right)\left(\frac{24}{50}\right) \approx 0.118$$

c)

$P(\text{none of the cards are spades}) = P(\text{no spade})P(\text{no spade})P(\text{no spade})$

$$= \left(\frac{39}{52}\right)\left(\frac{38}{51}\right)\left(\frac{37}{50}\right) \approx 0.414$$

d)

$$P(\text{at least one of the cards is an ace}) = 1 - P(\text{none of the cards are aces})$$
$$= 1 - \left[P(\text{no ace})P(\text{no ace})P(\text{no ace})\right]$$
$$= 1 - \left(\frac{48}{52}\right)\left(\frac{47}{51}\right)\left(\frac{46}{50}\right) \approx 0.217$$

31. Batteries.

Since batteries are not being replaced, use conditional probabilities throughout.

a)

$$P(\text{the first two batteries are good}) = P(\text{good})P(\text{good})$$
$$= \left(\frac{7}{12}\right)\left(\frac{6}{11}\right) \approx 0.318$$

b)

$$P(\text{at least one of the first three batteries works})$$
$$= 1 - P(\text{none of the first three batteries work})$$
$$= 1 - \left[P(\text{no good})P(\text{no good})P(\text{no good})\right]$$
$$= 1 - \left(\frac{5}{12}\right)\left(\frac{4}{11}\right)\left(\frac{3}{10}\right) \approx 0.955$$

c)

$$P(\text{the first four batteries are good}) = P(\text{good})P(\text{good})P(\text{good})P(\text{good})$$
$$= \left(\frac{7}{12}\right)\left(\frac{6}{11}\right)\left(\frac{5}{10}\right)\left(\frac{4}{9}\right) \approx 0.071$$

d)

$$P(\text{pick five to find one good})$$
$$= P(\text{not good})P(\text{not good})P(\text{not good})P(\text{not good})P(\text{good})$$
$$= \left(\frac{5}{12}\right)\left(\frac{4}{11}\right)\left(\frac{3}{10}\right)\left(\frac{2}{9}\right)\left(\frac{7}{8}\right) \approx 0.009$$

33. Eligibility.

a)

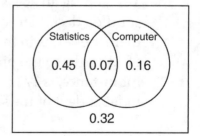

$$P(\text{eligible}) = P(\text{stats}) + P(\text{computer}) - P(\text{both})$$
$$= 0.52 + 0.23 - 0.07$$
$$= 0.68$$

68% of students are eligible for BioResearch, so
100 – 68 = 32% are ineligible.

From the Venn, the region outside the circles represents those students who have
taken neither course, and are therefore ineligible for BioResearch.

b)

$$P(\text{computer course} \mid \text{statistics}) = \frac{P(\text{computer and statistics})}{P(\text{statistics})} = \frac{0.07}{0.52} \approx 0.135$$

From the Venn, consider only the region inside the Statistics circle. The probability of having taken a computer course is 0.07 out of a total of 0.52 (the entire Statistics circle).

c) Taking the two courses are not disjoint events, since they have outcomes in common. In fact, 7% of juniors have taken both courses.

d) Taking the two courses are not independent events. The overall probability that a junior has taken a computer course is 0.23. The probability that a junior has taken a computer course given that he or she has taken a statistics course is 0.135. If taking the two courses were independent events, these probabilities would be the same.

35. Cell phones in the home.

Construct a Venn diagram of the possible outcomes.

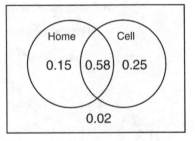

a)

$$P(\text{cell} \mid \text{home}) = \frac{P(\text{cell and home})}{P(\text{home})} = \frac{0.58}{0.73} \approx 0.795$$

From the Venn, consider only the region inside the Home circle. The probability that the person has a cell phone is 0.58 out of 0.73 (the entire Home circle).

b) Having a home phone and a cell phone are not independent events. 79.5% of people with home phones have cell phones. Overall, 83% of people have cell phones. If having a home phone and cell phone were independent events, these would be the same.

c) No, having a home phone and cell phone are not disjoint events. 58% of people have both.

37. Cards.

Yes, getting an ace is independent of the suit when drawing one card from a well shuffled deck. The overall probability of getting an ace is 4/52, or 1/13, since there are 4 aces in the deck. If you consider just one suit, there is only 1 ace out of 13 cards, so the probability of getting an ace given that the card is a diamond, for instance, is 1/13. Since the probabilities are the same, getting an ace is independent of the suit.

39. Unsafe food.

   a) Using the Venn diagram, the probability that a tested
   chicken was not contaminated with either kind of
   bacteria is 33%.

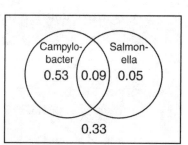

   b) Contamination with campylobacter and
   contamination with salmonella are not disjoint
   events, since 9% of chicken is contaminated with
   both.

   c) Contamination with campylobacter and contamination with salmonella may be
   independent events. The probability that a tested chicken is contaminated with
   campylobacter is 0.62. The probability that chicken contaminated with
   salmonella is also contaminated with campylobacter is 0.09/0.14 ≈ 0.64. If
   chicken is contaminated with salmonella, it is only slightly more likely to be
   contaminated with campylobacter than chicken in general. This difference could
   be attributed to expected variation due to sampling.

41. Men's health, again.

   Consider the two-way table from
   Exercise 23.

	Blood Pressure		
Cholesterol	High	OK	Total
High	0.11	0.21	0.32
OK	0.16	0.52	0.68
Total	0.27	0.73	1.00

   High blood pressure and high
   cholesterol are not independent events.
   28.8% of men with OK blood pressure
   have high cholesterol, while 40.7% of men with high blood pressure have high
   cholesterol. If having high blood pressure and high cholesterol were
   independent, these percentages would be the same.

43. Phone service.

   a) Since 25.4% of U.S. adults have only a cell phone, and
   1.8% have no phone at all, polling organizations can
   reach 100 – 25.4 – 1.8 = 72.8% of U.S. adults.

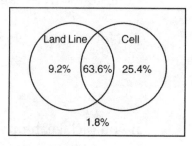

   b) Using the Venn diagram, about 72.8% of U.S. adults
   have a land line. The probability of a U.S. adults
   having a land line given that they have a cell phone is
   63.6/(63.6+25.4) or about 71.5%. It appears that
   having a cell phone and having a land line are independent, since the
   probabilities are roughly the same.

45. Gender.

   According to the poll, party affiliation is not independent of sex.
   Overall, (32+41)/186 ≈ 39.25% of the respondents were Democrats. Of the men,
   only 32/94 ≈ 34.04% were Democrats.

47. Luggage.

Organize using a tree diagram.

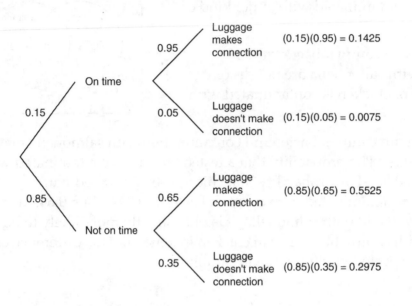

a) No, the flight leaving on time and the luggage making the connection are not independent events. The probability that the luggage makes the connection is dependent on whether or not the flight is on time. The probability is 0.95 if the flight is on time, and only 0.65 if it is not on time.

b) $P(\text{Luggage}) = P(\text{On time and Luggage}) + P(\text{Not on time and Luggage})$
$$= (0.15)(0.95) + (0.85)(0.65)$$
$$= 0.695$$

49. Late luggage.

Refer to the tree diagram constructed for Exercise 47.

$$P(\text{Not on time} \mid \text{No Lug.}) = \frac{P(\text{Not on time and No Luggage})}{P(\text{No Luggage})}$$

$$= \frac{(0.85)(0.35)}{(0.15)(0.05) + (0.85)(0.35)} \approx 0.975$$

If you pick Leah up at the Denver airport and her luggage is not there, the probability that her first flight was delayed is 0.975.

51. Absenteeism.

Organize the information in a tree diagram.

a) No, absenteeism is not independent of shift worked. The rate of absenteeism for the night shift is 2%, while the rate for the day shift is only 1%. If the two were independent, the percentages would be the same.

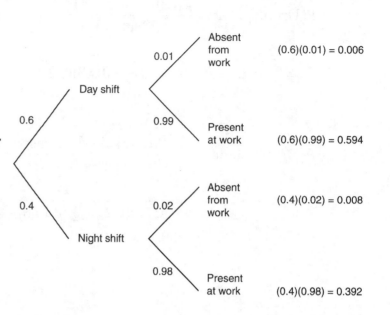

b)

$$P(\text{Absent})=P(\text{Day and Absent})+P(\text{Night and Absent})=(0.6)(0.01)+(0.4)(0.02)=0.014$$

The overall rate of absenteeism at this company is 1.4%.

53. Absenteeism, part II.

Refer to the tree diagram constructed for Exercise 51.

$$P(\text{Night} \mid \text{Absent})=\frac{P(\text{Night and Absent})}{P(\text{Absent})}=\frac{(0.4)(0.02)}{(0.6)(0.01)+(0.4)(0.02)}\approx 0.571$$

Approximately 57.1% of the company's absenteeism occurs on the night shift.

55. Drunks.

Organize the information into a tree diagram.

a) $P(\text{Detain} \mid \text{Not Drinking})=0.2$

b) $P(\text{Detain})$
$= P(\text{Drinking and Det.})$
$\quad +P(\text{Not Drinking and Det.})$
$= (0.12)(0.8)+(0.88)(0.2)$
$= 0.272$

c)

$$P(\text{Drunk} \mid \text{Det.}) = \frac{P(\text{Drunk and Det.})}{P(\text{Detain})}$$

$$= \frac{(0.12)(0.8)}{(0.12)(0.8) + (0.88)(0.2)}$$

$$\approx 0.353$$

d)

$$P(\text{Drunk} \mid \text{Release}) = \frac{P(\text{Drunk and Release})}{P(\text{Release})}$$

$$= \frac{(0.12)(0.2)}{(0.12)(0.2) + (0.88)(0.8)}$$

$$\approx 0.033$$

57. Dishwashers.

Organize the information in a tree diagram.

$$P(\text{Chuck} \mid \text{Break})$$

$$= \frac{P(\text{Chuck and Break})}{P(\text{Break})}$$

$$= \frac{(0.3)(0.03)}{(0.4)(0.01) + (0.3)(0.01) + (0.3)(0.03)}$$

$$\approx 0.563$$

If you hear a dish break, the probability that Chuck is on the job is approximately 0.563.

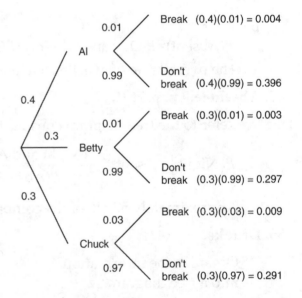

59. HIV Testing.

Organize the information in a tree diagram.

$$P(\text{No HIV} \mid \text{Test -}) = \frac{P(\text{No HIV and Test -})}{P(\text{Test -})}$$

$$= \frac{0.83725}{0.00045 + 0.83725}$$

$$\approx 0.999$$

The probability that a patient testing negative is truly free of HIV is about 99.9%.

Chapter 15 – Random Variables

Section 15.1

1. Oranges.

$\mu = E(\text{Oranges}) = 25(0.10) + 30(0.40) + 35(0.30) + 40(0.20) = 33$

The citrus farmer can expect an average of 33 oranges per tree.

Section 15.2

3. Oranges again.

$\sigma^2 = Var(\text{Oranges})$

$= (25-33)^2(0.10) + (30-33)^2(0.40) + (35-33)^2(0.30) + (40-33)^2(0.20) = 21$

$\sigma = SD(\text{Oranges}) = \sqrt{Var(\text{Oranges})} \approx \sqrt{21} = 4.58$

The standard deviation is 4.58 oranges per tree.

Section 15.3

5. Salary.

$E(\text{weekday} + \text{weekend}) = E(\text{weekday}) + E(\text{weekend}) = 1250 + 450 = \$1750$

To calculate the standard deviations, we must assume that the weekday and weekend salary expenses are independent.

$SD(\text{weekday} + \text{weekend}) = \sqrt{Var(\text{weekday}) + Var(\text{weekend})} = \sqrt{129^2 + 57^2} \approx \$141$

The total weekly salary has a mean of \$1750 and standard deviation of \$141.

Section 15.4

7. Battery.

Let X = the number of hours a calculator battery lasts

$z = \dfrac{x - \mu}{\sigma}$

$z = \dfrac{53 - 45}{5}$

$z = 1.6$

$P(X > 53) = P(z > 1.6) \approx 0.055$

According to the Normal model, the probability that a battery lasts more than 53 hours is approximately 0.055.

Chapter Exercises

9. Expected value.

a) $\mu = E(Y) = 10(0.3) + 20(0.5) + 30(0.2) = 19$

b) $\mu = E(Y) = 2(0.3) + 4(0.4) + 6(0.2) + 8(0.1) = 4.2$

11. Pick a card, any card.

a)

Win	$0	$5	$10	$30
P(amount won)	$\dfrac{26}{52}$	$\dfrac{13}{52}$	$\dfrac{12}{52}$	$\dfrac{1}{52}$

b) $\mu = E(\text{amount won}) = \$0\left(\dfrac{26}{52}\right) + \$5\left(\dfrac{13}{52}\right) + \$10\left(\dfrac{12}{52}\right) + \$30\left(\dfrac{1}{52}\right) \approx \$4.13$

c) Answers may vary. In the long run, the expected payoff of this game is $4.13 per play. Any amount less than $4.13 would be a reasonable amount to pay in order to play. Your decision should depend on how long you intend to play. If you are only going to play a few times, you should risk less.

13. Kids.

a)

Kids	1	2	3
P(Kids)	0.5	0.25	0.25

b) $\mu = E(\text{Kids}) = 1(0.5) + 2(0.25) + 3(0.25) = 1.75$ kids

c)

Boys	0	1	2	3
P(boys)	0.5	0.25	0.125	0.125

$\mu = E(\text{Boys}) = 0(0.5) + 1(0.25) + 2(0.125) + 3(0.125) = 0.875$ boys

15. Software.

Since the company can only get one contract, the probability that they will receive both is 0.

Profit	larger only $50,000	smaller only $20,000	both $70,000	neither $0
P(profit)	0.30	0.60	0	0.10

$\mu = E(\text{profit}) = \$50,000(0.30) + \$20,000(0.60) + \$70,000(0) + \$0(0.10)$
$= \$27,000$

17. Variation 1.

a)

$$\sigma^2 = Var(Y) = (10-19)^2(0.3) + (20-19)^2(0.5) + (30-19)^2(0.2) = 49$$
$$\sigma = SD(Y) = \sqrt{Var(Y)} = \sqrt{49} = 7$$

b)

$$\sigma^2 = Var(Y) = (2-4.2)^2(0.3) + (4-4.2)^2(0.4) + (6-4.2)^2(0.2) + (8-4.2)^2(0.1) = 3.56$$
$$\sigma = SD(Y) = \sqrt{Var(Y)} = \sqrt{3.56} \approx 1.89$$

19. Pick another card.

Answers may vary slightly (due to rounding of the mean)

$$\sigma^2 = Var(\text{Won}) = (0-4.13)^2\left(\frac{26}{52}\right) + (5-4.13)^2\left(\frac{13}{52}\right)$$
$$+ (10-4.13)^2\left(\frac{12}{52}\right) + (30-4.13)^2\left(\frac{1}{52}\right) \approx 29.5396$$
$$\sigma = SD(\text{Won}) = \sqrt{Var(\text{Won})} = \sqrt{29.5396} \approx \$5.44$$

21. Kids.

$$\sigma^2 = Var(\text{Kids}) = (1-1.75)^2(0.5) + (2-1.75)^2(0.25) + (3-1.75)^2(0.25) = 0.6875$$
$$\sigma = SD(\text{Kids}) = \sqrt{Var(\text{Kids})} = \sqrt{0.6875} \approx 0.83 \text{ kids}$$

23. Repairs.

a) $\mu = E(\text{Number of Repair Calls}) = 0(0.1) + 1(0.3) + 2(0.4) + 3(0.2) = 1.7 \text{ calls}$

b)

$$\sigma^2 = Var(\text{Calls}) = (0-1.7)^2(0.1) + (1-1.7)^2(0.3) + (2-1.7)^2(0.4) + (3-1.7)^2(0.2) = 0.81$$
$$\sigma = SD(\text{Calls}) = \sqrt{Var(\text{Calls})} = \sqrt{0.81} = 0.9 \text{ calls}$$

25. Defects.

The percentage of cars with *no* defects is 61%.

$$\mu = E(\text{Defects}) = 0(0.61) + 1(0.21) + 2(0.11) + 3(0.07) = 0.64 \text{ defects}$$

$$\sigma^2 = Var(\text{Defects}) = (0-0.64)^2(0.61) + (1-0.64)^2(0.21)$$
$$+ (2-0.64)^2(0.11) + (3-0.64)^2(0.07) \approx 0.8704$$
$$\sigma = SD(\text{Defects}) = \sqrt{Var(\text{Defects})} \approx \sqrt{0.8704} \approx 0.93 \text{ defects}$$

27. Cancelled flights.

a) $\mu = E(\text{gain}) = (-150)(0.20) + 100(0.80) = \$50$

b)

$$\sigma^2 = Var(\text{gain}) = (-150 - 50)^2 (0.20) + (100 - 50)^2 (0.80) = 10,000$$

$$\sigma = SD(\text{gain}) = \sqrt{Var(\text{gain})} \approx \sqrt{10,000} = \$100$$

29. Contest.

a) The two games are not independent. The probability that you win the second depends on whether or not you win the first.

b)

$$P(\text{losing both games}) = P(\text{losing the first}) \, P(\text{losing the second} \mid \text{first was lost})$$
$$= (0.6)(0.7) = 0.42$$

c)

$$P(\text{winning both games}) = P(\text{winning the first}) \, P(\text{winning the second} \mid \text{first was won})$$
$$= (0.4)(0.2) = 0.08$$

d)

X	0	1	2
$P(X = x)$	0.42	0.50	0.08

e)

$$\mu = E(X) = 0(0.42) + 1(0.50) + 2(0.08) = 0.66 \text{ games}$$

$$\sigma^2 = Var(X) = (0 - 0.66)^2 (0.42) + (1 - 0.66)^2 (0.50) + (2 - 0.66)^2 (0.08) = 0.3844$$

$$\sigma = SD(X) = \sqrt{Var(X)} = \sqrt{0.3844} \approx 0.62 \text{ games}$$

31. Batteries.

a)

# good	0	1	2
$P(\text{# good})$	$\left(\dfrac{3}{10}\right)\left(\dfrac{2}{9}\right) = \dfrac{6}{90}$	$\left(\dfrac{3}{10}\right)\left(\dfrac{7}{9}\right) + \left(\dfrac{7}{10}\right)\left(\dfrac{3}{9}\right) = \dfrac{42}{90}$	$\left(\dfrac{7}{10}\right)\left(\dfrac{6}{9}\right) = \dfrac{42}{90}$

b) $\mu = E(\text{number good}) = 0\left(\dfrac{6}{90}\right) + 1\left(\dfrac{42}{90}\right) + 2\left(\dfrac{42}{90}\right) = 1.4 \text{ batteries}$

c)

$$\sigma^2 = Var(\text{number good}) = (0-1.4)^2\left(\frac{6}{90}\right) + (1-1.4)^2\left(\frac{42}{90}\right) + (2-1.4)^2\left(\frac{42}{90}\right) \approx 0.3733$$

$$\sigma = SD(\text{number good}) = \sqrt{Var(\text{number good})} \approx \sqrt{0.3733} \approx 0.61 \text{ batteries.}$$

33. Random variables.

a)

$$\mu = E(3X) = 3(E(X)) = 3(10) = 30$$
$$\sigma = SD(3X) = 3(SD(X)) = 3(2) = 6$$

b)

$$\mu = E(Y+6) = E(Y) + 6 = 20 + 6 = 26$$
$$\sigma = SD(Y+6) = SD(Y) = 5$$

c)

$$\mu = E(X+Y)$$
$$= E(X) + E(Y) = 10 + 20 = 30$$
$$\sigma = SD(X+Y) = \sqrt{Var(X) + Var(Y)}$$
$$= \sqrt{2^2 + 5^2} \approx 5.39$$

d)

$$\mu = E(X-Y) = E(X) - E(Y)$$
$$= 10 - 20 = -10$$
$$\sigma = SD(X-Y) = \sqrt{Var(X) + Var(Y)}$$
$$= \sqrt{2^2 + 5^2} \approx 5.39$$

e)

$$\mu = E(X_1 + X_2) = E(X) + E(X) = 10 + 10 = 20$$
$$\sigma = SD(X_1 + X_2) = \sqrt{Var(X) + Var(X)}$$
$$= \sqrt{2^2 + 2^2} \approx 2.83$$

35. Random variables.

a)

$$\mu = E(0.8Y) = 0.8(E(Y)) = 0.8(300) = 240$$
$$\sigma = SD(0.8Y) = 0.8(SD(Y))$$
$$= 0.8(16) = 12.8$$

b)

$$\mu = E(2X-100) = 2(E(X)) - 100 = 140$$
$$\sigma = SD(2X-100)$$
$$= 2(SD(X)) = 2(12) = 24$$

c)

$$\mu = E(X+2Y) = E(X) + 2(E(Y))$$
$$= 120 + 2(300) = 720$$
$$\sigma = SD(X+2Y) = \sqrt{Var(X) + 2^2 Var(Y)}$$
$$= \sqrt{12^2 + 2^2(16^2)} \approx 34.18$$

d)

$$\mu = E(3X-Y) = 3(E(X)) - E(Y)$$
$$= 3(120) - 300 = 60$$
$$\sigma = SD(3X-Y)$$
$$= \sqrt{3^2 Var(X) + Var(Y)}$$
$$= \sqrt{3^2(12^2) + 16^2} \approx 39.40$$

e)

$$\mu = E(Y_1 + Y_2) = E(Y) + E(Y) = 300 + 300 = 600$$
$$\sigma = SD(Y_1 + Y_2) = \sqrt{Var(Y) + Var(Y)}$$
$$= \sqrt{16^2 + 16^2} \approx 22.63$$

37. Eggs.

a) $\mu = E(\text{Broken eggs in 3 doz.}) = 3(E(\text{Broken eggs in 1 doz.})) = 3(0.6) = 1.8$ eggs

b) $\sigma = SD(\text{Broken eggs in 3 dozen}) = \sqrt{0.5^2 + 0.5^2 + 0.5^2} \approx 0.87$ eggs

c) The cartons of eggs must be independent of each other.

39. Repair calls.

$\mu = E(\text{calls in 8 hours}) = 8(E(\text{calls in 1 hour}) = 8(1.7) = 13.6$ calls

$\sigma = SD(\text{calls in 8 hours}) = \sqrt{8(Var(\text{calls in 1 hour}))} = \sqrt{8(0.9)^2} \approx 2.55$ calls

This is only valid if the hours are independent of one another.

41. Tickets.

a)

$\mu = E(\text{tickets for 18 trucks})$
$= 18(E(\text{tickets for one truck})) = 18(1.3) = 23.4$ tickets

$\sigma = SD(\text{tickets for 18 trucks})$
$= \sqrt{18(Var(\text{tickets for one truck})} = \sqrt{18(0.7)^2} \approx 2.97$ tickets

b) We are assuming that trucks are ticketed independently.

43. Fire!

a) The standard deviation is large because the profits on insurance are highly variable. Although there will be many small gains, there will occasionally be large losses, when the insurance company has to pay a claim.

b)

$\mu = E(\text{two policies}) = 2(E(\text{one policy})) = 2(150) = \$300$

$\sigma = SD(\text{two policies}) = \sqrt{2(Var(\text{one policy}))} = \sqrt{2(6000^2)} \approx \$8,485.28$

c)

$\mu = E(10,000 \text{ policies}) = 10,000(E(\text{one policy})) = 10,000(150) = \$1,500,000$

$\sigma = SD(10,000 \text{ policies}) = \sqrt{10,000(Var(\text{one policy}))} = \sqrt{10,000(6000^2)} = \$600,000$

d) If the company sells 10,000 policies, they are likely to be successful. A profit of $0, is 2.5 standard deviations below the expected profit. This is unlikely to happen. However, if the company sells fewer policies, then the likelihood of turning a profit decreases. In an extreme case, where only two policies are sold, a profit of $0 is more likely, being only a small fraction of a standard deviation below the mean.

e) This analysis depends on each of the policies being independent from each other. This assumption of independence may be violated if there are many fire insurance claims as a result of a forest fire, or other natural disaster.

45. Cereal.

a) $E(\text{large bowl} - \text{small bowl}) = E(\text{large bowl}) - E(\text{small bowl}) = 2.5 - 1.5 = 1$ ounce

b) $\sigma = SD(\text{large} - \text{small}) = \sqrt{Var(\text{large}) + Var(\text{small})} = \sqrt{0.4^2 + 0.3^2} = 0.5$ ounces

c)

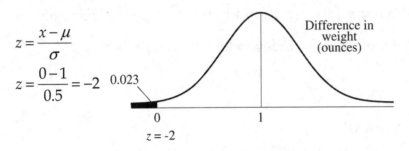

$$z = \frac{x - \mu}{\sigma}$$

$$z = \frac{0 - 1}{0.5} = -2 \quad 0.023$$

$$z = -2$$

The small bowl will contain more cereal than the large bowl when the difference between the amounts is less than 0. According to the Normal model, the probability of this occurring is approximately 0.023.

d)

$$\mu = E(\text{large} + \text{small}) = E(\text{large}) + E(\text{small}) = 2.5 + 1.5 = 4 \text{ ounce}$$

$$\sigma = SD(\text{large} + \text{small}) = \sqrt{Var(\text{large}) + Var(\text{small})} = \sqrt{0.4^2 + 0.3^2} = 0.5 \text{ ounces}$$

e)

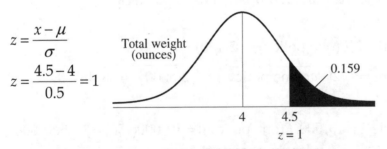

$$z = \frac{x - \mu}{\sigma}$$

$$z = \frac{4.5 - 4}{0.5} = 1$$

$$z = 1$$

According to the Normal model, the probability that the total weight of cereal in the two bowls is more than 4.5 ounces is approximately 0.159.

f)

$$\mu = E(\text{box} - \text{large} - \text{small}) = E(\text{box}) - E(\text{large}) - E(\text{small})$$

$$= 16.3 - 2.5 - 1.5 = 12.3 \text{ ounces}$$

$$\sigma = SD(\text{box} - \text{large} - \text{small}) = \sqrt{Var(\text{box}) + Var(\text{large}) + Var(\text{small})}$$

$$= \sqrt{0.2^2 + 0.3^2 + 0.4^2} \approx 0.54 \text{ ounces}$$

47. More cereal.

a)

$$\mu = E(\text{box} - \text{large} - \text{small})$$
$$= E(\text{box}) - E(\text{large}) - E(\text{small}) = 16.2 - 2.5 - 1.5 = 12.2 \text{ ounces}$$

b)

$$\sigma = SD(\text{box} - \text{large} - \text{small}) = \sqrt{Var(\text{box}) + Var(\text{large}) + Var(\text{small})}$$
$$= \sqrt{0.1^2 + 0.3^2 + 0.4^2} \approx 0.51 \text{ ounces}$$

c)

$$z = \frac{x - \mu}{\sigma}$$
$$z = \frac{13 - 12.2}{0.51}$$
$$z = 1.57$$

Weight left in box (ounces)

0.058

12.2   13

$z = 1.57$

According to the Normal model, the probability that the box contains more than 13 ounces is about 0.058.

49. Medley.

a)

$$\mu = E(\#1 + \#2 + \#3 + \#4) = E(\#1) + E(\#2) + E(\#3) + E(\#4)$$
$$= 50.72 + 55.51 + 49.43 + 44.91 = 200.57 \text{ seconds}$$
$$\sigma = SD(\#1 + \#2 + \#3 + \#4) = \sqrt{Var(\#1) + Var(\#2) + Var(\#3) + Var(\#4)}$$
$$= \sqrt{0.24^2 + 0.22^2 + 0.25^2 + 0.21^2} \approx 0.46 \text{ seconds}$$

b)

$$z = \frac{x - \mu}{\sigma}$$
$$z = \frac{199.48 - 200.57}{0.461}$$
$$z = -2.36$$

Total time (seconds)

0.009

199.48   200.57

$z = -2.36$

The team is not likely to swim faster than their best time. According to the Normal model, they are only expected to swim that fast or faster about 0.9% of the time.

51. Farmer's market.

a) Let A = price of a pound of apples, and let P = price of a pound of potatoes.

Profit $= 100A + 50P - 2$

b) $\mu = E(100A + 50P - 2) = 100(E(A)) + 50(E(P)) - 2 = 100(0.5) + 50(0.3) - 2 = \$63$

c)

$$\sigma = SD(100A + 50P - 2)$$

$$= \sqrt{100^2(Var(A)) + 50^2(Var(P))} = \sqrt{100^2(0.2^2) + 50^2(0.1^2)} \approx \$20.62$$

d) No assumptions are necessary to compute the mean. To compute the standard deviation, independent market prices must be assumed.

53. Coffee and doughnuts.

a)

$$\mu = E(\text{cups sold in 6 days}) = 6(E(\text{cups sold in 1 day})) = 6(320) = 1920 \text{ cups}$$

$$\sigma = SD(\text{cups sold in 6 days}) = \sqrt{6(Var(\text{cups sold in 1 day})} = \sqrt{6(20)^2} \approx 48.99 \text{ cups}$$

The distribution of total coffee sales for 6 days has distribution $N(1920,48.99)$.

$$z = \frac{x - \mu}{\sigma}$$

$$z = \frac{2000 - 1920}{48.99}$$

$$z = 1.633$$

According to the Normal model, the probability that he will sell more than 2000 cups of coffee in a week is approximately 0.051.

b) Let C = the number of cups of coffee sold. Let D = the number of doughnuts sold.

$$\mu = E(0.50C + 0.40D) = 0.50(E(C)) + 0.40(E(D)) = 0.50(320) + 0.40(150) = \$220$$

$$\sigma = SD(0.50C + 0.40D)$$

$$= \sqrt{0.50^2(Var(C)) + 0.40^2(Var(D))} = \sqrt{0.50^2(20^2) + 0.40^2(12^2)} \approx \$11.09$$

The day's profit can be modeled by $N(220,11.09)$. A day's profit of $300 is over 7 standard deviations above the mean. This is extremely unlikely. It would not be reasonable for the shop owner to expect the day's profit to exceed $300.

c) Consider the difference $D - 0.5C$. When this difference is greater than zero, the number of doughnuts sold is greater than half the number of cups of coffee sold.

$$\mu = E(D - 0.5C) = (E(D)) - 0.5(E(C)) = 150 - 0.5(320) = -\$10$$

$$\sigma = SD(D - 0.5C) = \sqrt{(Var(D)) + 0.5(Var(C))} = \sqrt{(12^2) + 0.5^2(20^2)} \approx \$15.62$$

The difference $D - 0.5C$ can be modeled by $N(-10, 15.62)$.

According to the Normal model, the probability that the shop owner will sell a doughnut to more than half of the coffee customers is approximately 0.26.

Chapter 16 – Probability Models

Section 16.1

1. Bernoulli.

a) These are not Bernoulli trials. The possible outcomes are 1, 2, 3, 4, 5, and 6. There are more than two possible outcomes.

b) These may be considered Bernoulli trials. There are only two possible outcomes, Type A and not Type A. Assuming the 120 donors are representative of the population, the probability of having Type A blood is 43%. The trials are not independent, because the population is finite, but the 120 donors represent less than 10% of all possible donors.

c) These are not Bernoulli trials. The probability of getting a heart changes as cards are dealt without replacement.

d) These are not Bernoulli trials. We are sampling without replacement, so the trials are not independent. Samples without replacement may be considered Bernoulli trials if the sample size is less than 10% of the population, but 500 is more than 10% of 3000.

e) These may be considered Bernoulli trials. There are only two possible outcomes, sealed properly and not sealed properly. The probability that a package is unsealed is constant, at about 10%, as long as the packages checked are a representative sample of all packages. Finally, the trials are not independent, since the total number of packages is finite, but the 24 packages checked probably represent less than 10% of the packages.

Section 16.2

3. Toasters.

Let X = the number of toasters that need repair.

The condition of the toasters can be considered Bernoulli trials. There are only two possible outcomes, needing to be sent back for repairs and not needing to be sent back for repairs. The probability that a toaster needs repair is constant, given as $p = 0.05$. The trials are not independent, since there are a finite number of toasters, but 20 toasters in each carton is less than 10% of all toasters produced.

The distribution of the number of repairs required follows $Binom(20, 0.05)$.

$$P(\text{exactly 3 toasters need repair}) = P(X = 3)$$
$$= {}_{20}C_3 (0.05)^3 (0.95)^{17}$$
$$\approx 0.0596$$

According to the Binomial model, the probability that exactly three toasters out of twenty require repair is 0.0596.

Section 16.3

5. **Toasters again.**

A binomial model and a normal model are both appropriate for modeling the number of toasters that need to be sent back for minor repair.

Let X = the number of toasters that need to be sent back.

The condition of the toasters can be considered Bernoulli trials, as verified in the previous exercise. The distribution of the number of repairs required follows *Binom*(10000, 0.05).

$E(X) = np = 10,000(0.05) = 500$ toasters.

$SD(X) = \sqrt{npq} = \sqrt{10000(0.05)(0.95)} \approx 21.79$ toasters.

Since $np = 500$ and $nq = 9,500$ are both greater than 10, *Binom*(10000, 0.05) may be approximated by the Normal model, $N(500, 21.79)$.

Using *Binom*(10000, 0.05):

$P(\text{fewer than 450 toasters}) = P(X < 450)$

$$= {}_{10000}C_0 (0.05)^0 (0.95)^{10000} + ... + {}_{10000}C_{449} (0.05)^{449} (0.95)^{9551}$$
$$\approx 0.009$$

According to the Binomial model, the probability that fewer than 450 of 10,000 toasters need repair is approximately 0.009.

Using $N(500, 21.79)$:

$z = \dfrac{x - \mu}{\sigma}$

$z = \dfrac{450 - 500}{21.79}$

$z = -2.29$

Number of toasters requiring repair out of 10,000

0.011

450   500

$z = -2.29$

$P(X < 450) \approx P(z < -2.29) \approx 0.011$

According to the Normal model, the probability that fewer than 450 of 10,000 toasters need repair is approximately 0.011.

Section 16.4

7. Sell!

$$E(X) = \lambda = 5$$

$$P(X = x) = \frac{e^{-\lambda}\lambda^x}{x!}$$

$$P(X = 3) = \frac{e^{-5}5^3}{3!} \approx 0.140$$

According to the Poisson model, the probability of the dealer selling 3 cars is 0.140.

Section 16.7

9. Telephone numbers.

a) The telephone numbers are distributed uniformly, since all numbers are equally likely to be selected.

b) According to the uniform distribution, the probability of choosing one of the numbers assigned the incubator is 200/10,000 = 0.02.

c) According to the uniform distribution, the probability of choosing a number above 9000 is 1000/10,000 = 0.10

11. Component lifetimes.

a) The mean of the exponential model is $\mu = \dfrac{1}{\lambda}$, so $\lambda = \dfrac{1}{\mu} = \dfrac{1}{3}$.

b) Let X = the number of years in the hard drive lifetime.

$$P(X \le 5) = 1 - e^{-\lambda t} = 1 - e^{-(1/3)(5)} \approx 0.811$$

According to the exponential model, the probability that hard drive lasts 5 years or less is approximately 0.811.

13. Simulating the model.

a) Answers will vary. A component is the simulation of the picture in one box of cereal. One possible way to model this component is to generate random digits 0-9. Let 0 and 1 represent Hope Solo and 2-9 a picture of another sports star. Each run will consist of generating random numbers until a 0 or 1 is generated. The response variable will be the number of digits generated until the first 0 or 1.

b) Answers will vary.

c) Answers will vary. To construct your simulated probability model, start by calculating the simulated probability that you get a picture of Hope Solo in the first box. This is the number of trials in which a 0 or 1 was generated first, divided by the total number of trials. Perform similar calculations for the simulated probability that you have to wait until the second box, the third box, etc.

d) Let X = the number of boxes opened until the first Hope Solo picture is found.

X	1	2	3	4	5	6	7	8	≥9
P(X)	0.20	$(0.80)(0.20)$ $= 0.16$	$(0.80)^2(0.20)$ $= 0.128$	$(0.80)^3(0.20)$ $= 0.1024$	0.082	0.066	0.052	0.042	0.168

e) Answers will vary.

15. Hope, again.

a) Answers will vary. A component is the simulation of the picture in one box of cereal. One possible way to model this component is to generate random digits 0-9. Let 0 and 1 represent Hope Solo and 2-9 a picture of another sports star. Each run will consist of generating five random numbers. The response variable will be the number of 0s and 1s in the five random numbers.

b) Answers will vary.

c) Answers will vary. To construct your simulated probability model, start by calculating the simulated probability that you get no pictures of Hope Solo in the five boxes. This is the number of trials in which neither 0 nor 1 were generated divided by the total number of trials. Perform similar calculations for the simulated probability that you would get one picture, 2 pictures, etc.

d) Let X = the number of Hope Solo pictures in 5 boxes.

X	0	1	2	3	4	5
P(X)	$(0.20)^0(0.80)^5$ ≈ 0.33	$_5C_1(0.20)^1(0.80)^4$ ≈ 0.41	$_5C_2(0.20)^2(0.80)^3$ ≈ 0.20	$_5C_3(0.20)^3(0.80)^2$ ≈ 0.05	$_5C_4(0.20)^4(0.80)^1$ ≈ 0.01	$(0.20)^5(0.80)^0$ ≈ 0.0

e) Answers will vary.

17. On time.

These departures cannot be considered Bernoulli trials. Departures from the same airport during a 2-hour period may not be independent. They all might be affected by weather and delays.

19. Hoops.

The player's shots may be considered Bernoulli trials. There are only two possible outcomes (make or miss), the probability of making a shot is constant (80%), and the shots are independent of one another (making, or missing, a shot does not affect the probability of making the next).

Let X = the number of shots until the first missed shot.
Let Y = the number of shots until the first made shot.

Since these problems deal with shooting until the first miss (or until the first made shot), a geometric model, either $Geom(0.8)$ or $Geom(0.2)$, is appropriate.

a) Use $Geom(0.2)$. $P(X = 5) = (0.8)^4(0.2) = 0.08192$ (Four shots made, followed by a miss.)

b) Use $Geom(0.8)$. $P(Y = 4) = (0.2)^3(0.8) = 0.0064$ (Three misses, then a made shot.)

c) Use $Geom(0.8)$. $P(Y = 1) + P(Y = 2) + P(Y = 3) = (0.8) + (0.2)(0.8) + (0.2)^2(0.8) = 0.992$

21. More hoops.

As determined in a previous exercise, the shots can be considered Bernoulli trials, and since the player is shooting until the first miss, $Geom(0.2)$ is the appropriate model.

$$E(X) = \frac{1}{p} = \frac{1}{0.2} = 5 \text{ shots}$$

The player is expected to take 5 shots until the first miss.

23. Customer center operator.

The calls can be considered Bernoulli trials. There are only two possible outcomes, taking the promotion, and not taking the promotion. The probability of success is constant at 5% (50% of the 10% Platinum cardholders.) The trials are not independent, since there are a finite number of cardholders, but this is a major credit card company, so we can assume we are selecting fewer than 10% of all cardholders. Since we are calling people until the first success, the model $Geom(0.05)$ may be used.

$$E(\text{calls}) = \frac{1}{p} = \frac{1}{0.05} = 20 \text{ calls.}$$ We expect it to take 20 calls to find the first cardholder to take the double miles promotion.

25. Blood.

These may be considered Bernoulli trials. There are only two possible outcomes, Type AB and not Type AB. Provided that the donors are representative of the population, the probability of having Type AB blood is constant at 4%. The trials are not independent, since the population is finite, but we are selecting fewer than 10% of all potential donors. Since we are selecting people until the first success, the model *Geom*(0.04) may be used.

Let X = the number of donors until the first Type AB donor is found.

a) $E(X) = \dfrac{1}{p} = \dfrac{1}{0.04} = 25$ people

We expect the 25th person to be the first Type AB donor.

b)

P(a Type AB donor among the first 5 people checked)
$$= P(X = 1) + P(X = 2) + P(X = 3) + P(X = 4) + P(X = 5)$$
$$= (0.04) + (0.96)(0.04) + (0.96)^2(0.04) + (0.96)^3(0.04) + (0.96)^4(0.04) \approx 0.185$$

c)

P(a Type AB donor among the first 6 people checked)
$$= P(X = 1) + P(X = 2) + P(X = 3) + P(X = 4) + P(X = 5) + P(X = 6)$$
$$= (0.04) + (0.96)(0.04) + (0.96)^2(0.04)$$
$$+ (0.96)^3(0.04) + (0.96)^4(0.04) + (0.96)^5(0.04) \approx 0.217$$

d) P(no Type AB donor before the 10th person checked) $= P(X > 9) = (0.96)^9 \approx 0.693$

This one is a bit tricky. There is no implication that we actually find a donor on the 10th trial. We only care that nine trials passed with no Type AB donor.

27. Coins and intuition.

a) Intuitively, we expect 50 heads.

b) $E(\text{heads}) = np = 100(0.5) = 50$ heads.

29. Lefties.

These may be considered Bernoulli trials. There are only two possible outcomes, left-handed and not left-handed. Since people are selected at random, the probability of being left-handed is constant at about 13%. The trials are not independent, since the population is finite, but a sample of 5 people is certainly fewer than 10% of all people.

Let X = the number of people checked until the first lefty is discovered.
Let Y = the number of lefties among n = 5.

a) Use *Geom*(0.13).

$P(\text{first lefty is the fifth person}) = P(X = 5) = (0.87)^4 (0.13) \approx 0.0745$

b) Use *Binom*(5, 0.13).

$P(\text{some lefties among the 5 people}) = 1 - P(\text{no lefties among the first 5 people})$
$$= 1 - P(Y = 0)$$
$$= 1 - {}_5C_0 (0.13)^0 (0.87)^5$$
$$\approx 0.502$$

c) Use *Geom*(0.13).

$P(\text{first lefty is second or third person}) = P(X = 2) + P(X = 3)$
$$= (0.87)(0.13) + (0.87)^2 (0.13) \approx 0.211$$

d) Use *Binom*(5, 0.13).

$P(\text{exactly 3 lefties in the group}) = P(Y = 3) = {}_5C_3 (0.13)^3 (0.87)^2 \approx 0.0166$

e) Use *Binom*(5, 0.13).

$P(\text{at least 3 lefties in the group})$
$= P(Y = 3) + P(Y = 4) + P(Y = 5)$
$= {}_5C_3 (0.13)^3 (0.87)^2 + {}_5C_4 (0.13)^4 (0.87)^1 + {}_5C_5 (0.13)^5 (0.87)^0$
≈ 0.0179

f) Use *Binom*(5, 0.13).

$P(\text{at most 3 lefties in the group}) = P(Y = 0) + P(Y = 1) + P(Y = 2) + P(Y = 3)$
$$= {}_5C_0 (0.13)^0 (0.87)^5 + {}_5C_1 (0.13)^1 (0.87)^4$$
$$+ {}_5C_2 (0.13)^2 (0.87)^3 + {}_5C_3 (0.13)^3 (0.87)^2$$
$$\approx 0.9987$$

31. Lefties redux.

a) In a previous exercise, we determined that the selection of lefties could be considered Bernoulli trials. Since our group consists of 5 people, use *Binom*(5, 0.13).

Let Y = the number of lefties among $n = 5$.

$E(Y) = np = 5(0.13) = 0.65$ lefties

b) $SD(Y) = \sqrt{npq} = \sqrt{5(0.13)(0.87)} \approx 0.75$ lefties

c) Use *Geom*(0.13). Let X = the number of people checked until the first lefty is discovered.

$$E(X) = \frac{1}{p} = \frac{1}{0.13} \approx 7.69 \text{ people}$$

33. Still more lefties.

a) In a previous exercise, we determined that the selection of lefties (and also righties) could be considered Bernoulli trials. Since our group consists of 12 people, and now we are considering the righties, use $Binom(12, 0.87)$.

Let $Y =$ the number of righties among $n = 12$.

$E(Y) = np = 12(0.87) = 10.44$ righties

$SD(Y) = \sqrt{npq} = \sqrt{12(0.87)(0.13)} \approx 1.16$ righties

b)
$$\begin{aligned} P(\text{not all righties}) &= 1 - P(\text{all righties}) \\ &= 1 - P(Y = 12) \\ &= 1 - {}_{12}C_{12}(0.87)^{12}(0.13)^0 \\ &\approx 0.812 \end{aligned}$$

c)
$$\begin{aligned} P(\text{no more than 10 righties}) &= P(Y \leq 10) \\ &= P(Y = 0) + P(Y = 1) + P(Y = 2) + \ldots + P(Y = 10) \\ &= {}_{12}C_0(0.87)^0(0.13)^{12} + \ldots + {}_{12}C_{10}(0.87)^{10}(0.13)^2 \\ &\approx 0.475 \end{aligned}$$

d)
$$P(\text{exactly six of each}) = P(Y = 6) = {}_{12}C_6(0.87)^6(0.13)^6 \approx 0.00193$$

e)
$$\begin{aligned} P(\text{majority righties}) &= P(Y \geq 7) \\ &= P(Y = 7) + P(Y = 8) + P(Y = 9) + \ldots + P(Y = 12) \\ &= {}_{12}C_7(0.87)^7(0.13)^5 + \ldots + {}_{12}C_{12}(0.87)^{12}(0.13)^0 \\ &\approx 0.998 \end{aligned}$$

35. Vision.

The vision tests can be considered Bernoulli trials. There are only two possible outcomes, nearsighted or not. The probability of any child being nearsighted is given as $p = 0.12$. Finally, since the population of children is finite, the trials are not independent. However, 169 is certainly less than 10% of all children, and we will assume that the children in this district are representative of all children in relation to nearsightedness. Use $Binom(169, 0.12)$.

$\mu = E(\text{nearsighted}) = np = 169(0.12) = 20.28$ children.

$\sigma = SD(\text{nearsighted}) = \sqrt{npq} = \sqrt{169(0.12)(0.88)} \approx 4.22$ children.

37. Tennis, anyone?

The first serves can be considered Bernoulli trials. There are only two possible outcomes, successful and unsuccessful. The probability of any first serve being good is given as $p = 0.70$. Finally, we are assuming that each serve is independent of the others. Since she is serving 6 times, use $Binom(6, 0.70)$.

Let $X =$ the number of successful serves in $n = 6$ first serves.

a)

$$P(\text{six serves in}) = P(X = 6)$$
$$= {}_6C_6 (0.70)^6 (0.30)^0$$
$$\approx 0.118$$

b)

$$P(\text{exactly four serves in})$$
$$= P(X = 4)$$
$$= {}_6C_4 (0.70)^4 (0.30)^2$$
$$\approx 0.324$$

c)

$$P(\text{at least four serves in}) = P(X = 4) + P(X = 5) + P(X = 6)$$
$$= {}_6C_4 (0.70)^4 (0.30)^2 + {}_6C_5 (0.70)^5 (0.30)^1 + {}_6C_6 (0.70)^6 (0.30)^0$$
$$\approx 0.744$$

d)

$$P(\text{no more than four serves in})$$
$$= P(X = 0) + P(X = 1) + P(X = 2) + P(X = 3) + P(X = 4)$$
$$= {}_6C_0 (0.70)^0 (0.30)^6 + {}_6C_1 (0.70)^1 (0.30)^5 + {}_6C_2 (0.70)^2 (0.30)^4$$
$$+ {}_6C_3 (0.70)^3 (0.30)^3 + {}_6C_4 (0.70)^4 (0.30)^2$$
$$\approx 0.580$$

39. And more tennis.

The first serves can be considered Bernoulli trials. There are only two possible outcomes, successful and unsuccessful. The probability of any first serve being good is given as $p = 0.70$. Finally, we are assuming that each serve is independent of the others. Since she is serving 80 times, use $Binom(80, 0.70)$.

Let $X =$ the number of successful serves in $n = 80$ first serves.

a) $E(X) = np = 80(0.70) = 56$ first serves in.

$SD(X) = \sqrt{npq} = \sqrt{80(0.70)(0.30)} \approx 4.10$ first serves in.

b) Since $np = 56$ and $nq = 24$ are both greater than 10, $Binom(80, 0.70)$ may be approximated by the Normal model, $N(56, 4.10)$.

c) According to the Normal model, in matches with 80 serves, she is expected to make between 51.9 and 60.1 first serves approximately 68% of the time, between 47.8 and 64.2 first serves approximately 95% of the time, and between 43.7 and 68.3 first serves approximately 99.7% of the time.

d) Using *Binom*(80, 0.70):

$$P(\text{at least 65 first serves}) = P(X \geq 65)$$
$$= P(X = 65) + P(X = 66) + \ldots + P(X = 80)$$
$$= {}_{80}C_{65}(0.70)^{65}(0.30)^{15} + \ldots + {}_{80}C_{80}(0.70)^{80}(0.30)^{0}$$
$$\approx 0.0161$$

According to the Binomial model, the probability that she makes at least 65 first serves out of 80 is approximately 0.0161.

Using *N*(56, 4.10):

$$z = \frac{x - \mu}{\sigma}$$
$$z = \frac{65 - 56}{4.10}$$
$$z \approx 2.195$$

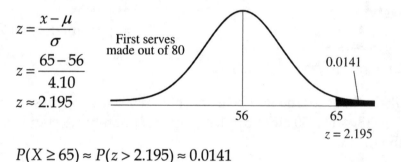

According to the Normal model, the probability that she makes at least 65 first serves out of 80 is approximately 0.0141.

$$P(X \geq 65) \approx P(z > 2.195) \approx 0.0141$$

41. Apples.

a) A binomial model and a normal model are both appropriate for modeling the number of cider apples that may come from the tree.

Let X = the number of cider apples found in the $n = 300$ apples from the tree.

The quality of the apples may be considered Bernoulli trials. There are only two possible outcomes, cider apple or not a cider apple. The probability that an apple must be used for a cider apple is constant, given as $p = 0.06$. The trials are not independent, since the population of apples is finite, but the apples on the tree are undoubtedly less than 10% of all the apples that the farmer has ever produced, so model with *Binom*(300, 0.06).

$$E(X) = np = 300(0.06) = 18 \text{ cider apples.}$$

$$SD(X) = \sqrt{npq} = \sqrt{300(0.06)(0.94)} \approx 4.11 \text{ cider apples.}$$

Since $np = 18$ and $nq = 282$ are both greater than 10, $Binom(300, 0.06)$ may be approximated by the Normal model, $N(18, 4.11)$.

b) Using $Binom(300, 0.06)$:

$$P(\text{at most 12 cider apples}) = P(X \le 12)$$
$$= P(X = 0) + \ldots + P(X = 12)$$
$$= {}_{300}C_0 (0.06)^0 (0.94)^{300} + \ldots + {}_{300}C_{12} (0.06)^{12} (0.94)^{282}$$
$$\approx 0.085$$

According to the Binomial model, the probability that no more than 12 cider apples come from the tree is approximately 0.085.

Using $N(18, 4.11)$:

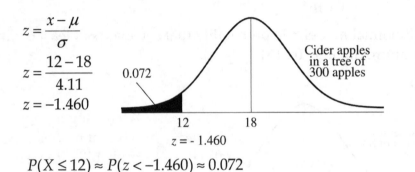

$$z = \frac{x - \mu}{\sigma}$$

$$z = \frac{12 - 18}{4.11}$$

$$z = -1.460$$

According to the Normal model, the probability that no more than 12 apples out of 300 are cider apples is approximately 0.072.

$$P(X \le 12) \approx P(z < -1.460) \approx 0.072$$

c) It is extremely unlikely that the tree will bear more than 50 cider apples. Using the Normal model, $N(18, 4.11)$, 50 cider apples is approximately 7.8 standard deviations above the mean.

43. Lefties again.

Let $X =$ the number of righties among a class of $n = 188$ students.

Using $Binom(188, 0.87)$:

These may be considered Bernoulli trials. There are only two possible outcomes, right-handed and not right-handed. The probability of being right-handed is assumed to be constant at about 87%. The trials are not independent, since the population is finite, but a sample of 188 students is certainly fewer than 10% of all people. Therefore, the number of righties in a class of 188 students may be modeled by $Binom(188, 0.87)$.

If there are 171 or more righties in the class, some righties have to use a left-handed desk.

$P(\text{at least 171 righties}) = P(X \geq 171)$
$$= P(X = 171) + \ldots + P(X = 188)$$
$$= {}_{188}C_{171}(0.87)^{171}(0.13)^{17} + \ldots + {}_{188}C_{188}(0.87)^{188}(0.13)^{0}$$
$$\approx 0.061$$

According to the binomial model, the probability that a right-handed student has to use a left-handed desk is approximately 0.061.

Using $N(163.56, 4.61)$:

$E(X) = np = 188(0.87) = 163.56$ righties.

$SD(X) = \sqrt{npq} = \sqrt{188(0.87)(0.13)} \approx 4.61$ righties.

Since $np = 163.56$ and $nq = 24.44$ are both greater than 10, $Binom(188, 0.87)$ may be approximated by the Normal model, $N(163.56, 4.61)$.

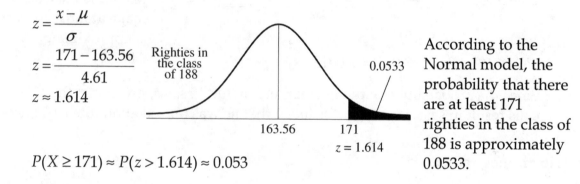

$z = \dfrac{x - \mu}{\sigma}$

$z = \dfrac{171 - 163.56}{4.61}$

$z \approx 1.614$

Righties in the class of 188

0.0533

163.56   171

$z = 1.614$

According to the Normal model, the probability that there are at least 171 righties in the class of 188 is approximately 0.0533.

$P(X \geq 171) \approx P(z > 1.614) \approx 0.053$

45. Annoying phone calls.

Let $X =$ the number of sales made after making $n = 200$ calls.

Using $Binom(200, 0.12)$:

These may be considered Bernoulli trials. There are only two possible outcomes, making a sale and not making a sale. The telemarketer was told that the probability of making a sale is constant at about $p = 0.12$. The trials are not independent, since the population is finite, but 200 calls is fewer than 10% of all calls. Therefore, the number of sales made after making 200 calls may be modeled by $Binom(200, 0.12)$.

$P(\text{at most 10}) = P(X \leq 10)$
$$= P(X = 0) + \ldots + P(X = 10)$$
$$= {}_{200}C_{0}(0.12)^{0}(0.88)^{200} + \ldots + {}_{200}C_{1900}(0.12)^{10}(0.88)^{190}$$
$$\approx 0.0006$$

According to the Binomial model, the probability that the telemarketer would make at most 10 sales is approximately 0.0006.

Using $N(24, 4.60)$:

$E(X) = np = 200(0.12) = 24$ sales.

$SD(X) = \sqrt{npq} = \sqrt{200(0.12)(0.88)} \approx 4.60$ sales.

Since $np = 24$ and $nq = 176$ are both greater than 10, $Binom(200, 0.12)$ may be approximated by the Normal model, $N(24, 4.60)$.

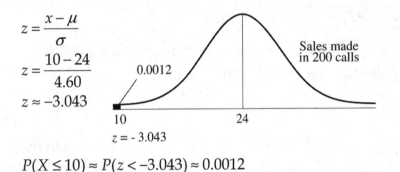

$z = \dfrac{x - \mu}{\sigma}$

$z = \dfrac{10 - 24}{4.60}$

$z \approx -3.043$

0.0012

Sales made in 200 calls

10   24

$z = -3.043$

$P(X \le 10) \approx P(z < -3.043) \approx 0.0012$

According to the Normal model, the probability that the telemarketer would make at most 10 sales is approximately 0.0012.

Since the probability that the telemarketer made 10 sales, given that the 12% of calls result in sales is so low, it is likely that he was misled about the true success rate.

47. Hurricanes, redux.

a)

$E(X) = \lambda = 2.45$

$P(\text{no hurricanes next year}) = \dfrac{e^{-2.45}(2.45)^0}{0!} \approx 0.0863$

b)

$P(\text{exactly one hurricane in next 2 years})$

$= P(\text{hurr. 1}^{st}\text{ yr})P(\text{no hurr. 2}^{nd}\text{ yr}) + P(\text{no hurr. 1}^{st}\text{ yr})P(\text{hurr. 2}^{nd}\text{ yr})$

$= \left(\dfrac{e^{-2.45}(2.45)^1}{1!} \right)\left(\dfrac{e^{-2.45}(2.45)^0}{0!} \right) + \left(\dfrac{e^{-2.45}(2.45)^0}{0!} \right)\left(\dfrac{e^{-2.45}(2.45)^1}{1!} \right)$

≈ 0.0365

49. TB again.

a) $E(X) = \lambda = np = 8000(0.0005) = 4\,\text{cases}.$

b)

$$P(\text{at least one new case}) = 1 - P(\text{no new cases})$$

$$= 1 - \frac{e^{-4}(4)^0}{0!} \approx 0.9817$$

51. Seatbelts II.

These stops may be considered Bernoulli trials. There are only two possible outcomes, belted or not belted. Police estimate that the probability that a driver is buckled is 80%. (The probability of not being buckled is therefore 20%.) Provided the drivers stopped are representative of all drivers, we can consider the probability constant. The trials are not independent, since the population of drivers is finite, but the police will not stop more than 10% of all drivers.

a) Let X = the number of cars stopped before finding a driver whose seat belt is not buckled. Use $Geom(0.2)$ to model the situation.

$$E(X) = \frac{1}{p} = \frac{1}{0.2} = 5 \text{ cars.}$$

b) $P(\text{First unbelted driver is in the sixth car}) = P(X = 6) = (0.8)^5(0.2) \approx 0.066$

c) $P(\text{The first ten drivers are wearing seatbelts}) = (0.8)^{10} \approx .107$

d) Let Y = the number of drivers wearing their seatbelts in 30 cars. Use $Binom(30, 0.8)$.

$$E(Y) = np = 30(0.8) = 24 \text{ drivers.}$$

$$SD(Y) = \sqrt{npq} = \sqrt{30(0.8)(0.2)} \approx 2.19 \text{ drivers.}$$

e) Let W = the number of drivers not wearing their seatbelts in 120 cars.

Using $Binom(120, 0.2)$:

$$P(\text{at least } 20) = P(W \geq 20)$$

$$= P(W = 20) + \ldots + P(W = 120)$$

$$= {}_{120}C_{20}(0.2)^{20}(0.8)^{100} + \ldots + {}_{120}C1_{20}(0.2)^{120}(0.8)^0$$

$$\approx 0.848$$

According to the Binomial model, the probability that at least 20 out of 120 drivers are not wearing their seatbelts is approximately 0.848.

Using $N(24, 4.38)$:

$$E(W) = np = 120(0.2) = 24 \text{ drivers.}$$

$$SD(W) = \sqrt{npq} = \sqrt{120(0.2)(0.8)} \approx 4.38 \text{ drivers.}$$

Since $np = 24$ and $nq = 96$ are both greater than 10, $Binom(120, 0.2)$ may be approximated by the Normal model, $N(24, 4.38)$.

$$z = \frac{w - \mu}{\sigma}$$

$$z = \frac{20 - 24}{4.38}$$

$$z \approx -0.913$$

$$P(W \geq 120) \approx P(z > -0.913) \approx 0.8194$$

According to the Normal model, the probability that at least 20 out of 120 drivers stopped are not wearing their seatbelts is approximately 0.8194.

53. ESP.

Choosing symbols may be considered Bernoulli trials. There are only two possible outcomes, correct or incorrect. Assuming that ESP does not exist, the probability of a correct identification from a randomized deck is constant, at $p = 0.20$. The trials are independent, as long as the deck is shuffled after each attempt. Since 100 trials will be performed, use $Binom(100, 0.2)$.

Let X = the number of symbols identified correctly out of 100 cards.

$E(X) = np = 100(0.2) = 20$ correct identifications.

$SD(X) = \sqrt{npq} = \sqrt{100(0.2)(0.8)} = 4$ correct identifications.

Answers may vary. In order be convincing, the "mind reader" would have to identify at least 32 out of 100 cards correctly, since 32 is three standard deviations above the mean. Identifying fewer cards than 32 could happen too often, simply due to chance.

55. Hot hand.

A streak like this is not unusual. The probability that he makes 4 in a row with a 55% free throw percentage is $(0.55)(0.55)(0.55)(0.55) \approx 0.09$. We can expect this to happen nearly one in ten times for every set of 4 shots that he makes. One out of ten times is not that unusual.

57. Hotter hand.

The shots may be considered Bernoulli trials. There are only two possible outcomes, make or miss. The probability of success is constant at 55%, and the shots are independent of one another. Therefore, we can model this situation with *Binom*(32, 0.55).

Let X = the number of free throws made out of 40.

$E(X) = np = 40(0.55) = 22$ free throws made.

$SD(X) = \sqrt{npq} = \sqrt{40(0.55)(0.45)} \approx 3.15$ free throws.

Answers may vary. The player's performance seems to have increased. 32 made free throws is $(32 - 22)/3.15 \approx 3.17$ standard deviations above the mean, an extraordinary feat, unless his free throw percentage has increased. This does NOT mean that the sneakers are responsible for the increase in free throw percentage. Some other variable may account for the increase. The player would need to set up a controlled experiment in order to determine what effect, if any, the sneakers had on his free throw percentage.

59. Web visitors.

a) The Poisson model because it is a good model to use when the data consists of counts of occurrences. The events must be independent and the mean number of occurrences stays constant.

b) Probability that in any one minute at least one purchase is made:

$$P(X = 1) + P(X = 2) + \ldots = \frac{e^{-3}3^1}{1!} + \frac{e^{-3}3^2}{2!} + \ldots = 0.9502$$

c) Probability that no one makes a purchase in the next 2 minutes:

$$P(X = 0) = \frac{e^{-6}6^0}{0!} = 0.0025$$

61. Web visitors, part 2.

a) The exponential model can be used to model the time between events especially when the number of arrivals of those events can be modeled by a Poisson model.

b) The mean time between purchases: 3 purchases are made per minute so each purchase is made every 1/3 minute.

c) Probability that the time to the next purchase will be between 1 and 2 minutes:

$e^{-3*1} - e^{-3*2} = 0.0473$

Review of Part IV – Randomness and Probability

1. **Quality Control.**

   a) $P(\text{defect}) = P(\text{cosm.}) + P(\text{func.}) - P(\text{cosm. and func.})$
   $$= 0.29 + 0.07 - 0.02 = 0.34$$

   Or, from the Venn: $0.27 + 0.02 + 0.05 = 0.34$

   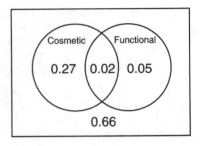

   b) $P(\text{cosm. and no func.})$

   $= P(\text{cosm.}) - P(\text{cosm. and func.})$
   $= 0.29 - 0.02 = 0.27$

   Or, from the Venn: 0.27 (region inside Cosmetic circle, outside Functional circle)

   c) $P(\text{func.} \mid \text{cosm.}) = \dfrac{P(\text{func. and cosm.})}{P(\text{cosm.})} = \dfrac{0.02}{0.29} \approx 0.069$

   From the Venn, consider only the region inside the Cosmetic circle. The probability that the car has a functional defect is 0.02 out of a total of 0.29 (the entire Cosmetic circle).

   d) The two kinds of defects are not disjoint events, since 2% of cars have both kinds.

   e) Approximately 6.9% of cars with cosmetic defects also have functional defects. Overall, the probability that a car has a cosmetic defect is 7%. The probabilities are estimates, so these are probably close enough to say that they two types of defects are independent.

3. **Airfares.**

   a) Let C = the price of a ticket to China
   Let F = the price of a ticket to France.

   Total price of airfare = $3C + 5F$

   b) $\mu = E(3C + 5F) = 3E(C) + 5E(F) = 3(1000) + 5(500) = \$5500$

   $\sigma = SD(3C + 5F) = \sqrt{3^2(Var(C)) + 5^2(Var(F))} = \sqrt{3^2(150^2) + 5^2(100^2)} \approx \$672.68$

   c) $\mu = E(C - F) = E(C) - E(F) = 1000 - 500 = \$500$

   $\sigma = SD(C - F) = \sqrt{Var(C) + Var(F)}$
   $$= \sqrt{150^2 + 100^2} \approx \$180.28$$

   d) No assumptions are necessary when calculating means. When calculating standard deviations, we must assume that ticket prices are independent of each other for different countries but all tickets to the same country are at the same price.

5. A game.

a) Let X = net amount won

X	$0	$2	- $2
$P(X)$	0.10	0.40	0.50

$$\mu = E(X) = 0(0.10) + 2(0.40) - 2(0.50) = -\$0.20$$

$$\sigma^2 = Var(X) = (0 - (-0.20))^2(0.10) + (2 - (-0.20))^2(0.40) + (-2 - (-0.20))^2(0.50) = 3.56$$

$$\sigma = SD(X) = \sqrt{Var(X)} = \sqrt{3.56} \approx \$1.89$$

b) $X + X$ = the total winnings for two plays.

$$\mu = E(X + X) = E(X) + E(X) = (-0.20) + (-0.20) = -\$0.40$$

$$\sigma = SD(X + X) = \sqrt{Var(X) + Var(X)}$$
$$= \sqrt{3.56 + 3.56} \approx \$2.67$$

7. Twins.

The selection of these women can be considered Bernoulli trials. There are two possible outcomes, twins or no twins. As long as the women selected are representative of the population of all pregnant women, then $p = 1/90$. (If the women selected are representative of the population of women taking Clomid, then $p = 1/10$.) The trials are not independent since the population of all women is finite, but 10 women are fewer than 10% of the population of women.

Let X = the number of twin births from $n = 10$ pregnant women.

Let Y = the number of twin births from $n = 10$ pregnant women taking Clomid.

a) Use *Binom*(10, 1/90)

$P(\text{at least one has twins}) = 1 - P(\text{none have twins})$
$$= 1 - P(X = 0)$$
$$= 1 - {}_{10}C_0 \left(\frac{1}{90}\right)^0 \left(\frac{89}{90}\right)^{10}$$
$$\approx 0.106$$

b) Use *Binom*(10, 1/10)

$P(\text{at least one has twins}) = 1 - P(\text{none have twins})$
$$= 1 - P(Y = 0)$$
$$= 1 - {}_{10}C_0 \left(\frac{1}{10}\right)^0 \left(\frac{9}{10}\right)^{10}$$
$$\approx 0.651$$

c) Use $Binom(5, 1/90)$ and $Binom(5, 1/10)$.

P(at least one has twins)

$= 1 - P$(no twins without Clomid)P(no twins with Clomid)

$$= 1 - \left({}_5C_0 \left(\frac{1}{90}\right)^0 \left(\frac{89}{90}\right)^5 \right)\left({}_5C_0 \left(\frac{1}{10}\right)^0 \left(\frac{9}{10}\right)^5 \right) \approx 0.442$$

9. More twins.

In the previous exercise, it was determined that these were Bernoulli trials. Use $Binom(5, 0.10)$.

Let X = the number of twin births from $n = 5$ pregnant women taking Clomid.

a)

P(none have twins) $= P(X = 0)$

$\qquad = {}_5C_0(0.1)^0(0.9)^5$

$\qquad \approx 0.590$

b)

P(one has twins) $= P(X = 1)$

$\qquad = {}_5C_1(0.1)^1(0.9)^4$

$\qquad \approx 0.328$

c)

P(at least three will have twins) $= P(X = 3) + P(X = 4) + P(X = 5)$

$\qquad = {}_5C_3(0.1)^3(0.9)^2 + {}_5C_4(0.1)^4(0.9)^1 + {}_5C_5(0.1)^5(0.9)^0$

$\qquad = 0.00856$

11. Twins, part III.

In a previous exercise, it was determined that these were Bernoulli trials. Use $Binom(152, 0.10)$.

Let X = the number of twin births from $n = 152$ pregnant women taking Clomid.

a) $E(X) = np = 152(0.10) = 15.2$ births.

$SD(X) = \sqrt{npq} = \sqrt{152(0.10)(0.90)} \approx 3.70$ births.

b) Since $np = 15.2$ and $nq = 136.8$ are both greater than 10, the Success/Failure condition is satisfied and $Binom(152, 0.10)$ may be approximated by $N(15.2, 3.70)$.

c) Using $Binom(152, 0.10)$:

P(no more than 10) $= P(X \le 10)$

$\qquad = P(X = 0) + \ldots + P(X = 10)$

$\qquad = {}_{152}C_0(0.10)^0(0.90)^{152} + \ldots + {}_{152}C_{10}(0.10)^{10}(0.90)^{142}$

$\qquad \approx 0.097$

According to the Binomial model, the probability that no more than 10 women would have twins is approximately 0.097.

Using *N*(15.2, 3.70):

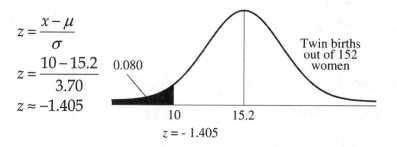

$$z = \frac{x - \mu}{\sigma}$$

$$z = \frac{10 - 15.2}{3.70}$$

$$z \approx -1.405$$

0.080

Twin births out of 152 women

10   15.2

z = - 1.405

$$P(X \le 10) \approx P(z < -1.405) \approx 0.080$$

According to the Normal model, the probability that no more than 10 women would have twins is approximately 0.080.

13. Language.

Assuming that the freshman composition class consists of 25 randomly selected people, these may be considered Bernoulli trials. There are only two possible outcomes, having the specified language center or not having the specified language center. The probabilities of the specified language centers are constant at 80%, 10%, or 10%, for right, left, and two-sided language center, respectively. The trials are not independent, since the population of people is finite, but we will select fewer than 10% of all people.

a) Let $L =$ the number of people with left-brain language control from $n = 25$ people.

Use *Binom*(25, 0.80).

$$P(\text{no more than } 15) = P(L \le 15)$$
$$= P(L = 0) + \ldots + P(L = 15)$$
$$= {}_{25}C_0 (0.80)^0 (0.20)^{25} + \ldots + {}_{25}C_{15} (0.80)^{15} (0.20)^{10}$$
$$\approx 0.0173$$

According to the Binomial model, the probability that no more than 15 students in a class of 25 will have left-brain language centers is approximately 0.0173.

b) Let $T =$ the number of people with two-sided language control from $n = 5$ people.

Use *Binom*(5, 0.10).

$$P(\text{none have two-sided language control}) = P(T = 0)$$
$$= {}_5C_0 (0.10)^0 (0.90)^5$$
$$\approx 0.590$$

c) Use Binomial models:

$E(\text{left}) = np_L = 1200(0.80) = 960$ people

$E(\text{right}) = np_R = 1200(0.10) = 120$ people

$E(\text{two} - \text{sided}) = np_T = 1200(0.10) = 120$ people

d) Let R = the number of people with right-brain language control.

$E(R) = np_R = 1200(0.10) = 120$ people

$SD(R) = \sqrt{np_R q_R} = \sqrt{1200(0.10)(0.90)} \approx 10.39$ people.

e) Since $np_R = 120$ and $nq_R = 1080$ are both greater than 10, the Normal model, $N(120, 10.39)$, may be used to approximate $Binom(1200, 0.10)$. According to the Normal model, about 68% of randomly selected groups of 1200 people could be expected to have

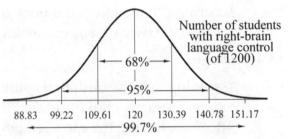

Number of students with right-brain language control (of 1200)

88.83   99.22   109.61   120   130.39   140.78   151.17

between 109.61 and 130.39 people with right-brain language control. About 95% of randomly selected groups of 1200 people could be expected to have between 99.22 and 140.78 people with right-brain language control. About 99.7% of randomly selected groups of 1200 people could be expected to have between 88.83 and 151.17 people with right-brain language control.

15. Beanstalks.

a) The greater standard deviation for men's heights indicates that men's heights are more variable than women's heights.

b) Admission to a Beanstalk Club is based upon extraordinary height for both men and women, but men are slightly more likely to qualify. The qualifying height for women is about 2.4 SDs above the mean height of women, while the qualifying height for men is about 1.75 SDS above the mean height for men.

c) Let M = the height of a randomly selected man from $N(69.1, 2.8)$.

Let W = the height of a randomly selected woman from $N(64.0, 2.5)$.

$M - W$ = the difference in height of a randomly selected man and woman.

d) $E(M - W) = E(M) - E(W) = 69.1 - 64.0 = 5.1$ inches

e) $SD(M - W) = \sqrt{Var(M) + Var(W)} = \sqrt{2.8^2 + 2.5^2} \approx 3.75$ inches

f) Since each distribution is described by a Normal model, the distribution of the difference in height between a randomly selected man and woman is $N(5.1, 3.75)$.

$$z = \frac{y - \mu}{\sigma}$$

$$z = \frac{0 - 5.1}{\sqrt{2.8^2 + 2.5^2}}$$

$$z \approx -1.359$$

According to the Normal model, the probability that a randomly selected man is taller than a randomly selected woman is approximately 0.913.

g) If people chose spouses independent of height, we would expect 91.3% of married couples to consist of a taller husband and shorter wife. The 92% that was seen in the survey is close to 91.3%, and the difference may be due to natural sampling variability. Unless this survey is very large, there is not sufficient evidence of association between height and choice of partner.

17. Multiple choice.

Guessing at questions can be considered Bernoulli trials. There are only two possible outcomes, correct or incorrect. If you are guessing, the probability of success is $p = 0.25$, and the questions are independent. Use $Binom(50, 0.25)$ to model the number of correct guesses on the test.

a) Let $X =$ the number of correct guesses.
$$P(\text{at least } 30 \text{ of } 50 \text{ correct}) = P(X \geq 30)$$
$$= P(X = 30) + \ldots + P(X = 50)$$
$$= {}_{50}C_{30}(0.25)^{30}(0.75)^{20} + \ldots + {}_{50}C_{30}(0.25)^{50}(0.75)^{0}$$
$$\approx 0.00000016$$

You are **very** unlikely to pass by guessing on every question.

b) Use $Binom(50, 0.70)$.
$$P(\text{at least } 30 \text{ of } 50 \text{ correct}) = P(X \geq 30)$$
$$= P(X = 30) + \ldots + P(X = 50)$$
$$= {}_{50}C_{30}(0.70)^{30}(0.30)^{20} + \ldots + {}_{50}C_{50}(0.70)^{50}(0.30)^{0}$$
$$\approx 0.952$$

According to the Binomial model, your chances of passing are about 95.2%.

c) Use $Geom(0.70)$.
$$P(\text{first correct on third question}) = (0.30)^2(0.70) = 0.063$$

19. Insurance.

The company is expected to pay $100,000 only 2.6% of the time, while always gaining $520 from every policy sold. When they pay, they actually only pay $99,480.

$$E(\text{profit}) = \$520(0.974) - \$99,480(0.026) = -\$2,080.$$

The expected profit is actually a **loss** of $2,080 per policy. The company had better raise its premiums if it hopes to stay in business.

21. Passing stats.

Organize the information in a tree diagram.

a)

$$P(\text{Passing Statistics})$$
$$= P(\text{Scedastic and Pass})$$
$$\quad + P(\text{Kurtosis and Pass})$$
$$\approx 0.4667 + 0.25$$
$$\approx 0.717$$

Pass   (70/120)(0.8) ≈ 0.4667

0.8

Scedastic

70/120

0.2

Fail   (70/120)(0.2) ≈ 0.1167

Pass   (50/120)(0.6) = 0.25

50/120

0.6

Kurtosis

0.4

Fail   (50/120)(0.4) ≈ 0.1667

b)

$$P(\text{Kurtosis} \mid \text{Fail}) = \frac{P(\text{Kurtosis and Fail})}{P(\text{Fail})} \approx \frac{0.1667}{0.1167 + 0.1667} \approx 0.588$$

23. Random variables.

a) $\mu = E(X + 50)$
$\quad = E(X) + 50 = 50 + 50 = 100$
$\sigma = SD(X + 50) = SD(X) = 8$

b) $\mu = E(10Y) = 10E(Y)$
$\quad = 10(100) = 1000$
$\sigma = SD(10Y) = 10SD(Y) = 60$

c)

$$\mu = E(X + 0.5Y) = E(X) + 0.5E(Y)$$
$$= 50 + 0.5(100) = 100$$
$$\sigma = SD(X + 0.5Y) = \sqrt{Var(X) + 0.5^2 Var(Y)}$$
$$= \sqrt{8^2 + 0.5^2(6^2)} \approx 8.54$$

d)

$$\mu = E(X - Y) = E(X) - E(Y) = 50 - 100 = -50$$
$$\sigma = SD(X - Y) = \sqrt{Var(X) + Var(Y)}$$
$$= \sqrt{8^2 + 6^2} = 10$$

e)

$$\mu = E(X_1 + X_2) = E(X) + E(X) = 50 + 50 = 100$$

$$\sigma = SD(X_1 + X_2) = \sqrt{Var(X) + Var(X)} = \sqrt{8^2 + 8^2} \approx 11.31$$

25. Youth survey.

a) Many boys play computer games and use email, so the probabilities can total more than 100%. There is no evidence that there is a mistake in the report.

b) Playing computer games and using email are not mutually exclusive. If they were, the probabilities would total 100% or less.

c) Emailing friends and being a boy or girl are not independent. 76% of girls emailed friends in the last week, but only 65% of boys emailed. If emailing were independent of being a boy or girl, the probabilities would be the same.

d) Let X = the number of students chosen until the first student is found who does not use the Internet. Use $Geom(0.07)$. $P(X = 5) = (0.93)^4 (0.07) \approx 0.0524$.

27. Travel to Kyrgyzstan.

a) Spending an average of 4237 soms per day, you can stay about $\dfrac{90{,}000}{4237} \approx 21$ days.

b) Assuming that your daily spending is independent, the standard deviation is the square root of the sum of the variances for 21 days.

$$\sigma = \sqrt{21(360)^2} \approx 1649.73 \text{ soms}$$

c) The standard deviation in your total expenditures is about 1650 soms, so if you don't think you will exceed your expectation by more than 2 standard deviations, bring an extra 3300 soms. This gives you a cushion of about 157 soms for each of the 21 days.

29. Home sweet home.

Since the homes are randomly selected, these can be considered Bernoulli trials. There are only two possible outcomes, owning the home or not owning the home. The probability of any randomly selected resident home being owned by the current resident is 0.66. The trials are not independent, since the population is finite, but as long as the city has more than 8200 homes, we are not sampling more than 10% of the population. The Binomial model, $Binom(820, 0.66)$, can be used to model the number of homeowners among the 820 homes surveyed. Let H = the number of homeowners found in $n = 820$ homes.

$E(H) = np = 820(0.66) = 541.2$ homes.

$SD(H) = \sqrt{npq} = \sqrt{820(0.66)(0.34)} \approx 13.56$ homes.

The 523 homeowners found in the candidate's survey represent a number of homeowners that is only about 1.34 standard deviations below the expected number of homeowners. It is not particularly unusual to be 1.34 standard deviations below the mean. There is little support for the candidate's claim of a low level of home ownership.

31. Who's the boss?

a) $P(\text{first three owned by women}) = (0.30)^3 \approx 0.027$

b) $P(\text{none of the first four are owned by women}) = (0.70)^4 \approx 0.240$

c) $P(\text{sixth firm called is owned by women} \mid \text{none of the first five were}) = 0.30$

Since the firms are chosen randomly, the fact that the first five firms were owned by men has no bearing on the ownership of the sixth firm.

33. When to stop?

a) Since there are only two outcomes, 6 or not 6, the probability of getting a 6 is 1/6, and the trials are independent, these are Bernoulli trials. Use *Geom*(1/6).

$$\mu = \frac{1}{p} = \frac{1}{\frac{1}{6}} = 6 \text{ rolls}$$

b) If 6's are not allowed, the mean of each die roll is $\dfrac{1+2+3+4+5}{5} = 3$. You would expect to get 15 if you rolled 5 times.

c) $P(\text{5 rolls without a 6}) = \left(\dfrac{5}{6}\right)^5 \approx 0.402$

35. Technology on campus.

a)

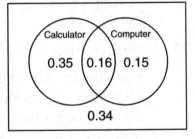

$P(\text{neither tech.})$
$= 1 - P(\text{either tech.})$
$= 1 - \left[P(\text{calculator}) + P(\text{computer}) - P(\text{both}) \right]$
$= 1 - [0.51 + 0.31 - 0.16] = 0.34$

Or, from the Venn: 0.34 (the region outside both circles) This is MUCH easier.

34% of students use neither type of technology.

b) $P(\text{calc. and no comp.}) = P(\text{calc.}) - P(\text{calc. and comp.}) = 0.51 - 0.16 = 0.35$

Or, from the Venn: 0.35 (region inside the Calculator circle, outside the Computer circle)

35% of students use calculators, but not computers.

c) $P(\text{computer} \mid \text{calculator}) = \dfrac{P(\text{comp. and calc.})}{P(\text{calc.})} = \dfrac{0.16}{0.51} \approx 0.314$

About 31.4% of calculator users have computer assignments.

d) The percentage of computer users overall is 31%, while 31.4% of calculator users were computer users. These are very close. There is no indication of an association between computer use and calculator use.

37. O-rings.

a) A Poisson model would be used.

b) If the probability of failure for one O-ring is 0.01, then the mean number of failures for 10 O-rings is $E(X) = \lambda = np = 10(0.01) = 0.1$ O-ring. We are able to calculate this because the Poisson model scales according to sample size.

c) $P(\text{one failed O-ring}) = \dfrac{e^{-0.1}(0.1)^1}{1!} \approx 0.090$

d)

$P(\text{at least one failed O-ring}) = 1 - P(\text{no failures})$

$$= 1 - \dfrac{e^{-0.1}(0.1)^0}{0!} \approx 0.095$$

39. Socks.

Since we are sampling without replacement, use conditional probabilities throughout.

a) $P(2 \text{ blue}) = \left(\dfrac{4}{12}\right)\left(\dfrac{3}{11}\right) = \dfrac{12}{132} = \dfrac{1}{11}$

b) $P(\text{no grey}) = \left(\dfrac{7}{12}\right)\left(\dfrac{6}{11}\right) = \dfrac{42}{132} = \dfrac{7}{22}$

c) $P(\text{at least one black}) = 1 - P(\text{no black}) = 1 - \left(\dfrac{9}{12}\right)\left(\dfrac{8}{11}\right) = \dfrac{60}{132} = \dfrac{5}{11}$

d) $P(\text{green}) = 0$ (There aren't any green socks in the drawer.)

e) $P(\text{match}) = P(2 \text{ blue}) + P(2 \text{ grey}) + P(2 \text{ black})$

$$= \left(\dfrac{4}{12}\right)\left(\dfrac{3}{11}\right) + \left(\dfrac{5}{12}\right)\left(\dfrac{4}{11}\right) + \left(\dfrac{3}{12}\right)\left(\dfrac{2}{11}\right) = \dfrac{19}{66}$$

41. The Drake equation.

a) $N \cdot f_p$ represents the number of stars in the Milky Way Galaxy expected to have planets.

b) $N \cdot f_p \cdot n_e \cdot f_i$ represents the number of planets in the Milky Way Galaxy expected to have intelligent life.

c) $f_l \cdot f_i$ is the probability that a planet has a suitable environment and has intelligent life.

d) $f_l = P(\text{life} \mid \text{suitable environment})$. This is the probability that life develops, if a planet has a suitable environment.

$f_i = P(\text{intelligence} \mid \text{life})$. This is the probability that the life develops intelligence, if a planet already has life.

$f_c = P(\text{communication} \mid \text{intelligence})$. This is the probability that radio communication develops, if a planet already has intelligent life.

43. Pregnant?

Organize the information in a tree diagram.

$P(\text{pregnant} \mid + \text{ test})$

$= \dfrac{P(\text{preg. and } + \text{ test})}{P(+ \text{ test})}$

$= \dfrac{0.686}{0.686 + 0.006}$

≈ 0.991

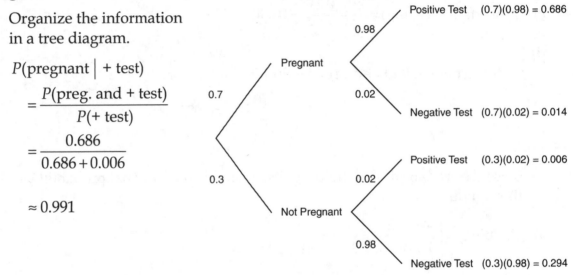

Chapter 17 – Sampling Distribution Models

Section 17.1

1. Website.

a) Since the sample is drawn at random, and assuming that 200 investors is a small portion of their customers, the sampling distribution for the proportion of 200 investors that use smartphones will be unimodal and symmetric (roughly Normal).

b) The center of the sampling distribution of the sample proportion is 0.36.

c) The standard deviation of the sample proportion is $\sqrt{\dfrac{pq}{n}} = \sqrt{\dfrac{(0.36)(0.64)}{200}} \approx 0.034$.

3. Send money.

All of the histograms are centered around $p = 0.05$. As n gets larger, the shape of the histograms get more unimodal and symmetric, approaching a Normal model, while the variability in the sample proportions decreases.

Section 17.2

5. Marriage.

The data come from a random sample, so the randomization condition is met. We don't know the exact value of p, we can estimate it with $\hat{p}$.

$n\hat{p} = (1500)(0.27) = 405$, and $n\hat{q} = (1500)(0.73) = 1095$. So, there are well over 10 successes and 10 failures, meeting the Success/Failure Condition. Since there are more than $(10)(1500) = 15,000$ adults in the United States, the 10% Condition is met. A Normal model is appropriate for the sampling distribution of the sample proportion.

7. Send more money.

a) The histogram for $n = 200$ looks quite unimodal and symmetric. We should be able to use the Normal model.

b) The Success/Failure condition requires np and nq to both be at least 10, which is not satisfied until $n = 200$ for $p = 0.05$. The theory supports the choice in part a.

Section 17.3

9. Sample maximum.

a) A Normal model is not appropriate for the sampling distribution of the sample maximum. The histogram is skewed strongly to the left.

b) No. The 95% rule is based on the Normal model, and the Normal model is not appropriate here.

Section 17.4

11. Market research.

a) The standard deviation of the sample proportion is $\sqrt{\dfrac{pq}{n}} = \sqrt{\dfrac{(0.15)(0.85)}{100}} \approx 0.0357$.

b) To reduce the standard deviation by half, she needs a sample 4 times as large, or 400 people.

13. Tips.

a) Since the distribution of tips is skewed to the right, we can't use the Normal model to determine the probability that a given party will tip at least $20.

b) No. A sample of 4 parties is probably not a large enough sample for the CLT to allow us to use the Normal model to estimate the distribution of averages.

c) A sample of 10 parties may not be large enough to allow the use of a Normal model to describe the distribution of averages. It would be risky to attempt to estimate the probability that his next 10 parties tip an average of $15. However, since the distribution of tips has $\mu = \$9.60$, with standard deviation $\sigma = \$5.40$, we still know that the mean of the sampling distribution model is $\mu_{\bar{y}} = \$9.60$ with standard deviation $SD(\bar{y}) = \dfrac{5.40}{\sqrt{10}} \approx \$1.71$.

We don't know the exact shape of the distribution, but we can still assess the likelihood of specific means. A mean tip of $15 is over 3 standard deviations above the expected mean tip for 10 parties. That's not very likely to happen.

15. More tips.

a) **Randomization condition:** Assume that the tips from 40 parties can be considered a representative sample of all tips.

Independence assumption: It is reasonable to think that the tips are mutually independent, unless the service is particularly good or bad during this weekend.

10% condition: The tips of 40 parties certainly represent less than 10% of all tips.

Large Enough Sample condition: The sample of 40 parties is large enough.

The mean tip is $\mu = \$9.60$, with standard deviation $\sigma = \$5.40$. Since the conditions are satisfied, the CLT allows us to model the sampling distribution of $\bar{y}$ with a Normal model, with $\mu_{\bar{y}} = \$9.60$ and standard deviation

$$SD(\bar{y}) = \dfrac{5.40}{\sqrt{40}} \approx \$0.8538.$$

In order to earn at least $500, the waiter would have to average $\dfrac{500}{40} = \$12.50$ per party.

According to the Normal model, the probability that the waiter earns at least $500 in tips in a weekend is approximately 0.0003.

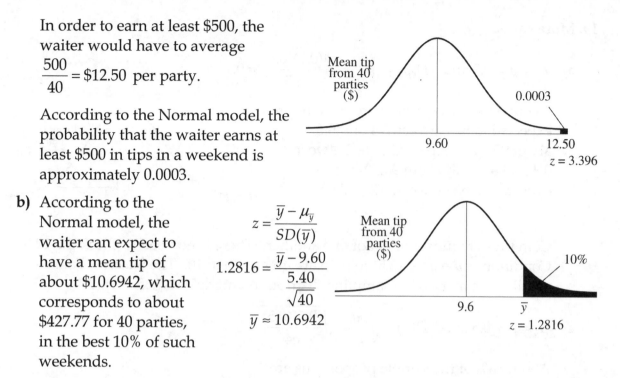

b) According to the Normal model, the waiter can expect to have a mean tip of about $10.6942, which corresponds to about $427.77 for 40 parties, in the best 10% of such weekends.

$$z = \frac{\bar{y} - \mu_{\bar{y}}}{SD(\bar{y})}$$

$$1.2816 = \frac{\bar{y} - 9.60}{\dfrac{5.40}{\sqrt{40}}}$$

$$\bar{y} \approx 10.6942$$

17. Coin tosses.

a) The histogram of these proportions is expected to be symmetric, but **not** because of the Central Limit Theorem. The sample of 16 coin flips is not large. The distribution of these proportions is expected to be symmetric because the probability that the coin lands heads is the same as the probability that the coin lands tails.

b) The histogram is expected to have its center at 0.5, the probability that the coin lands heads.

c) The standard deviation of data displayed in this histogram should be approximately equal to the standard deviation of the sampling distribution model, $\sqrt{\dfrac{pq}{n}} = \sqrt{\dfrac{(0.5)(0.5)}{16}} = 0.125$.

d) The expected number of heads, $np = 16(0.5) = 8$, which is less than 10. The Success/Failure condition is not met. The Normal model is not appropriate in this case.

19. More coins.

a) $\mu_{\hat{p}} = p = 0.5$ and $SD(\hat{p}) = \sqrt{\dfrac{pq}{n}} = \sqrt{\dfrac{(0.5)(0.5)}{25}} = 0.1$

About 68% of the sample proportions are expected to be between 0.4 and 0.6, about 95% are expected to be between 0.3 and 0.7, and about 99.7% are expected to be between 0.2 and 0.8.

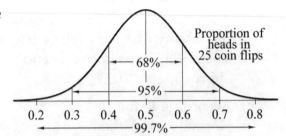

b) Coin flips are independent of one another. There is no need to check the 10% Condition. $np = nq = 12.5$, so both are greater than 10. The Success/Failure condition is met, so the sampling distribution model is $N(0.5, 0.1)$.

c) $\mu_{\hat{p}} = p = 0.5$ and $SD(\hat{p}) = \sqrt{\dfrac{pq}{n}} = \sqrt{\dfrac{(0.5)(0.5)}{64}} = 0.0625$

About 68% of the sample proportions are expected to be between 0.4375 and 0.5625, about 95% are expected to be between 0.375 and 0.625, and about 99.7% are expected to be between 0.3125 and 0.6875.

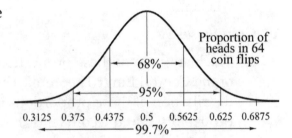

Coin flips are independent of one another, and $np = nq = 32$, so both are greater than 10. The Success/Failure condition is met, so the sampling distribution model is $N(0.5, 0.0625)$.

d) As the number of tosses increases, the sampling distribution model will still be Normal and centered at 0.5, but the standard deviation will decrease. The sampling distribution model will be less spread out.

21. Just (un)lucky.

For 200 flips, the sampling distribution model is Normal with $\mu_{\hat{p}} = p = 0.5$ and

$SD(\hat{p}) = \sqrt{\dfrac{pq}{n}} = \sqrt{\dfrac{(0.5)(0.5)}{200}} \approx 0.0354$. Her sample proportion of $\hat{p} = 0.42$ is about 2.26 standard deviations below the expected proportion, which is unusual, but not extraordinary. According to the Normal model, we expect sample proportions this low or lower about 1.2% of the time.

Chapter Exercises

23. Speeding.

a) $\mu_{\hat{p}} = p = 0.70$

$$SD(\hat{p}) = \sqrt{\frac{pq}{n}} = \sqrt{\frac{(0.7)(0.3)}{80}} \approx 0.051 \, .$$

About 68% of the sample proportions are expected to be between 0.649 and 0.751, about 95% are expected to be between 0.598 and 0.802, and about 99.7% are expected to be between 0.547 and 0.853.

b) **Randomization condition:** The sample may not be representative. If the flow of traffic is very fast, the speed of the other cars around may have some effect on the speed of each driver. Likewise, if traffic is slow, the police may find a smaller proportion of speeders than they expect.

10% condition: 80 cars represent less than 10% of all cars

Success/Failure condition: $np = 56$ and $nq = 24$ are both greater than 10.

The Normal model may not be appropriate. Use caution. (And don't speed!)

25. Vision.

a) **Randomization condition:** Assume that the 170 children are a representative sample of all children.

10% condition: A sample of this size is less than 10% of all children.

Success/Failure condition: $np = 20.4$ and $nq = 149.6$ are both greater than 10.

Therefore, the sampling distribution model for the proportion of 170 children

who are nearsighted is $N\left(0.12, \sqrt{\dfrac{(0.12)(0.88)}{170}} \right)$ or $N(0.12, 0.025)$.

b) The Normal model is to the right.

c) They might expect that the proportion of nearsighted students to be within 2 standard deviations of the mean. According to the Normal model, this means they might expect

between 7% and 17% of the students to be nearsighted, or between about 12 and 29 students.

27. Loans.

a) $\mu_{\hat{p}} = p = 7\%$

$$SD(\hat{p}) = \sqrt{\frac{pq}{n}} = \sqrt{\frac{(0.07)(0.93)}{200}} \approx 0.018 \approx 1.8\%$$

b) **Randomization condition:** Assume that the 200 people are a representative sample of all loan recipients.

10% condition: A sample of this size is less than 10% of all loan recipients.

Success/Failure condition: $np = 14$ and $nq = 186$ are both greater than 10.

Therefore, the sampling distribution model for the proportion of 200 loan recipients who will not make payments on time is $N(0.07, 0.018)$.

c) According to the Normal model, the probability that over 10% of these clients will not make timely payments is approximately 0.048.

$$z = \frac{\hat{p} - \mu_{\hat{p}}}{\sqrt{\frac{pq}{n}}}$$

$$z = \frac{0.10 - 0.07}{\sqrt{\frac{(0.07)(0.93)}{200}}}$$

$$z \approx 1.663$$

29. Back to school? 2013.

Randomization condition: We are considering random samples of 400 students who took the ACT.

10% Condition: 400 students is less than 10% of all college students.

Success/Failure condition: $np = 288.8$ and $nq = 111.2$ are both greater than 10.

Therefore, the sampling distribution model for $\hat{p}$ is Normal, with:

$$\mu_{\hat{p}} = p = 0.722$$

$$SD(\hat{p}) = \sqrt{\frac{pq}{n}} = \sqrt{\frac{(0.722)(0.278)}{400}} \approx 0.0224$$

According to the sampling distribution model, about 68% of the colleges are expected to have retention rates between 0.700 and 0.744, about 95% of the colleges are expected to have retention

rates between 0.677 and 0.767, and about 99.7% of the colleges are expected to have retention rates between 0.655 and 0.789. However, the conditions for the use of this model may not be met. We should be cautious about making any conclusions based on this model.

31. Back to school, again.

Provided that the students at this college are typical, the sampling distribution model for the retention rate, $\hat{p}$, is Normal with $\mu_{\hat{p}} = p = 0.722$ and standard

deviation $SD(\hat{p}) = \sqrt{\dfrac{pq}{n}} = \sqrt{\dfrac{(0.722)(0.278)}{603}} \approx 0.0182$

This college has a right to brag about their retention rate. $522/603 = 86.6\%$ is over 7 standard deviations above the expected rate of 72.2%.

33. Polling.

Randomization condition: We must assume that the 400 voters were polled randomly.

10% condition: 400 voters polled represent less than 10% of potential voters.

Success/Failure condition: $np = 208$ and $nq = 192$ are both greater than 10.

Therefore, the sampling distribution model for $\hat{p}$ is Normal, with:

$\mu_{\hat{p}} = p = 0.52$ $\qquad SD(\hat{p}) = \sqrt{\dfrac{pq}{n}} = \sqrt{\dfrac{(0.52)(0.48)}{400}} \approx 0.025$

According to the Normal model, the probability that the newspaper's sample will lead them to predict defeat (that is, predict budget support below 50%) is approximately 0.212.

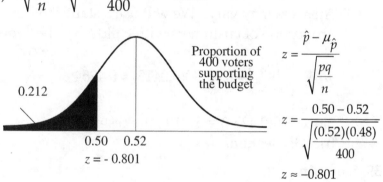

$z = \dfrac{\hat{p} - \mu_{\hat{p}}}{\sqrt{\dfrac{pq}{n}}}$

$z = \dfrac{0.50 - 0.52}{\sqrt{\dfrac{(0.52)(0.48)}{400}}}$

$z \approx -0.801$

35. Gaydar.

Randomization condition: We must assume that the 100 photographs were randomized before they were shown to the women.

10% condition: This study doesn't involve a sample, just randomization of a set of photos. The 10% condition does not apply.

Success/Failure condition: $np = 65$ and $nq = 35$ are both greater than 10.

Therefore, the sampling distribution model for $\hat{p}$ is Normal, with:

$\mu_{\hat{p}} = p = 0.65$ $\qquad SD(\hat{p}) = \sqrt{\dfrac{pq}{n}} = \sqrt{\dfrac{(0.65)(0.35)}{100}} \approx 0.048$

According to the Normal model, the probability of correctly classifying 80 of 100 men is approximately 0.0008.

$$z = \frac{\hat{p} - \mu_{\hat{p}}}{\sqrt{\frac{pq}{n}}}$$

$$z = \frac{0.80 - 0.65}{\sqrt{\frac{(0.65)(0.35)}{100}}}$$

$z = 3.145$ $z \approx 3.145$

37. "No Children" section.

Randomization condition: We will assume that the 120 customers (to fill the restaurant to capacity) are representative of all customers.

10% condition: 120 customers represent less than 10% of all potential customers.

Success/Failure condition: $np = 36$ and $nq = 84$ are both greater than 10.

Therefore, the sampling distribution model for $\hat{p}$ is Normal, with:

$$\mu_{\hat{p}} = p = 0.30 \qquad SD(\hat{p}) = \sqrt{\frac{pq}{n}} = \sqrt{\frac{(0.30)(0.70)}{120}} \approx 0.042$$

Answers may vary. We will use 3 standard deviations above the expected proportion of customers with children to be "very sure".

$$\mu_{\hat{p}} + 3\left(\sqrt{\frac{pq}{n}}\right) \approx 0.30 + 3(0.0418) \approx 0.4254$$

Since $120(0.4254) = 51.048$, the restaurant needs about 51 seats in the family-friendly section.

39. Sampling.

a) The sampling distribution model for the sample mean is $N\left(\mu, \frac{\sigma}{\sqrt{n}}\right)$.

b) If we choose a larger sample, the mean of the sampling distribution model will remain the same, but the standard deviation will be smaller.

41. Waist size.

a) The distribution of waist size of 250 men is unimodal and slightly skewed to the right. A typical waist size is approximately 36 inches, and the standard deviation in waist sizes is approximately 4 inches.

b) All of the histograms show distributions of sample means centered near 36 inches. As n gets larger the histograms approach the Normal model in shape, and the variability in the sample means decreases. The histograms are fairly Normal by the time the sample reaches size 5.

43. Waist size revisited.

a)

n	Observed mean	Theoretical mean	Observed st. dev.	Theoretical standard deviation
2	36.314	36.33	2.855	$4.019 / \sqrt{2} \approx 2.842$
5	36.314	36.33	1.805	$4.019 / \sqrt{5} \approx 1.797$
10	36.341	36.33	1.276	$4.019 / \sqrt{10} \approx 1.271$
20	36.339	36.33	0.895	$4.019 / \sqrt{20} \approx 0.899$

b) The observed values are all very close to the theoretical values.

c) For samples as small as 5, the sampling distribution of sample means is unimodal and symmetric. The Normal model would be appropriate.

d) The distribution of the original data is nearly unimodal and symmetric, so it doesn't take a very large sample size for the distribution of sample means to be approximately Normal.

45. GPAs.

Randomization condition: Assume that the students are randomly assigned to seminars.

Independence assumption: It is reasonable to think that GPAs for randomly selected students are mutually independent.

10% condition: The 25 students in the seminar certainly represent less than 10% of the population of students.

Large Enough Sample condition: The distribution of GPAs is roughly unimodal and symmetric, so the sample of 25 students is large enough.

The mean GPA for the freshmen was $\mu = 3.4$, with standard deviation $\sigma = 0.35$. Since the conditions are met, the Central Limit Theorem tells us that we can model the sampling distribution of the mean GPA with a Normal model, with $\mu_{\bar{y}} = 3.4$ and standard deviation $SD(\bar{y}) = \dfrac{0.35}{\sqrt{25}} \approx 0.07$.

The sampling distribution model for the sample mean GPA is approximately $N(3.4, 0.07)$.

47. Lucky spot?

a) Smaller outlets have more variability than the larger outlets, just as the Central Limit Theorem predicts.

b) If the lottery is truly random, all outlets are equally likely to sell winning tickets.

49. Pregnancy.

a)

$$z = \frac{y - \mu}{\sigma}$$

$$z = \frac{270 - 266}{16}$$

$$z = 0.25$$

$$z = \frac{y - \mu}{\sigma}$$

$$z = \frac{280 - 266}{16}$$

$$z = 0.875$$

According to the Normal model, approximately 21.1% of all pregnancies are expected to last between 270 and 280 days.

b)

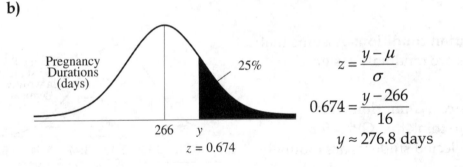

$$z = \frac{y - \mu}{\sigma}$$

$$0.674 = \frac{y - 266}{16}$$

$$y \approx 276.8 \text{ days}$$

According to the Normal model, the longest 25% of pregnancies are expected to last approximately 276.8 days or more.

c) **Randomization condition:** Assume that the 60 women the doctor is treating can be considered a representative sample of all pregnant women.

Independence assumption: It is reasonable to think that the durations of the patients' pregnancies are mutually independent.

10% condition: The 60 women that the doctor is treating certainly represent less than 10% of the population of all women.

Large Enough Sample condition: The sample of 60 women is large enough. In this case, any sample would be large enough, since the distribution of pregnancies is Normal.

The mean duration of the pregnancies was $\mu = 266$ days, with standard deviation $\sigma = 16$ days. Since the distribution of pregnancy durations is Normal, we can model the sampling distribution of the mean pregnancy duration with a Normal model, with $\mu_{\bar{y}} = 266$ days and standard deviation

$$SD(\bar{y}) = \frac{16}{\sqrt{60}} \approx 2.07 \text{ days}.$$

d) According to the Normal model, with mean 266 days and standard deviation 2.07 days, the probability that the mean pregnancy duration is less than 260 days is 0.002.

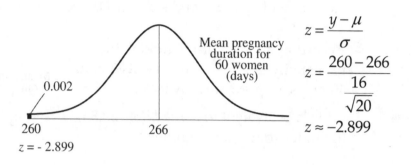

$$z = \frac{y - \mu}{\sigma}$$

$$z = \frac{260 - 266}{\frac{16}{\sqrt{20}}}$$

$$z \approx -2.899$$

51. Pregnant again.

a) The distribution of pregnancy durations may be skewed to the left since there are more premature births than very long pregnancies. Modern practice of medicine stops pregnancies at about 2 weeks past normal due date by inducing labor or performing a Caesarean section.

b) We can no longer answer the questions posed in parts a and b. The Normal model is not appropriate for skewed distributions. The answer to part c is still valid. The Central Limit Theorem guarantees that the sampling distribution model is Normal when the sample size is large.

53. Dice and dollars.

a) Let X = the number of dollars won in one play.

$$\mu = E(X) = 0\left(\frac{3}{6}\right) + 1\left(\frac{2}{6}\right) + 10\left(\frac{1}{6}\right) = \$2$$

$$\sigma^2 = Var(X) = (0-2)^2\left(\frac{3}{6}\right) + (1-2)^2\left(\frac{2}{6}\right) + (10-2)^2\left(\frac{1}{6}\right) = 13$$

$$\sigma = SD(X) = \sqrt{Var(X)} = \sqrt{13} \approx \$3.61$$

b) $X + X$ = the total winnings for two plays.

$$\mu = E(X + X) = E(X) + E(X) = 2 + 2 = \$4$$

$$\sigma = SD(X + X) = \sqrt{Var(X) + Var(X)}$$
$$= \sqrt{13 + 13} \approx \$5.10$$

c) In order to win at least \$100 in 40 plays, you must average at least $\dfrac{100}{40} = \$2.50$ per play.

The expected value of the winnings is $\mu = \$2$, with standard deviation $\sigma = \$3.61$. Rolling a die is random and the outcomes are mutually independent, so the Central Limit Theorem guarantees that the sampling distribution model is Normal with $\mu_{\bar{x}} = \$2$ and standard deviation $SD(\bar{x}) = \dfrac{\$3.61}{\sqrt{40}} \approx \$0.571$.

According to the Normal model, the probability that you win at least \$100 in 40 plays is approximately 0.191.

(This is equivalent to using $N(80, 22.83)$ to model your total winnings.)

Mean winnings from 40 plays (\$)

0.191

2   2.50

$z = 0.876$

55. AP Stats 2013.

a) $\mu = 5(0.126) + 4(0.202) + 3(0.250) + 2(0.188) + 1(0.234) = 2.798$

$$\sigma = \sqrt{\begin{aligned}&(5-2.798)^2(0.126) + (4-2.798)^2(0.202) + (3-2.798)^2(0.250)\\ &+ (2-2.798)^2(0.188) + (1-2.798)^2(0.234)\end{aligned}} \approx 1.338$$

b) The distribution of scores for 40 randomly selected students would not follow a Normal model. The distribution would resemble the population, mostly uniform for scores 1 – 4, with about half as many 5s.

c) **Randomization condition:** The scores are selected randomly.

Independence assumption: It is reasonable to think that the randomly selected scores are independent of one another.

10% condition: The 40 scores represent less than 10% of all scores.

Large Enough Sample condition: A sample of 40 scores is large enough.

Since the conditions are satisfied, the sampling distribution model for the mean of 40 randomly selected AP Stat scores is Normal, with $\mu_{\bar{y}} = \mu \approx 2.798$ and standard deviation $SD(\bar{y}) = \dfrac{\sigma}{\sqrt{n}} = \dfrac{1.338}{\sqrt{40}} \approx 0.2116$.

57. AP Stats 2013, again.

Since the teacher considers his 63 students "typical", and 63 is less than 10% of all students, the sampling distribution model for the mean AP Stat score for 63 students is Normal, with mean $\mu_{\bar{y}} = \mu \approx 2.798$ and standard deviation

$$SD(\bar{y}) = \frac{\sigma}{\sqrt{n}} = \frac{1.338}{\sqrt{63}} \approx 0.1686 .$$

$$z = \frac{\bar{y} - \mu_{\bar{y}}}{SD(\bar{y})}$$

$$z = \frac{3 - 2.798}{\frac{1.338}{\sqrt{63}}}$$

$$z = 1.198$$

According to the sampling distribution model, the probability that the class of 63 students achieves an average of 3 on the AP Stat exam is about 11.5%.

59. Fuel economy again.

a) **Randomization condition:** Assume that the 150 cars can be considered a representative sample of all cars of this type.

Independence assumption: It is reasonable to think that the mileages for these cars are mutually independent.

10% condition: The 150 cars in the lot certainly represent less than 10% of all midsize cars.

Large Enough Sample condition: A sample of 150 cars is large enough.

The mean mileage was $\mu = 24$ mpg, with standard deviation $\sigma = 5.5$ mpg. Since the conditions are met, the CLT allows us to model the sampling distribution of the $\bar{y}$ with a Normal model, with $\mu_{\bar{y}} = 24$ mpg and standard deviation

$$SD(\bar{y}) = \frac{5.5}{\sqrt{150}} \approx 0.45 \text{ mpg}.$$

b) According to the Normal model, the probability that $\bar{y}$ is between 25 and 27 mpg is approximately 0.0131. (27 mpg is about 6.67 standard deviations above the mean, well off the scale of the picture.)

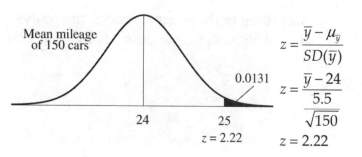

$$z = \frac{\bar{y} - \mu_{\bar{y}}}{SD(\bar{y})}$$

$$z = \frac{\bar{y} - 24}{\frac{5.5}{\sqrt{150}}}$$

$$z = 2.22$$

c) According to the Normal model, there is only a 5% chance that the mean gas mileage is greater than approximately 24.7 mpg.

$$z = \frac{\bar{y} - \mu_{\bar{y}}}{SD(\bar{y})}$$

$$1.645 = \frac{\bar{y} - 24}{\frac{5.5}{\sqrt{150}}}$$

$$\bar{y} \approx 24.7$$

61. IQs.

a) According to the Normal model, the probability that the IQ of a student from East State is at least 125 is approximately 0.734.

b) First, we will need to generate a model for the difference in IQ between the two schools. Since we are choosing at random, it is reasonable to believe that the students' IQs are independent, which allows us to calculate the standard deviation of the difference.

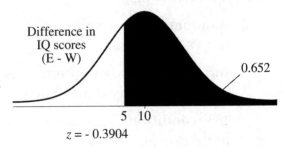

$$\mu = E(E - W) = E(E) - E(W) = 130 - 120 = 10$$
$$\sigma = SD(E - W) = \sqrt{Var(E) + Var(W)}$$
$$= \sqrt{8^2 + 10^2} \approx 12.806$$

Since both distributions are Normal, the distribution of the difference is $N(10, 12.806)$.

According to the Normal model, the probability that the IQ of a student at ESU is at least 5 points higher than a student at WSU is approximately 0.652.

c) **Randomization condition:** Students are randomly sampled from WSU.

Independence assumption: It is reasonable to think that the IQs are mutually independent.

10% condition: The 3 students certainly represent less than 10% of students.

Large Enough Sample condition: The distribution of IQs is Normal, so the distribution of sample means of samples of any size will be Normal, so a sample of 3 students is large enough.

The mean IQ is $\mu_w = 120$, with standard deviation $\sigma_w = 10$. Since the distribution IQs is Normal, we can model the sampling distribution of $\overline{w}$ with a Normal model, with $\mu_{\overline{w}} = 120$ with standard deviation

$$SD(\overline{w}) = \frac{10}{\sqrt{3}} \approx 5.7735.$$

Mean IQ of 3 WSU students

0.193

120   125

$z = 0.8660$

According to the Normal model, the probability that the mean IQ of the 3 WSU students is above 125 is approximately 0.193.

d) As in part c, the sampling distribution of $\overline{e}$, the mean IQ of 3 ESU students, can be modeled with a Normal model, with $\mu_{\overline{e}} = 130$ with standard deviation

$$SD(\overline{e}) = \frac{8}{\sqrt{3}} \approx 4.6188.$$

The distribution of the difference in mean IQ is Normal, with the following parameters:

$$\mu_{\overline{e}-\overline{w}} = E(\overline{e} - \overline{w}) = E(\overline{e}) - E(\overline{w}) = 130 - 120 = 10$$
$$\sigma_{\overline{e}-\overline{w}} = SD(\overline{e} - \overline{w}) = \sqrt{Var(\overline{e}) + Var(\overline{w})}$$
$$= \sqrt{\left(\frac{10}{\sqrt{3}}\right)^2 + \left(\frac{8}{\sqrt{3}}\right)^2} \approx 7.3937$$

According to the Normal model, the probability that the mean IQ of 3 ESU students is at least 5 points higher than the mean IQ of 3 WSU students is approximately 0.751.

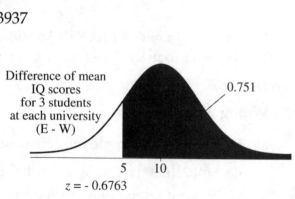

Difference of mean IQ scores for 3 students at each university (E - W)

0.751

5   10

$z = -0.6763$

Chapter 18 – Confidence Intervals for Proportions

Section 18.1

1. **Lying about age.**

   a) This means that 49% of the 799 teens in the sample said they have misrepresented their age online. This is our best estimate of p, the proportion of all U.S. teens who would say they have done so.

   b) $SE(\hat{p}) = \sqrt{\dfrac{(0.49)(0.51)}{799}} \approx 0.018$

   c) Because we don't know p, we use $\hat{p}$ to estimate the standard deviation of the sampling distribution. So the standard error is our estimate of the amount of variation in the sample proportion we expect to see from sample to sample when we ask 799 teens whether they've misrepresented their age online.

Section 18.2

3. **Lying about age again.**

   a) We are 95% confident that, if we were to ask all U.S. teens whether they have misrepresented their age online, between 45.6% and 52.5% of them would say they have.

   b) If we were to collect many random samples of 799 teens, about 95% of the confidence intervals we construct would contain the proportion of all U.S. teens who admit to misrepresenting their age online.

Section 18.3

5. **Wrong direction.**

   a) $ME = z^* \times SE(\hat{p}) = z^* \times \sqrt{\dfrac{\hat{p}\hat{q}}{n}} = 1.645 \times \sqrt{\dfrac{(0.66)(0.34)}{402}} \approx 0.039$ or 3.9%.

   b) The margin of error for 95% confidence would be larger. The critical value of 1.96 is greater than the 1.645 needed for 90% confidence.

Section 18.4

7. **Wrong direction again.**

   a) The sample is a simple random sample. Both $n\hat{p} = (402)(0.66) = 265.32$ and $n\hat{q} = (402)(0.34) = 136.68$ are at least ten. The sample size of 402 is less than 10% of the adult population of Wisconsin. Conditions for the confidence interval are met.

b)

$$ME = z^* \sqrt{\frac{\hat{p}\hat{q}}{n}}$$

$$0.02 = 1.645 \sqrt{\frac{(0.66)(0.34)}{n}}$$

$$n = \frac{(1.645)^2 (0.66)(0.34)}{(0.02)^2}$$

$$n \approx 1519 \, \text{people}$$

In order to estimate the proportion to within 2% with 90% confidence, we would need a sample of at least 1519 people. All decimals in the final answer must be rounded up, to the next person.

(For a more cautious answer, let $\hat{p} = \hat{q} = 0.5$. This method results in a required sample of 1692 people.)

Chapter Exercises

9. Margin of error.

The newscaster believes the true proportion of voters with a certain opinion is within 4% of the estimate, with some degree of confidence, perhaps 95% confidence.

11. Conditions.

   a) *Population* – all cars; *sample* – 134 cars actually stopped at the checkpoint; p – proportion of all cars with safety problems; $\hat{p}$ – proportion of cars in the sample that actually have safety problems (10.4%).
   Randomization condition: This sample is not random, so hopefully the cars stopped are representative of cars in the area.
   10% condition: The 134 cars stopped represent a small fraction of all cars, certainly less than 10%.
   Success/Failure condition: $n\hat{p} = 14$ and $n\hat{q} = 120$ are both greater than 10, so the sample is large enough.
   A one-proportion z-interval can be created for the proportion of all cars in the area with safety problems.

   b) *Population* – the general public; *sample* – 602 viewers that logged on to the Web site;
   p –proportion of the general public that support prayer in school; $\hat{p}$ - proportion of viewers that logged on to the Web site and voted that support prayer in schools (81.1%).
   Randomization condition: This sample is not random, but biased by voluntary response.
   It would be very unwise to attempt to use this sample to infer anything about the opinion of the general public related to school prayer.

c) *Population* – parents at the school; *sample* – 380 parents who returned surveys; p – proportion of all parents in favor of uniforms; $\hat{p}$ – proportion of those who responded that are in favor of uniforms (60%).
Randomization condition: This sample is not random, but rather biased by nonresponse. There may be lurking variables that affect the opinions of parents who return surveys (and the children who deliver them!).
It would be very unwise to attempt to use this sample to infer anything about the opinion of the parents about uniforms.

d) *Population* – all freshmen enrollees at the college (not just one year); *sample* – 1632 freshmen during the specified year; p – proportion of all students who will graduate on time; $\hat{p}$ – proportion of students from that year who graduate on time (85.05%).
Randomization condition: This sample is not random, but this year's freshmen class is probably representative of freshman classes in other years.
10% condition: The 1632 students in that years freshmen class represent less than 10% of all possible students.
Success/Failure condition: $n\hat{p} = 1388$ and $n\hat{q} = 244$ are both greater than 10, so the sample is large enough.
A one-proportion z-interval can be created for the proportion of freshmen that graduate on time from this college.

13. Conclusions.

a) Not correct. This statement implies certainty. There is no level of confidence in the statement.

b) Not correct. Different samples will give different results. Many fewer than 95% of samples are expected to have *exactly* 88% on-time orders.

c) Not correct. A confidence interval should say something about the unknown population proportion, not the sample proportion in different samples.

d) Not correct. We *know* that 88% of the orders arrived on time. There is no need to make an interval for the sample proportion.

e) Not correct. The interval should be about the proportion of on-time orders, not the days.

15. Confidence intervals.

a) False. For a given sample size, higher confidence means a *larger* margin of error.

b) True. Larger samples lead to smaller standard errors, which lead to smaller margins of error.

c) True. Larger samples are less variable, which makes us more confident that a given confidence interval succeeds in catching the population proportion.

d) False. The margin of error decreases as the square root of the sample size increases. Halving the margin of error requires a sample four times as large as the original.

17. Cars.

We are 90% confident that between 29.9% and 47.0% of cars are made in Japan.

19. Mislabeled seafood.

a) $\hat{p} \pm z^* \sqrt{\dfrac{\hat{p}\hat{q}}{n}} = \left(\dfrac{42}{190}\right) \pm 1.960 \sqrt{\dfrac{\left(\frac{42}{190}\right)\left(\frac{148}{190}\right)}{190}} = (0.162, 0.280)$

b) We are 95% confident that between 16.2% and 28.0% of all seafood packages sold in these three states are mislabeled.

c) The size of the population is irrelevant. If *Consumer Reports* had a random sample, 95% of intervals generated by studies like this are expected to capture the true proportion of seafood packages that are mislabeled.

21. Baseball fans.

a) $ME = z^* \times SE(\hat{p}) = z^* \times \sqrt{\dfrac{\hat{p}\hat{q}}{n}} = 1.645 \times \sqrt{\dfrac{(0.48)(0.52)}{1006}} \approx 0.026$

b) We're 90% confident that this poll's estimate is within 2.6% of the true proportion of people who are baseball fans.

c) The margin of error for 99% confidence would be larger. To be more certain, we must be less precise.

d) $ME = z^* \times SE(\hat{p}) = z^* \times \sqrt{\dfrac{\hat{p}\hat{q}}{n}} = 2.576 \times \sqrt{\dfrac{(0.48)(0.52)}{1006}} \approx 0.041$

e) Smaller margins of error involve less confidence. The narrower the confidence interval, the less likely we are to believe that we have succeeded in capturing the true proportion of people who are baseball fans.

23. Contributions please.

a) **Randomization condition:** Letters were sent to a random sample of 100,000 potential donors.
10% condition: We assume that the potential donor list has more than 1,000,000 names.
Success/Failure condition: $n\hat{p} = 4781$ and $n\hat{q} = 95{,}219$ are both much greater than 10, so the sample is large enough.

$\hat{p} \pm z^* \sqrt{\dfrac{\hat{p}\hat{q}}{n}} = \left(\dfrac{4781}{100{,}000}\right) \pm 1.960 \sqrt{\dfrac{\left(\frac{4781}{100{,}000}\right)\left(\frac{95{,}219}{100{,}000}\right)}{100{,}000}} = (0.0465, 0.0491)$

We are 95% confident that the between 4.65% and 4.91% of potential donors would donate.

b) The confidence interval gives the set of plausible values with 95% confidence. Since 5% is above the interval, it seems to be a bit optimistic.

25. Teenage drivers.

a) Randomization condition: The insurance company randomly selected 582 accidents.
10% condition: 582 accidents represent less than 10% of all accidents.
Success/Failure condition: $n\hat{p} = 91$ and $n\hat{q} = 491$ are both greater than 10, so the sample is large enough.

Since the conditions are met, we can use a one-proportion z-interval to estimate the percentage of accidents involving teenagers.

$$\hat{p} \pm z^* \sqrt{\frac{\hat{p}\hat{q}}{n}} = \left(\frac{91}{582}\right) \pm 1.960 \sqrt{\frac{\left(\frac{91}{582}\right)\left(\frac{491}{582}\right)}{582}} = (12.7\%, 18.6\%)$$

b) We are 95% confident that between 12.7% and 18.6% of all accidents involve teenagers.

c) About 95% of random samples of size 582 will produce intervals that contain the true proportion of accidents involving teenagers.

d) Our confidence interval contradicts the assertion of the politician. The figure quoted by the politician, 1 out of every 5, or 20%, is above the interval.

27. Safe food.

The grocer can conclude nothing about the opinions of all his customers from this survey. Those customers who bothered to fill out the survey represent a voluntary response sample, consisting of people who felt strongly one way or another about irradiated food. The random condition was not met.

29. Death penalty, again.

a) There may be response bias based on the wording of the question.

b) $\hat{p} \pm z^* \sqrt{\frac{\hat{p}\hat{q}}{n}} = (0.61) \pm 1.960 \sqrt{\frac{(0.61)(0.39)}{1020}} = (58\%, 64\%)$

c) The margin of error based on the pooled sample is smaller, since the sample size is larger.

31. Rickets.

a) **Randomization condition:** The 2700 children were chosen at random.
10% condition: 2700 children are less than 10% of all English children.
Success/Failure condition: $n\hat{p} = (2700)(0.20) = 540$ and $n\hat{q} = (2700)(0.80) = 2160$ are both greater than 10, so the sample is large enough.

Since the conditions are met, we can use a one-proportion z-interval to estimate the proportion of the English children with vitamin D deficiency.

$$\hat{p} \pm z^* \sqrt{\frac{\hat{p}\hat{q}}{n}} = (0.20) \pm 2.326 \sqrt{\frac{(0.20)(0.80)}{2700}} = (18.2\%, 21.8\%)$$

b) We are 98% confident that between 18.2% and 21.8% of English children are deficient in vitamin D.

c) About 98% of random samples of size 2700 will produce confidence intervals that contain the true proportion of English children that are deficient in vitamin D.

33. Privacy or Security?

a) The confidence interval will be wider. The sample size is probably about one-sixth (17%) of the sample size of for all adults, so we would expect the confidence interval to be about two and a half times as wide.

b) The second poll's margin of error should be slightly wider. There are fewer "young" people (13%) in the sample than seniors (17%).

35. Deer ticks.

a) **Independence assumption:** Deer ticks are parasites. A deer carrying the parasite may spread it to others. Ticks may not be distributed evenly throughout the population.
Randomization condition: The sample is not random and may not represent all deer.
10% condition: 153 deer are less than 10% of all deer.
Success/Failure condition: $n\hat{p} = 32$ and $n\hat{q} = 121$ are both greater than 10, so the sample is large enough.

The conditions are not satisfied, so we should use caution when a one-proportion z-interval is used to estimate the proportion of deer carrying ticks.

$$\hat{p} \pm z^* \sqrt{\frac{\hat{p}\hat{q}}{n}} = \left(\frac{32}{153}\right) \pm 1.645 \sqrt{\frac{\left(\frac{32}{153}\right)\left(\frac{121}{153}\right)}{153}} = (15.5\%, 26.3\%)$$

We are 90% confident that between 15.5% and 26.3% of deer have ticks.

b) In order to cut the margin of error in half, they must sample 4 times as many deer.

4(153) = 612 deer.

c) The incidence of deer ticks is not plausibly independent, and the sample may not be representative of all deer, since females and young deer are usually not hunted.

37. Graduation.

a)

$$ME = z^* \sqrt{\frac{\hat{p}\hat{q}}{n}}$$

$$0.06 = 1.645 \sqrt{\frac{(0.25)(0.75)}{n}}$$

$$n = \frac{(1.645)^2 (0.25)(0.75)}{(0.06)^2}$$

$$n \approx 141 \text{ people}$$

In order to estimate the proportion of non-graduates in the 25-to 30-year-old age group to within 6% with 90% confidence, we would need a sample of at least 141 people. All decimals in the final answer must be rounded up, to the next person.

(For a more cautious answer, let $\hat{p} = \hat{q} = 0.5$. This method results in a required sample of 188 people.)

b)

$$ME = z^* \sqrt{\frac{\hat{p}\hat{q}}{n}}$$

$$0.04 = 1.645 \sqrt{\frac{(0.25)(0.75)}{n}}$$

$$n = \frac{(1.645)^2 (0.25)(0.75)}{(0.04)^2}$$

$$n \approx 318 \text{ people}$$

In order to estimate the proportion of non-graduates in the 25-to 30-year-old age group to within 4% with 90% confidence, we would need a sample of at least 318 people. All decimals in the final answer must be rounded up, to the next person.

(For a more cautious answer, let $\hat{p} = \hat{q} = 0.5$. This method results in a required sample of 423 people.)

Alternatively, the margin of error is now 2/3 of the original, so the sample size must be increased by a factor of 9/4. 141(9/4) ≈ 318 people.

c)

$$ME = z^* \sqrt{\frac{\hat{p}\hat{q}}{n}}$$

$$0.03 = 1.645 \sqrt{\frac{(0.25)(0.75)}{n}}$$

$$n = \frac{(1.645)^2 (0.25)(0.75)}{(0.03)^2}$$

$$n \approx 564 \text{ people}$$

In order estimate the proportion of non-graduates in the 25-to 30-year-old age group to within 3% with 90% confidence, we would need a sample of at least 564 people. All decimals in the final answer must be rounded up, to the next person.

(For a more cautious answer, let $\hat{p} = \hat{q} = 0.5$. This method results in a required sample of 752 people.)

Alternatively, the margin of error is now half that of the original, so the sample size must be increased by a factor of 4. $141(4) \approx 564$ people.

39. Graduation, again.

$$ME = z^* \sqrt{\frac{\hat{p}\hat{q}}{n}}$$

$$0.02 = 1.960 \sqrt{\frac{(0.25)(0.75)}{n}}$$

$$n = \frac{(1.960)^2 (0.25)(0.75)}{(0.02)^2}$$

$$n \approx 1801 \text{ people}$$

In order to estimate the proportion of non-graduates in the 25-to 30-year-old age group to within 2% with 95% confidence, we would need a sample of at least 1801 people. All decimals in the final answer must be rounded up, to the next person.

(For a more cautious answer, let $\hat{p} = \hat{q} = 0.5$. This method results in a required sample of 2401 people.)

41. Pilot study.

$$ME = z^* \sqrt{\frac{\hat{p}\hat{q}}{n}}$$

$$0.03 = 1.645 \sqrt{\frac{(0.15)(0.85)}{n}}$$

$$n = \frac{(1.645)^2 (0.15)(0.85)}{(0.03)^2}$$

$$n \approx 384 \text{ cars}$$

Use $\hat{p} = \frac{9}{60} = 0.15$ from the pilot study as an estimate.

In order to estimate the percentage of cars with faulty emissions systems to within 3% with 90% confidence, the state's environmental agency will need a sample of at least 384 cars. All decimals in the final answer must be rounded up, to the next car.

43. Approval rating.

$$ME = z^* \sqrt{\frac{\hat{p}\hat{q}}{n}}$$

$$0.025 = z^* \sqrt{\frac{(0.65)(0.35)}{972}}$$

$$z^* = \frac{0.025}{\sqrt{\dfrac{(0.65)(0.35)}{972}}}$$

$$z^* \approx 1.634$$

Since $z^* \approx 1.634$, which is close to 1.645, the pollsters were probably using 90% confidence. The slight difference in the z^* values is due to rounding of the governor's approval rating.

Chapter 19 – Testing Hypotheses about Proportions

Section 19.1

1. **Better than aspirin?**

   a) The new drug is not more effective than aspirin, and reduces the risk of heart attack by 44%. ($p = 0.44$)

   b) The new drug is more effective than aspirin, and reduces the risk of heart attack by more than 44%. ($p > 0.44$)

Section 19.2

3. **Better than aspirin 2?**

   a) Since the *P*-value of 0.28 is greater than 0.05, fail to reject the null hypothesis. There is not sufficient evidence to conclude that the new drug is better than aspirin.

   b) Since the *P*-value of 0.004 is less than 0.05, reject the null hypothesis. There is evidence that the new drug is more effective than aspirin.

Section 19.3

5. **Hispanic origin.**

   a) H_0 : The proportion of people in the county that are of Hispanic or Latino origin is 0.16. ($p = 0.16$)

   H_A : The proportion of people in the county that are of Hispanic or Latino origin is different from 0.16. ($p \neq 0.16$)

   b) **Randomization condition:** The 437 county residents were a random sample of all county residents.
   10% condition: 437 is likely to be less than 10% of all county residents.

   Success/Failure condition: $np_0 = (437)(0.16) = 69.92$ and $nq_0 = (437)(0.84) = 367.08$ are both greater than 10, so the sample is large enough.

   Since the conditions are met, we will model the sampling distribution of $\hat{p}$ with a Normal model and perform a one-proportion *z*-test.

c) $\hat{p} = \dfrac{44}{437} = 0.101$

$SD(\hat{p}) = \sqrt{\dfrac{p_0 q_0}{n}} = \sqrt{\dfrac{(0.16)(0.84)}{437}} \approx 0.018$

$z = \dfrac{\hat{p} - p_0}{SD(\hat{p})} = \dfrac{0.101 - 0.16}{0.018} = -3.28$

P-value $= 2 \cdot P(z < -3.28) = 0.001$

d) Since the P-value = 0.001 is so low, reject the null hypothesis. There is evidence that the Hispanic/Latino population in this county differs from that as the nation as a whole. These data suggest that the proportion of Hispanic/Latino residents is, in fact, lower than the national proportion.

Section 19.4

7. Psychic again (you should have seen this coming).

The alternative hypothesis would be one-sided, because the only evidence that would support the friend's claim is guessing more than 25% of the suits correctly.

Section 19.5

9. Bad medicine.

a) The drug may not be approved for use, and people would miss out on a beneficial product.

b) The drug will go into production and people will suffer the side effect.

Chapter Exercises

11. Hypotheses.

a) H_0 : The governor's "negatives" are 30%. ($p = 0.30$)
H_A : The governor's "negatives" are less than 30%. ($p < 0.30$)

b) H_0 : The proportion of heads is 50%. ($p = 0.50$)
H_A : The proportion of heads is not 50%. ($p \neq 0.50$)

c) H_0 : The proportion of people who quit smoking is 20%. ($p = 0.20$)
H_A : The proportion of people who quit smoking is greater than 20%. ($p > 0.20$)

13. Negatives.

Statement d is the correct interpretation of a P-value.

15. Relief.

It is *not* reasonable to conclude that the new formula and the old one are equally effective. Furthermore, our inability to make that conclusion has nothing to do with the *P*-value. We can not prove the null hypothesis (that the new formula and the old formula are equally effective), but can only fail to find evidence that would cause us to reject it. All we can say about this *P*-value is that there is a 27% chance of seeing the observed effectiveness from natural sampling variation if the new formula and the old one are equally effective.

17. He cheats?

a) Two losses in a row aren't convincing. There is a 25% chance of losing twice in a row, and that is not unusual.

b) If the process is fair, three losses in a row can be expected to happen about 12.5% of the time. $(0.5)(0.5)(0.5) = 0.125$.

c) Three losses in a row is still not a convincing occurrence. We'd expect that to happen about once every eight times we tossed a coin three times.

d) Answers may vary. Maybe 5 times would be convincing. The chances of 5 losses in a row are only 1 in 32, which seems unusual.

19. Smartphones.

1) Null and alternative hypotheses should involve p, not $\hat{p}$.

2) The question is about *failing* to meet the goal. H_A should be $p < 0.96$.

3) The student failed to check $nq_0 = (200)(0.04) = 8$. Since $nq_0 < 10$, the Success/Failure condition is violated. Similarly, the 10% Condition is not verified.

4) $SD(\hat{p}) = \sqrt{\dfrac{pq}{n}} = \sqrt{\dfrac{(0.96)(0.04)}{200}} \approx 0.014$. The student used $\hat{p}$ and $\hat{q}$.

5) Value of z is incorrect. The correct value is $z = \dfrac{0.94 - 0.96}{0.014} \approx -1.43$.

6) *P*-value is incorrect. $P = P(z < -1.43) = 0.076$

7) For the *P*-value given, an incorrect conclusion is drawn. A *P*-value of 0.12 provides no evidence that the new system has failed to meet the goal. The correct conclusion for the corrected *P*-value is: Since the *P*-value of 0.076 is fairly low, there is weak evidence that the new system has failed to meet the goal.

21. Dowsing.

a) H_0 : The percentage of successful wells drilled by the dowser is 30%. ($p = 0.30$)
H_A : The percentage of successful wells drilled is greater than 30%. ($p > 0.30$)

b) **Independence assumption:** There is no reason to think that finding water in one well will affect the probability that water is found in another, unless the wells are close enough to be fed by the same underground water source.
Randomization condition: This sample is not random, so hopefully the customers you check with are representative of all of the dowser's customers.
10% condition: The 80 customers sampled may be considered less than 10% of all possible customers.
Success/Failure condition: $np_0 = (80)(0.30) = 24$ and $nq_0 = (80)(0.70) = 56$ are both greater than 10, so the sample is large enough.

c) The sample of customers may not be representative of all customers, so we will proceed cautiously. A Normal model can be used to model the sampling distribution of the proportion, with $\mu_{\hat{p}} = p_0 = 0.30$ and

$$SD(\hat{p}) = \sqrt{\frac{p_0 q_0}{n}} = \sqrt{\frac{(0.30)(0.70)}{80}} \approx 0.0512 \,.$$

We can perform a one-proportion z-test. The observed proportion of successful wells is $\hat{p} = \dfrac{27}{80} = 0.3375$.

$$z = \frac{\hat{p} - p_0}{\sqrt{\dfrac{pq}{n}}}$$

$$z = \frac{0.3375 - 0.30}{\sqrt{\dfrac{(0.30)(0.70)}{80}}}$$

$$z \approx 0.73$$

$P = 0.232$

0.3   0.3375
$z = 0.73$

d) If his dowsing has the same success rate as standard drilling methods, there is more than a 23% chance of seeing results as good as those of the dowser, or better, by natural sampling variation.

e) With a P-value of 0.232, we fail to reject the null hypothesis. There is no evidence to suggest that the dowser has a success rate any higher than 30%.

23. Absentees.

a) H_0 : The percentage of students in 2000 with perfect attendance the previous month is 34% ($p = 0.34$)
H_A : The percentage of students in 2000 with perfect attendance the previous month is different from 34% ($p \neq 0.34$)

b) **Randomization condition:** Although not specifically stated, we can assume that the National Center for Educational Statistics used random sampling.
10% condition: The 8302 students are less than 10% of all students.
Success/Failure condition: $np_0 = (8302)(0.34) = 2822.68$ and $nq_0 = (8302)(0.66) = 5479.32$ are both greater than 10, so the sample is large enough.

c) Since the conditions for inference are met, a Normal model can be used to model the sampling distribution of the proportion, with $\mu_{\hat{p}} = p_0 = 0.34$ and

$$SD(\hat{p}) = \sqrt{\frac{p_0 q_0}{n}} = \sqrt{\frac{(0.34)(0.66)}{8302}} \approx 0.0052$$

We can perform a two-tailed one-proportion z-test. The observed proportion of perfect attendees is $\hat{p} = 0.33$.

d) With a *P*-value of 0.0544, we reject the null hypothesis. There is some evidence to suggest that the percentage of students with perfect attendance in the previous month has changed in 2000.

$$z = \frac{\hat{p} - p_0}{\sqrt{\dfrac{pq}{n}}}$$

$$z = \frac{0.33 - 0.34}{\sqrt{\dfrac{(0.34)(0.66)}{8302}}}$$

$$z \approx -1.923$$

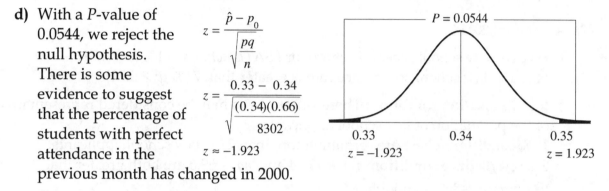

$P = 0.0544$

0.33   0.34   0.35

$z = -1.923$ $z = 1.923$

e) This result is not meaningful. A difference this small, although statistically significant, is of little practical significance.

25. Contributions, please, part II.

a) H_0 : The contribution rate is 5% ($p = 0.05$)
H_A : The contribution rate is less than 5% ($p < 0.05$)

b) **Randomization condition:** Potential donors were randomly selected.
10% condition: We will assume the entire mailing list has over 1,000,000 names.
Success/Failure condition: $np_0 = 5000$ and $nq_0 = 95,000$ are both greater than 10, so the sample is large enough.

The conditions have been satisfied, so a Normal model can be used to model the sampling distribution of the proportion, with $\mu_{\hat{p}} = p_0 = 0.05$ and

$$SD(\hat{p}) = \sqrt{\frac{p_0 q_0}{n}} = \sqrt{\frac{(0.05)(0.95)}{100,000}} \approx 0.0007 \, .$$

We can perform a one-proportion z-test. The observed contribution rate is $\hat{p} = \dfrac{4781}{100,000} = 0.04781$.

c) Since the *P*-value = 0.0006 is low, we reject the null hypothesis. There is strong evidence that contribution rate for all potential donors is lower than 5%.

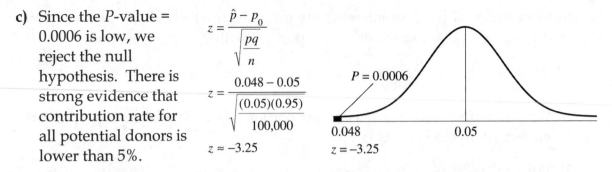

$$z = \frac{\hat{p} - p_0}{\sqrt{\dfrac{pq}{n}}}$$

$$z = \frac{0.048 - 0.05}{\sqrt{\dfrac{(0.05)(0.95)}{100{,}000}}}$$

$$z \approx -3.25$$

$P = 0.0006$

0.048 0.05
$z = -3.25$

27. Law school 2011.

a) H_0 : The law school acceptance rate for *LSATisfaction* is 71% ($p = 0.71$)
H_A : The law school acceptance rate is greater than 71% ($p > 0.71$)

b) **Randomization condition:** These 240 students may be considered representative of the population of law school applicants.
10% condition: There are certainly more than 2400 law school applicants.
Success/Failure condition: $np_0 = 170.4$ and $nq_0 = 69.6$ are both greater than 10, so the sample is large enough.

The conditions have been satisfied, so a Normal model can be used to model the sampling distribution of the proportion, with $\mu_{\hat{p}} = p_0 = 0.71$ and

$$SD(\hat{p}) = \sqrt{\frac{pq}{n}} = \sqrt{\frac{(0.71)(0.29)}{240}} \approx 0.0293 \,.$$

We can perform a one-proportion z-test. The observed success rate is
$$\hat{p} = \frac{174}{240} = 0.725 \,.$$

$$z = \frac{\hat{p} - p_0}{\sqrt{\dfrac{pq}{n}}}$$

$$z = \frac{0.725 - 0.71}{\sqrt{\dfrac{(0.71)(0.29)}{240}}}$$

$$z \approx 0.5121$$

$P = 0.304$

0.71 0.725
$z = 0.5121$

c) Since the *P*-value = 0.304 is high, we fail to reject the null hypothesis. There is no evidence that the law school acceptance rate is higher for *LSATisfaction* applicants than for applicants nationwide.

29. Pollution.

H_0 : The percentage of cars with faulty emissions is 20%. ($p = 0.20$)
H_A : The percentage of cars with faulty emissions is greater than 20%. ($p > 0.20$)

Two conditions are not satisfied. 22 is greater than 10% of the population of 150 cars, and $np_0 = (22)(0.20) = 4.4$, which is not greater than 10. It's not advisable to proceed with a test.

31. Twins.

H_0 : The percentage of twin births to teenage girls is 3%. ($p = 0.03$)
H_A : The percentage of twin births to teenage girls differs from 3%. ($p \neq 0.03$)

Independence assumption: One mother having twins will not affect another. Observations are plausibly independent.

Randomization condition: This sample may not be random, but it is reasonable to think that this hospital has a representative sample of teen mothers, with regards to twin births.

10% condition: The sample of 469 teenage mothers is less than 10% of all such mothers.

Success/Failure condition: $np_0 = (469)(0.03) = 14.07$ and $nq_0 = (469)(0.97) = 454.93$ are both greater than 10, so the sample is large enough.

The conditions have been satisfied, so a Normal model can be used to model the sampling distribution of the proportion, with $\mu_{\hat{p}} = p_0 = 0.03$ and

$$SD(\hat{p}) = \sqrt{\frac{p_0 q_0}{n}} = \sqrt{\frac{(0.03)(0.97)}{469}} \approx 0.0079 \,.$$

We can perform a one-proportion z-test. The observed proportion of twin births to teenage mothers is $\hat{p} = \dfrac{7}{469} \approx 0.015$.

Since the P-value = 0.0556 is fairly low, we reject the null hypothesis. There is some evidence that the proportion of twin births for teenage mothers at this large

$$z = \frac{\hat{p} - p_0}{\sqrt{\dfrac{pq}{n}}}$$

$$z = \frac{0.015 - 0.03}{\sqrt{\dfrac{(0.03)(0.97)}{469}}}$$

$$z \approx -1.91$$

city hospital is lower than the proportion of twin births for all mothers.

33. WebZine.

H_0 : The percentage of readers interested in an online edition is 25%. ($p = 0.25$)
H_A : The percentage of readers interested is greater than 25%. ($p > 0.25$)

Randomization condition: The magazine conducted an SRS of 500 current readers.
10% condition: 500 readers are less than 10% of all potential subscribers.
Success/Failure condition: $np_0 = (500)(0.25) = 125$ and $nq_0 = (500)(0.75) = 375$ are both greater than 10, so the sample is large enough.

The conditions have been satisfied, so a Normal model can be used to model the sampling distribution of the proportion, with $\mu_{\hat{p}} = p_0 = 0.25$ and

$$SD(\hat{p}) = \sqrt{\frac{p_0 q_0}{n}} = \sqrt{\frac{(0.25)(0.75)}{500}} \approx 0.0194 .$$

We can perform a one-proportion z-test. The observed proportion of interested readers is $\hat{p} = \dfrac{137}{500} = 0.274$.

Since the P-value = 0.1076 is high, we fail to reject the null hypothesis. There is little evidence to suggest that the proportion of interested readers is greater than 25%. The magazine should not publish the online edition.

$$z = \frac{\hat{p} - p_0}{\sqrt{\dfrac{pq}{n}}}$$

$$z = \frac{0.274 - 0.25}{\sqrt{\dfrac{(0.25)(0.75)}{500}}}$$

$$z \approx 1.24$$

35. Women executives.

H_0 : The proportion of female executives is similar to the overall proportion of female employees at the company. ($p = 0.40$)
H_A : The proportion of female executives is lower than the overall proportion of female employees at the company. ($p < 0.40$)

Independence assumption: It is reasonable to think that executives at this company were chosen independently.
Randomization condition: The executives were not chosen randomly, but it is reasonable to think of these executives as representative of all potential executives over many years.
10% condition: 43 executives are less than 10% of all executives at the company.
Success/Failure condition: $np_0 = (43)(0.40) = 17.2$ and $nq_0 = (43)(0.60) = 25.8$ are both greater than 10, so the sample is large enough.

The conditions have been satisfied, so a Normal model can be used to model the sampling distribution of the proportion, with $\mu_{\hat{p}} = p_0 = 0.40$ and

$$SD(\hat{p}) = \sqrt{\frac{p_0 q_0}{n}} = \sqrt{\frac{(0.40)(0.60)}{43}} \approx 0.0747 \, .$$

Perform a one-proportion z-test. The observed proportion is $\hat{p} = \dfrac{13}{43} \approx 0.302$.

Since the *P*-value = 0.0955 is high, we fail to reject the null hypothesis. There is little evidence to suggest proportion of female executives is any different from the overall proportion of 40% female employees at the company.

$$z = \frac{\hat{p} - p_0}{\sqrt{\dfrac{pq}{n}}}$$

$$z = \frac{0.302 - 0.40}{\sqrt{\dfrac{(0.40)(0.60)}{43}}}$$

$$z \approx -1.31$$

$P = 0.0955$

0.302   0.40

$z = -1.31$

37. Dropouts.

H_0 : The proportion of dropouts at this high school is similar to 10.3%, the proportion of dropouts nationally. ($p = 0.103$)
H_A : The proportion of dropouts at this high school is greater than 10.3%, the proportion of dropouts nationally. ($p > 0.103$)

Independence assumption /Randomization condition: Assume that the students at this high school are representative of all students nationally. This is really what we are testing. The dropout rate at this high school has traditionally been close to the national rate. If we reject the null hypothesis, we will have evidence that the dropout rate at this high school is no longer close to the national rate.
10% condition: 1782 students are less than 10% of all students nationally.
Success/Failure condition: $np_0 = (1782)(0.103) = 183.546$ and $nq_0 = (1782)(0.897) = 1598.454$ are both greater than 10, so the sample is large enough.

The conditions have been satisfied, so a Normal model can be used to model the sampling distribution of the proportion, $\mu_{\hat{p}} = p_0 = 0.103$ and

$$SD(\hat{p}) = \sqrt{\frac{p_0 q_0}{n}} = \sqrt{\frac{(0.103)(0.897)}{1782}} \approx 0.0072 \, .$$

We can perform a one-proportion z-test. The observed proportion of dropouts is $\hat{p} = \dfrac{210}{1782} \approx 0.117845$.

Since the *P*-value = 0.02 is low, we reject the null hypothesis. There is evidence that the dropout rate at this high school is higher than 10.3%.

$$z = \frac{\hat{p} - p_0}{\sqrt{\dfrac{pq}{n}}}$$

$$z = \frac{0.117845 - 0.109}{\sqrt{\dfrac{(0.103)(0.897)}{1782}}}$$

$z \approx 2.06$

39. Lost luggage.

H_0 : The proportion of lost luggage returned the next day is 90%. ($p = 0.90$)
H_A : The proportion of lost luggage returned is lower than 90%. ($p < 0.90$)

Independence assumption: It is reasonable to think that the people surveyed were independent with regards to their luggage woes.
Randomization condition: Although not stated, we will hope that the survey was conducted randomly, or at least that these air travelers are representative of all air travelers for that airline.

10% condition: 122 air travelers are less than 10% of all air travelers on the airline.
Success/Failure condition: $np_0 = (122)(0.90) = 109.8$ and $nq_0 = (122)(0.10) = 12.2$ are both greater than 10, so the sample is large enough.

The conditions have been satisfied, so a Normal model can be used to model the sampling distribution of the proportion, with $\mu_{\hat{p}} = p_0 = 0.90$ and

$$SD(\hat{p}) = \sqrt{\frac{p_0 q_0}{n}} = \sqrt{\frac{(0.90)(0.10)}{122}} \approx 0.0272 \, .$$

We can perform a one-proportion *z*-test. The observed proportion of lost luggage is $\hat{p} = \dfrac{103}{122} \approx 0.844 \, .$

Since the *P*-value = 0.0201 is low, we reject the null hypothesis. There is evidence that the proportion of lost luggage returned the next day is lower than the 90% claimed by the airline.

$$z = \frac{\hat{p} - p_0}{\sqrt{\dfrac{pq}{n}}}$$

$$z = \frac{0.844 - 0.90}{\sqrt{\dfrac{(0.90)(0.10)}{122}}}$$

$z \approx -2.05$

41. John Wayne.

a) H_0 : The death rate from cancer for people working on the film was similar to that predicted by cancer experts, 30 out of 220.

H_A : The death rate from cancer for people working on the film was higher than the rate predicted by cancer experts.

The conditions for inference are not met, since this is not a random sample. We will assume that the cancer rates for people working on the film are similar to those predicted by the cancer experts, and a Normal model can be used to model the sampling distribution of the rate, with $\mu_{\hat{p}} = p_0 = 30/220$ and

$$SD(\hat{p}) = \sqrt{\frac{p_0 q_0}{n}} = \sqrt{\frac{\left(\frac{30}{220}\right)\left(\frac{190}{220}\right)}{220}} \approx 0.0231 \,.$$

We can perform a one-proportion z-test. The observed cancer rate is

$$\hat{p} = \frac{46}{220} \approx 0.209 \,.$$

$$z = \frac{\hat{p} - p_0}{SD(\hat{p})}$$

Since the P-value = 0.0008 is very low, we reject the null hypothesis. There is strong evidence that the cancer rate is higher than expected among the workers on the film.

$$z = \frac{\frac{46}{220} - \frac{30}{220}}{\sqrt{\frac{\left(\frac{30}{220}\right)\left(\frac{190}{220}\right)}{220}}}$$

$$z = 3.14$$

b) This does not prove that exposure to radiation may increase the risk of cancer. This group of people may be atypical for reasons that have nothing to do with the radiation.

Chapter 20 – Inferences About Means

Section 20.1

1. **Salmon.**

a) The shipment of 4 salmon has $SD(\bar{y}) = \dfrac{\sigma}{\sqrt{n}} = \dfrac{2}{\sqrt{4}} = 1\,\text{pound}.$

The shipment of 16 salmon has $SD(\bar{y}) = \dfrac{\sigma}{\sqrt{n}} = \dfrac{2}{\sqrt{16}} = 0.5\,\text{pounds}.$

The shipment of 100 salmon has $SD(\bar{y}) = \dfrac{\sigma}{\sqrt{n}} = \dfrac{2}{\sqrt{100}} = 0.2\,\text{pounds}.$

b) The Normal model would better characterize the shipping weight of the pallets than the shipping weight of the boxes, since the pallets contain a large number of salmon. The Central Limit Theorem tells us that the distribution of means (and therefore totals) approaches the Normal model, regardless of the underlying distribution. As samples get larger, the approximation gets better.

Section 20.2

3. **t-models, part I.**

a) 1.74   b) 2.37

5. **t-models, part III.**

As the number of degrees of freedom increases, the shape and center of t-models do not change. The spread of t-models decreases as the number of degrees of freedom increases, and the shape of the distribution becomes closer to Normal.

Section 20.3

7. **Home sales.**

a) The estimates of home value losses must be independent. This is verified using the Randomization condition, since the houses were randomly sampled. The distribution of home value losses must be Normal. A histogram of home value losses in the sample would be checked to verify this, using the Nearly Normal condition. Even if the histogram is not unimodal and symmetric, the sample size of 36 should allow for some small departures from Normality.

b) $\bar{y} \pm t^*_{n-1}\left(\dfrac{s}{\sqrt{n}}\right) = 9560 \pm t^*_{35}\left(\dfrac{1500}{\sqrt{36}}\right) \approx (9052.50,\ 10067.50)$

c) We are 95% confident that average home value loss is between \$9052.50 and \$10,067.50.

Section 20.4

9. *t*-models, again.

    a) 0.0524   **b)** 0.0889

11. **Home prices.**

H_0: The mean loss in home value has not decreased from the mean loss in home value in 2010 of \$10,200. $(\mu = 10,200)$

H_A: The mean loss in home value has decreased significantly from the mean loss in home value in 2010 of \$10,200. $(\mu < 10,200)$

Randomization condition: The home value losses were collected from 36 randomly selected homes.

Nearly Normal condition: We don't have the data, so we can't make a scatterplot, but with a sample size of 36 homes, the Central Limit Theorem should allow us to use a Normal model to describe the sampling distribution of mean home value losses.

In 2010, home value losses had a mean of \$10,200 and we will use the sample standard deviation of \$1500. Since the conditions for inference are satisfied, we can model the sampling distribution of the mean home value loss with a

Student's *t* model, with 36 – 1 = 35 degrees of freedom, $t_{35}\left(10200, \dfrac{1500}{\sqrt{36}}\right)$. We

will perform a one-sample *t*-test.

$$t = \frac{\bar{y} - \mu_0}{\dfrac{\sigma}{\sqrt{n}}}$$

$$t = \frac{9560 - 10200}{\dfrac{1500}{\sqrt{36}}}$$

$$t = -2.56$$

$P = 0.0075$

$t = -2.56$

Since the *P*-value = 0.0075 is low, we reject the null hypothesis. There is evidence that the loss of home values in this community has decreased.

Section 20.5

13. Jelly.

A preliminary calculation with $z^* = 2.576$ gives an approximate sample size of 27. Refining the approximation with $df = n-1 = 27-1 = 26$ and $t_{26}^* = 2.779$ gives an approximate sample size of 31. The consumer advocate should collect a sample of at least 31 jelly jars to achieve a margin of error of no more than 2 grams.

$$ME = z^* \left(\frac{s}{\sqrt{n}} \right) \qquad ME = t_{n-1}^* \left(\frac{s}{\sqrt{n}} \right)$$

$$2 = 2.576 \left(\frac{4}{\sqrt{n}} \right) \qquad 2 = 2.779 \left(\frac{4}{\sqrt{n}} \right)$$

$$n = \frac{(2.576)^2 (4)^2}{(2)^2} \qquad n = \frac{(2.779)^2 (4)^2}{(2)^2}$$

$$n \approx 27 \qquad\qquad n \approx 31$$

Chapter Exercises

15. Cattle.

a) Not correct. A confidence interval is for the mean weight gain of the population of all cows. It says nothing about individual cows. This interpretation also appears to imply that there is something special about the interval that was generated, when this interval is actually one of many that could have been generated, depending on the cows that were chosen for the sample.

b) Not correct. A confidence interval is for the mean weight gain of the population of all cows, not individual cows.

c) Not correct. We don't need a confidence interval about the average weight gain for cows in this study. We are certain that the mean weight gain of the cows in this study is 56 pounds. Confidence intervals are for the mean weight gain of the population of all cows.

d) Not correct. This statement implies that the average weight gain varies. It doesn't. We just don't know what it is, and we are trying to find it. The average weight gain is either between 45 and 67 pounds, or it isn't.

e) Not correct. This statement implies that there is something special about our interval, when this interval is actually one of many that could have been generated, depending on the cows that were chosen for the sample. The correct interpretation is that 95% of samples of this size will produce an interval that will contain the mean weight gain of the population of all cows.

17. Meal plan.

a) Not correct. The confidence interval is not about the individual students.

b) Not correct. The confidence interval is not about individual students in the sample. In fact, we know exactly what these students spent, so there is no need to estimate.

c) Not correct. We know that the mean cost for students in this sample was $1467.

d) Not correct. A confidence interval is not about other sample means.

e) This is the correct interpretation of a confidence interval. It estimates a population parameter.

19. Pulse rates.

a) We are 95% confident the interval 70.9 to 74.5 beats per minute contains the true mean heart rate.

b) The width of the interval is about 74.5 – 70.9 = 3.6 beats per minute. The margin of error is half of that, about 1.8 beats per minute.

c) The margin of error would have been larger. More confidence requires a larger critical value of *t*, which increases the margin of error.

21. CEO compensation.

We should be hesitant to trust this confidence interval, since the conditions for inference are not met. The distribution is highly skewed and there is an outlier.

23. Normal temperature.

a) **Randomization condition:** The adults were randomly selected.
Nearly Normal condition: The sample of 52 adults is large, and the histogram shows no serious skewness, outliers, or multiple modes.

The people in the sample had a mean temperature of 98.2846° and a standard deviation in temperature of 0.682379°. Since the conditions are satisfied, the sampling distribution of the mean can be modeled by a Student's *t* model, with 52 – 1 = 51 degrees of freedom. We will use a one-sample *t*-interval with 98% confidence for the mean body temperature.
(By hand, use $t^*_{50} \approx 2.403$ from the table.)

b) $\bar{y} \pm t^*_{n-1}\left(\dfrac{s}{\sqrt{n}}\right) = 98.2846 \pm t^*_{51}\left(\dfrac{0.682379}{\sqrt{52}}\right) \approx (98.06,\ 98.51)$

c) We are 98% confident that the interval 98.06°F to 98.51°F contains the true mean body temperature for adults. (If you calculated the interval by hand, using $t^*_{50} \approx 2.403$ from the table, your interval may be slightly different than intervals calculated using technology. With the rounding used here, they are identical. Even if they aren't, it's not a big deal.)

d) 98% of all random samples of size 52 will produce intervals that contain the true mean body temperature of adults.

e) Since the interval is below the body temperature of 98.6°F, there is strong evidence that the true mean body temperature of adults is lower than 98.6°F.

25. Normal temperatures, part II.

a) The 90% confidence interval would be narrower than the 98% confidence interval. We can be more precise with our interval when we are less confident.

b) The 98% confidence interval has a greater chance of containing the true mean body temperature of adults than the 90% confidence interval, but the 98% confidence interval is less precise (wider) than the 90% confidence interval.

c) The 98% confidence interval would be narrower if the sample size were increased from 52 people to 500 people. The smaller standard error would result in a smaller margin of error.

d) Our sample of 52 people gave us a 98% confidence interval with a margin of error of $(98.51 - 98.05)/2 = 0.225°F$. In order to get a margin of error of 0.1, less than half of that, we need a sample over 4 times as large. It should be safe to use $t^*_{100} \approx 2.364$ from the table, since the sample will need to be larger than 101. Or we could use $z^* \approx 2.326$, since we expect the sample to be large. We need a sample of about 252 people in order to estimate the mean body temperature of adults to within 0.1°F.

$$ME = t^*_{n-1}\left(\frac{s}{\sqrt{n}}\right)$$

$$0.1 = 2.326\left(\frac{0.682379}{\sqrt{n}}\right)$$

$$n = \frac{(2.326)^2 (0.682379)^2}{(0.1)^2}$$

$$n \approx 252$$

27. Speed of Light.

a) $\bar{y} \pm t^*_{n-1}\left(\dfrac{s}{\sqrt{n}}\right) = 756.22 \pm t^*_{22}\left(\dfrac{107.12}{\sqrt{23}}\right) \approx (709.9,\ 802.5)$

b) We are 95% confident that the interval 299,709.9 to 299,802.5 km/sec contains the speed of light.

c) We have assumed that the measurements are independent of each other and that the distribution of the population of all possible measurements is Normal. The assumption of independence seems reasonable, but it might be a good idea to look at a display of the measurements made by Michelson to verify that the Nearly Normal Condition is satisfied.

29. Flight on time 2013.

a) **Randomization condition:** Since there is no time trend, the monthly on-time departure rates should be independent. This is not a random sample, but should be representative.
Nearly Normal condition: The histogram looks unimodal, and slightly skewed to the left. Since the sample size is 238, this should not be of concern.

b) The on-time departure rates in the sample had a mean of 78.02%, and a standard deviation in of 5.045%. Since the conditions have been satisfied, construct a one-sample t-interval, with $238 - 1 = 237$ degrees of freedom, at 90% confidence.

$$\bar{y} \pm t_{n-1}^* \left(\frac{s}{\sqrt{n}} \right) = 78.02 \pm t_{237}^* \left(\frac{5.045}{\sqrt{238}} \right) \approx (77.48, 78.56)$$

c) We are 90% confident that the interval from 77.48% to 78.56% contains the true mean monthly percentage of on-time flight departures.

31. Farmed salmon, second look.

The 95% confidence interval lies entirely above the 0.08 ppm limit. This is evidence that mirex contamination is too high and consistent with rejecting the null hypothesis. We used an upper-tail test, so the P-value should be smaller than $\frac{1}{2}(1 - 0.95) = 0.025$, and it was.

33. Pizza.

If in fact the mean cholesterol of pizza eaters does not indicate a health risk, then only 7 out of every 100 samples would be expected to have mean cholesterol as high or higher than the mean cholesterol observed in the sample.

35. Fuel economy 2014 revisited.

a) Randomization condition: The 35 cars were not selected randomly. We will have to assume that they are representative of all 2014 automobiles. **Nearly Normal condition:** The distribution doesn't appear to be unimodal and symmetric, but the sample size is reasonably large.

The mileages in the sample had a mean of 23.5143 mpg, and a standard deviation in of 6.0603 mpg. Since the conditions have been satisfied, construct a one-sample t-interval, with $35 - 1 = 34$ degrees of freedom, at 95% confidence.

$$\bar{y} \pm t_{n-1}^* \left(\frac{s}{\sqrt{n}} \right) = 23.5143 \pm t_{34}^* \left(\frac{6.0603}{\sqrt{35}} \right) \approx (21.43, 25.60)$$

We are 95% confident that the interval from 21.43 to 25.60 contains the true mean mileage of 2014 automobiles.

b) The data is a mix of small, mid-size, and large vehicles. Without knowing how the data were selected, we are cautious about generalizing to all 2014 cars.

37. Marriage.

a) H_0: The mean age at which American men first marry is 23.3 years. $(\mu = 23.3)$

H_A: The mean age is greater than 23.3 years. $(\mu > 23.3)$

b) **Randomization condition:** The 40 men were selected randomly.
Nearly Normal condition: The population of ages of men at first marriage is likely to be skewed to the right. It is much more likely that there are men who marry for the first time at an older age than at an age that is very young. We should examine the distribution of the sample to check for serious skewness and outliers, but with a large sample of 40 men, it should be safe to proceed.

c) Since the conditions for inference are satisfied, we can model the sampling distribution of the mean age of men at first marriage with $N\left(23.3, \dfrac{\sigma}{\sqrt{n}}\right)$. Since we do not know σ, the standard deviation of the population, $\sigma(\bar{y})$ will be estimated by $SE(\bar{y}) = \dfrac{s}{\sqrt{n}}$, and we will use a Student's t model, with $40 - 1 = 39$ degrees of freedom, $t_{39}\left(23.3, \dfrac{5.3}{\sqrt{40}}\right)$.

d) The mean age at first marriage in the sample was 24.2 years, with a standard deviation in age of 5.3 years. Use a one-sample t-test, modeling the sampling distribution of $\bar{y}$ with $t_{39}\left(23.3, \dfrac{5.3}{\sqrt{40}}\right)$. The P-value is 0.1447.

$$t = \frac{\bar{y} - \mu_0}{SE(\bar{y})}$$

$$t = \frac{24.2 - 23.3}{\dfrac{5.3}{\sqrt{40}}}$$

$$t \approx 1.07$$

P = 0.1447

23.3 24.2
$t = 1.07$

t_{39}

e) If the mean age at first marriage is still 23.3 years, there is a 14.5% chance of getting a sample mean of 24.2 years or older simply from natural sampling variation.

f) Since the P-value = 0.1447 is high, we fail to reject the null hypothesis. There is no evidence to suggest that the mean age of men at first marriage has changed from 23.3 years, the mean in 1960.

39. Ruffles.

a) **Randomization condition:** The 6 bags were not selected at random, but it is reasonable to think that these bags are representative of all bags of chips.
10% condition: 6 bags are less than 10% of all bags of chips.
Nearly Normal condition: The histogram of the weights of chips in the sample is nearly normal.

b) $\bar{y} \approx 28.78$ grams, $s \approx 0.40$ grams

c) Since the conditions for inference have been satisfied, use a one-sample t-interval, with $6 - 1 = 5$ degrees of freedom, at 95% confidence.

$$\bar{y} \pm t^*_{n-1}\left(\frac{s}{\sqrt{n}}\right) = 28.78 \pm t^*_5\left(\frac{0.40}{\sqrt{6}}\right) \approx (28.36, 29.21)$$

d) We are 95% confident that the mean weight of the contents of Ruffles bags is between 28.36 and 29.21 grams.

e) Since the interval is above the stated weight of 28.3 grams, there is evidence that the company is filling the bags to more than the stated weight, on average.

41. Popcorn.

H_0: The mean proportion of unpopped kernels is 10%. $(\mu = 10)$

H_A: The mean proportion of unpopped kernels is lower than 10%. $(\mu < 10)$

Randomization condition: The 8 bags were randomly selected.
Nearly Normal condition: The histogram of the percentage of unpopped kernels is unimodal and roughly symmetric.

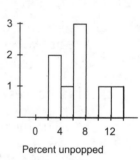

The bags in the sample had a mean percentage of unpopped kernels of 6.775 percent and a standard deviation in percentage of unpopped kernels of 3.637 percent. Since the conditions for inference are satisfied, we can model the sampling distribution of the mean percentage of unpopped kernels with a Student's t model, with $8 - 1 = 7$ degrees of freedom, $t_7\left(6.775, \dfrac{3.637}{\sqrt{8}}\right)$.

We will perform a one-sample t-test.

Since the *P*-value = 0.0203 is low, we reject the null hypothesis. There is evidence to suggest the mean percentage of unpopped kernels is less than 10% at this setting.

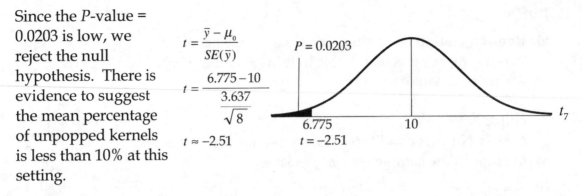

$$t = \frac{\bar{y} - \mu_0}{SE(\bar{y})}$$

$$t = \frac{6.775 - 10}{\frac{3.637}{\sqrt{8}}}$$

$$t \approx -2.51$$

$P = 0.0203$

6.775   10

$t = -2.51$

t_7

43. Chips ahoy.

a) **Randomization condition:** The bags of cookies were randomly selected.
   Nearly Normal condition: The Normal probability plot is reasonably straight, and the histogram of the number of chips per bag is unimodal and symmetric.

b) The bags in the sample had a mean number of chips of 1238.19, and a standard deviation of 94.282 chips. Since the conditions for inference have been satisfied, use a one-sample *t*-interval, with 16 – 1 = 15 degrees of freedom, at 95% confidence.

$$\bar{y} \pm t^*_{n-1}\left(\frac{s}{\sqrt{n}}\right) = 1238.19 \pm t^*_{15}\left(\frac{94.282}{\sqrt{16}}\right) \approx (1187.9, \ 1288.4)$$

We are 95% confident that the mean number of chips in an 18-ounce bag of Chips Ahoy cookies is between 1187.9 and 1288.4.

c) H₀: The mean number of chips per bag is 1000. $(\mu = 1000)$

   Hₐ: The mean number of chips per bag is greater than 1000. $(\mu > 1000)$

   Since the confidence interval is well above 1000, there is strong evidence that the mean number of chips per bag is well above 1000.

   However, since the "1000 Chip Challenge" is about individual bags, not means, the claim made by Nabisco may not be true. If the mean was around 1188 chips, the low end of our confidence interval, and the standard deviation of the population was about 94 chips, our best estimate obtained from our sample, a bag containing 1000 chips would be about 2 standard deviations below the mean. This is not likely to happen, but not an outrageous occurrence. These data do not provide evidence that the "1000 Chip Challenge" is true.

45. Maze.

a) **Independence assumption:** It is reasonable to think that the rats' times will be independent, as long as the times are for different rats.

Nearly Normal condition: There is an outlier in both the Normal probability plot and the histogram that should probably be eliminated before continuing the test. One rat took a long time to complete the maze.

b) H_0: The mean time for rats to complete this maze is 60 seconds. $(\mu = 60)$

H_A: The mean time for rats to complete this maze is not 60 seconds. $(\mu \neq 60)$

The rats in the sample finished the maze with a mean time of 52.21 seconds and a standard deviation in times of 13.5646 seconds. Since the conditions for inference are satisfied, we can model the sampling distribution of the mean time in which rats complete the maze with a Student's t model, with

21 – 1 = 20 degrees of freedom, $t_{20}\left(60, \dfrac{13.5646}{\sqrt{21}}\right)$. We will perform a one-sample

t-test.

Since the P-value = 0.0160 is low, we reject the null hypothesis. There is evidence that the mean time required for rats to finish the maze is

$$t = \frac{\bar{y} - \mu_0}{SE(\bar{y})}$$

$$t = \frac{52.21 - 60}{\dfrac{13.56}{\sqrt{21}}}$$

$$t \approx -2.63$$

not 60 seconds. Our evidence suggests the mean time is less than 60 seconds.

c) Without the outlier, the rats in the sample finished the maze with a mean time of 50.13 seconds and standard deviation in times of 9.90 seconds. Since the conditions for inference are satisfied, we can model the sampling distribution of the mean time in which rats complete the maze with a Student's t model, with

20 – 1 = 19 degrees of freedom, $t_{19}\left(60, \dfrac{9.90407}{\sqrt{20}}\right)$. We will use a one-sample t-test.

This test results in a value of $t = -4.46$, and a two-sided P-value = 0.0003. Since the P-value is low, we reject the null hypothesis. There is evidence that the mean time required for rats to finish the maze is not 60 seconds. Our evidence suggests that the mean time is actually less than 60 seconds.

d) According to both tests, there is evidence that the mean time required for rats to complete the maze is different than 60 seconds. The maze does not meet the "one-minute average" requirement. It should be noted that the test without the outlier is the appropriate test. The one slow rat made the mean time required seem much higher than it probably was.

47 Golf Drives III 2013.

a) $\bar{y} \pm t^{*}_{n-1}\left(\dfrac{s}{\sqrt{n}}\right) = 289.24 \pm t^{*}_{154}\left(\dfrac{11.18}{\sqrt{155}}\right) \approx (287.47,\ 291.01)$

b) These data are not a random sample of golfers. The top professionals are not representative of all golfers and were not selected at random. We might consider the 2013 data to represent the population of all professional golfers, past, present, and future.

c) The data are means for each golfer, so they are less variable than if we looked at separate drives, and inference is invalid.

Chapter 21 – More About Tests and Intervals

Section 21.1

1. **Parameters and hypotheses.**

   a) Let $p =$ probability of winning on a slot machine.
   $H_0 : p = 0.01$ vs. $H_A : p \neq 0.01$

   b) Let $\mu =$ mean spending per customer this year.
   $H_0 : \mu = \$35.32$ vs. $H_A : \mu \neq \$35.32$

   c) Let $p =$ proportion of patients cured by the new drug.
   $H_0 : p = 0.3$ vs. $H_A : p \neq 0.3$

   d) Let $p =$ proportion of clients now using the website.
   $H_0 : p = 0.4$ vs. $H_A : p \neq 0.4$

Section 21.2

3. **P-values.**

   a) False. A low P-value provides evidence for rejecting the null hypothesis.

   b) False. It results in rejecting the null hypothesis, but does not prove that it is false.

   c) False. A high P-value shows that the data are consistent with the null hypothesis but does not prove that the null hypothesis is true.

   d) False. Whether a P-value provides enough evidence to reject the null hypothesis depends on the risk of a type I error that one is willing to assume (the α level).

5. **Hypotheses.**

   a) Two-sided. Let $p =$ proportion of accounting reports that are on time.
   $H_0 :$ 40% of accounting reports were on time this year. $(p = 0.40)$
   $H_A :$ The percentage of reports submitted on time is not 40%. $(p \neq 0.40)$

   b) One-sided. Let $\mu =$ mean click-through rate.
   $H_0 :$ The mean click-through rate for the new website is 5.4 minutes. $(\mu = 5.4)$
   $H_A :$ The mean click-through rate is less than 5.4 minutes. $(\mu < 5.4)$

   c) One-sided. Let $p =$ percentage of employees enrolled in at least one wellness class.
   $H_0 :$ 42% of employees are enrolled in at least one wellness class. $(p = 0.42)$
   $H_A :$ The percentage of employees enrolled in at least one wellness class is greater than 42%. $(p > 0.42)$

d) One-sided. Let p = percentage of voters who will vote for the candidate.

H_0 : 50% of voters will vote for the candidate. ($p = 0.50$)

H_A : More than 50% of voters will vote for the candidate. ($p > 0.50$)

Section 21.3

7. Alpha true and false.

a) True.

b) False. The alpha level is set independently and does not depend on the sample size.

c) False. The P-value would have to be less than 0.01 to reject the null hypothesis.

d) False. It simply means we do not have enough evidence at that alpha level to reject the null hypothesis.

Section 21.4

9. Critical values.

a) $z = \pm 1.96$

b) $z = 1.645$

c) $t = \pm 2.03$

d) $z = 2.33$ (n is not relevant for critical values of z)

e) $z = -2.33$

Section 21.5

11. Errors.

a) Type I error. The actual value is not greater than 0.3 but they rejected the null hypothesis.

b) No error. The actual value is 0.50, which was not rejected.

c) Type II error. The actual value was 55.3 points, which is greater than 52.5.

d) Type II error. The null hypothesis was not rejected, but it was false. The true relief rate was greater than 0.25.

Chapter Exercises

13. One sided or two?

a) Two sided. Let p be the percentage of students who prefer Diet Coke.

H_0 : 50% of students prefer Diet Coke. ($p = 0.50$)

H_A : The percentage of students who prefer Diet Coke is not 50%. ($p \neq 0.50$)

b) One sided. Let p be the percentage of teenagers who prefer the new formulation.

H_0 : 50% of students prefer the new formulation. ($p = 0.50$)
H_A : More than 50% of students prefer the new formulation. ($p > 0.50$)

c) One sided. Let p be the percentage of people who plan to vote for the override.

H_0 : 2/3 of the residents intend to vote for the override. ($p = 2/3$)
H_A : More than 2/3 of the residents intend to vote for the override. ($p > 2/3$)

d) Two sided. Let p be the percentage of days the market goes up.

H_0 : The market goes up on 50% of days. ($p = 0.50$)
H_A : The percentage of days the market goes up is not 50%. ($p \neq 0.50$)

15. P-value.

If the effectiveness of the new poison ivy treatment is the same as the effectiveness of the old treatment, the chance of observing an effectiveness this large or larger in a sample of the same size is 4.7% by natural sampling variation alone.

17. Alpha.

Since the null hypothesis was rejected at $\alpha = 0.05$, the P-value for the researcher's test must have been less than 0.05. He would have made the same decision at $\alpha = 0.10$, since the P-value must also be less than 0.10. We can't be certain whether or not he would have made the same decision at $\alpha = 0.01$, since we only know that the P-value was less than 0.05. It may have been less than 0.01, but we can't be sure.

19. Significant?

a) If 98% of children have really been vaccinated, there is practically no chance of observing 97.4% of children (in a sample of 13,000) vaccinated by natural sampling variation alone.

b) We conclude that the proportion of children who have been vaccinated is below 98%, but a 95% confidence interval would show that the true proportion is between 97.1% and 97.7%. Most likely a decrease from 98% to 97.7% would not be considered important. The 98% figure was probably an approximate figure anyway. However, if the 98% figure was not as estimate, and with 1,000,000 kids per year vaccinated, even 0.1% represents 1,000 kids, so this may be important.

21. Groceries.

a) **Randomization condition:** We will assume that the Yahoo survey was conducted randomly.
10% condition: 2400 is less than 10% of all men.
Success/Failure condition: $n\hat{p} = 1224$ and $n\hat{q} = 1176$ are both greater than 10, so the sample is large enough.

Since the conditions are met, we can use a one-proportion z-interval to estimate the percentage of men who identify themselves as the primary grocery shopper in their household.

$$\hat{p} \pm z^* \sqrt{\frac{\hat{p}\hat{q}}{n}} = \left(\frac{1224}{2400}\right) \pm 2.326 \sqrt{\frac{\left(\frac{1224}{2400}\right)\left(\frac{1176}{2400}\right)}{2400}} = (48.6\%, 53.4\%)$$

We are 98% confident that between 48.6% and 53.4% of all men identify themselves as the primary grocery shopper in their household.

b) Since 45% is not in the interval, there is strong evidence that more than 45% of all men identify themselves as the primary grocery shopper in their household.

c) The significance level of this test is $\alpha = 0.01$. It's an upper tail test based on a 98% confidence interval.

23. Approval 2014.

a) **Randomization condition:** The adults were randomly selected.
10% condition: 1500 adults represent less than 10% of all adults.
Success/Failure condition: $n\hat{p} = (1500)(0.40) = 600$ and $n\hat{q} = (1500)(0.60) = 900$ are both greater than 10, so the sample is large enough.

Since the conditions are met, we can use a one-proportion z-interval to estimate Barack Obama's approval rating.

$$\hat{p} \pm z^* \sqrt{\frac{\hat{p}\hat{q}}{n}} = (0.40) \pm 1.960 \sqrt{\frac{(0.40)(0.60)}{1500}} = (0.375, 0.425)$$

We are 95% confident that Barack Obama's approval rating is between 37.5% and 42.5%.

b) Since 45% is not within the interval, this is not a plausible value for Barack Obama's approval rating. There is evidence against the null hypothesis.

25. Dogs.

a) We cannot construct a confidence interval for the rate of occurrence of early hip dysplasia among 6-month old puppies because only 5 of 42 puppies were found with early hip dysplasia. The Success/Failure condition is not satisfied.

b) **Independence assumption:** If hip dysplasia is hereditary, and puppies brought to the vaccination clinic are from the same litter, independence might be an issue.
Randomization condition: The veterinarian considers the 42 puppies to be a random sample of all puppies.
10% condition: 42 puppies represent less than 10% of all puppies.
Success/Failure condition: As previously mentioned, this condition is not met, since there aren't at least 10 puppies with hip dysplasia in the sample.

Since the other conditions are met, we can construct a one-proportion plus-four z-interval to estimate the percentage of puppies with early hip dysplasia.

$$\tilde{p} = \frac{y+2}{n+4} = \frac{7}{46} = 0.152$$

$$\tilde{p} \pm z^* \sqrt{\frac{\tilde{p}\tilde{q}}{\tilde{n}}} = (0.152) \pm 1.96 \sqrt{\frac{(0.152)(0.848)}{46}} = (4.8\%, 25.6\%)$$

We are 95% confident that between 4.8% and 25.6% of puppies have early hip dysplasia.

27. Loans.

a) The bank has made a Type II error. The person was not a good credit risk, and the bank failed to notice this.

b) The bank has made a Type I error. The person was a good credit risk, and the bank was convinced that he/she was not.

c) By making it easier to get a loan, the bank has reduced the alpha level. It takes less evidence to grant the person the loan.

d) The risk of Type I error is decreased and the risk of Type II error has increased.

29. Second loan.

a) Power is the probability that the bank denies a loan that could not have been repaid.

b) To increase power, the bank could raise the cutoff score.

c) If the bank raised the cutoff score, a larger number of trustworthy people would be denied credit, and the bank would lose the opportunity to collect the interest on these loans.

31. Homeowners 2013.

a) The null hypothesis is that the level of home ownership does not rise. The alternative hypothesis is that it rises.

b) In this context, a Type I error is when the city concludes that home ownership is on the rise, but in fact, the tax breaks don't help.

c) In this context, a Type II error is when the city abandons the tax breaks, thinking they don't help, when in fact they were helping.

d) A Type I error causes the city to forego tax revenue, while a Type II error withdraws help from those who might have otherwise been able to buy a house.

e) The power of the test is the city's ability to detect an actual increase in home ownership.

33. Testing cars.

H₀ : The shop is meeting the emissions standards.
Hₐ : The shop is not meeting the emissions standards.

a) Type I error is when the regulators decide that the shop is not meeting standards when they actually are meeting the standards.

b) Type II error is when the regulators certify the shop when they are not meeting the standards.

c) Type I would be more serious to the shop owners. They would lose their certification, even though they are meeting the standards.

d) Type II would be more serious to environmentalists. Shops are allowed to operate, even though they are allowing polluting cars to operate.

35. Cars again.

a) The power of the test is the probability of detecting that the shop is not meeting standards when they are not.

b) The power of the test will be greater when 40 cars are tested. A larger sample size increases the power of the test.

c) The power of the test will be greater when the level of significance is 10%. There is a greater chance that the null hypothesis will be rejected.

d) The power of the test will be greater when the shop is out of compliance "a lot". Larger problems are easier to detect.

37. Equal opportunity?

H₀ : The company is not discriminating against minorities.
Hₐ : The company is discriminating against minorities.

a) This is a one-tailed test. They wouldn't sue if "too many" minorities were hired.

b) Type I error would be deciding that the company is discriminating against minorities when they are not discriminating.

c) Type II error would be deciding that the company is not discriminating against minorities when they actually are discriminating.

d) The power of the test is the probability that discrimination is detected when it is actually occurring.

e) The power of the test will increase when the level of significance is increased from 0.01 to 0.05.

f) The power of the test is lower when the lawsuit is based on 37 employees instead of 87. Lower sample size leads to less power.

39. Software for learning.

a) The test is one-tailed. We are testing to see if an increase in average score is associated with the software.

b) H_0 : The average score does not change following the use of software. ($\mu = 105$)
H_A : The average score increases following the use of the software. ($\mu > 105$)

c) The professor makes a Type I error if he buys the software when the average score has not actually increased.

d) The professor makes a Type II error if he doesn't buy the software when the average has actually increased.

e) The power of the test is the probability of buying the software when the average score has actually increased.

41. Software, part II.

a) H_0 : The average score does not change following the use of software. ($\mu = 105$)
H_A : The average score increases following the use of the software. ($\mu > 105$)

Randomization condition: This year's class of 203 students is probably representative of all stats students.
Nearly Normal condition: We don't have the scores from the 203 individuals, so we can't check a plot of the data. However, with a sample this large, the Central Limit Theorem allows us to model the sampling distribution of the means with a t-distribution.

The mean score was 108 points, with a standard deviation of 8.7 points. Since the conditions for inference are satisfied, we can model the sampling distribution of the mean score with a Student's t model, with $203 - 1 = 202$ degrees of freedom, $t_{202}\left(108, \dfrac{8.7}{\sqrt{203}}\right)$. We will perform a one-sample t-test.

The value of t is approximately 4.91, which results in a P-value of less than 0.0001, so we reject the null hypothesis. There is strong evidence that the mean score has increased since use of the software program was implemented. As long as the professor feels confident that this class of stats students is representative of all potential students, then he should buy the program.

$$t = \frac{\bar{y} - \mu_0}{\dfrac{\sigma}{\sqrt{n}}}$$

$$t = \frac{108 - 105}{\dfrac{8.7}{\sqrt{203}}}$$

$$t = 4.91$$

If you used a 95% confidence interval to assess the effectiveness of the program: $\bar{y} \pm t^*_{n-1}\left(\dfrac{s}{\sqrt{n}}\right) = 108 \pm t^*_{202}\left(\dfrac{8.7}{\sqrt{203}}\right) \approx (106.8,\ 109.2)$

We are 95% confident that the mean score is between 106.8 and 109.2. Since 105 is above the interval, this provides evidence that the mean score has increased following the implementation of the software program.

b) The mean score on the exam only increased by 1 to 4 points. This small difference might not be enough to be worth the cost of the program.

43. TV safety.

a) This is an upper-tail test. We hope to show that the TV stand will hold 500 or more pounds easily.

b) The inspectors will commit a Type I error if they decide the TV stands are safe when they are not.

c) The inspectors will commit a Type II error if they decide the TVs are unsafe when they are actually safe.

45. TV safety, revisited.

a) To decrease the likelihood of producing an unsafe TV stand, they should decrease α. This lower the chance of making a Type I error.

b) The power of the test is the ability of the inspectors to determine that the TV stand is safe when it is actually capable of holding more than 500 pounds.

c) The company could increase the power of the test by lowering the standard deviation by testing more stands. This could prove costly and would require more time to test. They could also increase α, but then they will commit Type I errors with greater frequency, approving stands that cannot hold 500 pounds of more. Finally, they could require that TV stands have a higher weight capacity than 500 pounds as the standard. Again, that might prove costly, since they would be rejecting many more stands that were safe.

47. Two coins.

a) The alternative hypothesis is that your coin produces 30% heads.

b) Reject the null hypothesis if the coin comes up tails.   Otherwise, fail to reject.

c) There is a 10% chance that the coin comes up tails if the null hypothesis is true, so alpha is 10%.

d) Power is our ability to detect the 30% coin. That coin will come up tails 70% of the time. That's the power of our test.

e) To increase the power and lower the probability of Type I error at the same time, simply flip the coin more times.

49. Hoops.

H_0 : The player's foul-shot percentage is only 60%. ($p = 0.60$)
H_A : The player's foul-shot percentage is better than 60%. ($p > 0.60$)

a) The player's shots can be considered Bernoulli trials. There are only two possible outcomes, make the shot and miss the shot. The probability of making any shot is constant at $p = 0.60$. Assume that the shots are independent of each other. Use *Binom*(10, 0.60).

Let X = the number of shots made out of $n = 10$.

$$P(\text{makes at least 9 out of 10}) = P(X \geq 9)$$
$$= P(X = 9) + P(X = 10)$$
$$= {}_{10}C_9(0.60)^9(0.40)^1 + {}_{10}C_{10}(0.60)^{10}(0.40)^0$$
$$\approx 0.0464$$

b) The coach made a Type I error.

c) The power of the test can be calculated for specific values of the new probability of success. Each true value of p has a power calculation associated with it. In this case, we are finding the power of the test to detect an 80% foul-shooter. Use *Binom*(10, 0.80).

Let X = the number of shots made out of $n = 10$.

$$P(\text{makes at least 9 out of 10}) = P(X \geq 9)$$
$$= P(X = 9) + P(X = 10)$$
$$= {}_{10}C_9(0.80)^9(0.20)^1 + {}_{10}C_{10}(0.80)^{10}(0.20)^0 \approx 0.376$$

The power of the test to detect an increase in foul-shot percentage from 60% to 80% is about 37.6%.

d) The power of the test to detect improvement in foul-shooting can be increased by increasing the number of shots, or by keeping the number of shots at 10 but increasing the level of significance by declaring that 8, 9, or 10 shots made will convince the coach that the player has improved. In other words, the coach can increase the power of the test by lowering the standard of proof.

Review of Part V – From the Data at Hand to the World at Large

1. Crohn's disease.

H_0: The relapse rate for those using Omega-3 fatty acids is the same as the relapse rate for those who are not. $\left(p_{Omega-3} = p_{Not}\ \text{ or }\ p_{Omega-3} - p_{Not} = 0\right)$

H_A: The relapse rate for those using Omega-3 fatty acids is lower than the relapse rate for those who are not. $\left(p_{Omega-3} < p_{Not}\ \text{ or }\ p_{Omega-3} - p_{Not} < 0\right)$

3. Hamsters.

a) **Randomization condition:** Assume these litters are representative of all litters. **Nearly Normal condition:** We don't have the actual data, so we can't look at a graphical display. However, since the sample size is large, the Central Limit Theorem guarantees that the distribution of averages will be approximately Normal, as long as there are no outliers.

The litters in the sample had a mean size of 7.72 baby hamsters and a standard deviation of 2.5 baby hamsters. Since the conditions are satisfied, the sampling distribution of the mean can be modeled by a Student's t- model, with $47 - 1 = 46$ degrees of freedom. We will use a one-sample t-interval with 90% confidence for the mean number of baby hamsters per litter.

$$\bar{y} \pm t^*_{n-1}\left(\frac{s}{\sqrt{n}}\right) = 7.72 \pm t^*_{46}\left(\frac{2.5}{\sqrt{47}}\right) \approx (7.11,\ 8.33)$$

We are 90% confident that the mean number of baby hamsters per litter is between 7.11 and 8.33.

b) A 98% confidence interval would have a larger margin of error. Higher levels of confidence come at the price of less precision in the estimate.

c) A quick estimate using z gives us a sample size of about 25 litters. Using this estimate, $t^*_{24} = 2.064$ at 95% confidence. We need a sample of about 27 litters in order to estimate the number of baby hamsters per litter to within 1 baby hamster.

$$ME = t^*_{24}\left(\frac{s}{\sqrt{n}}\right)$$

$$1 = 2.064\left(\frac{2.5}{\sqrt{n}}\right)$$

$$n = \frac{(2.064)^2 (2.5)^2}{(1)^2}$$

$$n \approx 27$$

5. Leaky gas tanks.

a) H_0: The proportion of leaky gas tanks is 40%. ($p = 0.40$)
H_A: The proportion of leaky gas tanks is less than 40%. ($p < 0.40$)

b) **Randomization condition:** A random sample of 27 service stations in California was taken.
10% condition: 27 stations are less than 10% of all service stations in California.
Success/Failure condition: $np = (27)(0.40) = 10.8$ and $nq = (27)(0.60) = 16.2$ are both greater than 10, so the sample is large enough.

c) Since the conditions have been satisfied, a Normal model can be used to model the sampling distribution of the proportion, with $\mu_{\hat{p}} = p = 0.40$ and

$$\sigma(\hat{p}) = \sqrt{\frac{pq}{n}} = \sqrt{\frac{(0.40)(0.60)}{27}} \approx 0.09428 \, .$$ We can perform a one-proportion z-test.

The observed proportion of leaky gas tanks is

$$\hat{p} = \frac{7}{27} \approx 0.2593 \, .$$

$$z = \frac{\hat{p} - p_0}{\sqrt{\dfrac{pq}{n}}}$$

$$z = \frac{0.2593 - 0.40}{\sqrt{\dfrac{(0.40)(0.60)}{27}}}$$

$$z \approx -1.49$$

$P = 0.0677$

0.2593   0.4

$z = -1.49$

d) Since the *P*-value = 0.0677 is relatively high, we fail to reject the null hypothesis. There is little evidence that the proportion of leaky gas tanks is less than 40%. The new program doesn't appear to be effective in decreasing the proportion of leaky gas tanks.

e) If the program actually works, we haven't done anything *wrong*. Our methods are correct. Statistically speaking, we have committed a Type II error.

f) In order to decrease the probability of making this type of error, we could lower our standards of proof, by raising the level of significance. This will increase the power of the test to detect a decrease in the proportion of leaky gas tanks. Another way to decrease the probability that we make a Type II error is to sample more service stations. This will decrease the variation in the sample proportion, making our results more reliable.

g) Increasing the level of significance is advantageous, since it decreases the probability of making a Type II error, and increases the power of the test. However, it also increases the probability that a Type I error is made, in this case, thinking that the program is effective when it really is not effective. Increasing the sample size decreases the probability of making a Type II error and increases power, but can be costly and time-consuming.

7. Scrabble.

a) The researcher believes that the true proportion of As is within 10% of the estimated 54%, namely, between 44% and 64%.

b) A large margin of error is usually associated with a small sample, but the sample consisted of "many" hands. The margin of error is large because the standard error of the sample is large. This occurs because the true proportion of As in a hand is close to 50%, the most difficult proportion to predict.

c) This provides no evidence that the simulation is faulty. The true proportion of As is contained in the confidence interval. The researcher's results are consistent with 63% As.

9. Net-Newsers.

a) The Pew Research Foundation believes that the true proportion of people who obtain news from the Internet is between 11% and 15%.

b) The smaller sample size in the cell sample would result in a larger standard error. This would make the margin of error larger, as well.

c) $\hat{p} \pm z^* \sqrt{\dfrac{\hat{p}\hat{q}}{n}} = (0.82) \pm 1.960 \sqrt{\dfrac{(0.82)(0.18)}{470}} = (78.5\%, 85.5\%)$

We are 95% confident that between 78.5% and 85.5% of Net-Newsers get news during the course of the day.

d) The sample of 470 Net-Newsers is smaller than either of the earlier samples. This results in a larger margin of error.

11. Bimodal.

a) The *sample's* distribution (NOT the *sampling* distribution), is expected to look more and more like the distribution of the population, in this case, bimodal.

b) The expected value of the sample's mean is expected to be μ, the population mean, regardless of sample size.

c) The variability of the sample mean, $\sigma(\bar{y})$, is $\dfrac{\sigma}{\sqrt{n}}$, the population standard deviation divided by the square root of the sample size, regardless of the sample size.

d) As the sample size increases, the sampling distribution model becomes closer and closer to a Normal model.

13. Archery.

a) $\mu_{\hat{p}} = p = 0.80$ $\sigma(\hat{p}) = \sqrt{\dfrac{pq}{n}} = \sqrt{\dfrac{(0.80)(0.20)}{200}} \approx 0.028$

b) $np = (200)(0.80) = 160$ and $nq = (200)(0.20) = 40$ are both greater than 10, so the Normal model is appropriate.

c) The Normal model of the sampling distribution of the proportion of bull's-eyes she makes out of 200 is at the right.

Proportion of shots made out of 200

Approximately 68% of the time, we expect her to hit the bull's-eye on between 77.2% and 82.8% of her shots. Approximately 95% of the time, we expect her to hit the bull's-eye on between 74.4% and 85.6% of her shots. Approximately 99.7% of the time, we expect her to hit the bull's-eye on between 71.6% and 88.4% of her shots.

d) According to the Normal model, the probability that she hits the bull's-eye in at least 85% of her 200 shots is approximately 0.039.

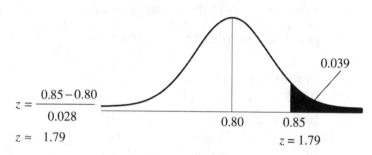

$$z = \frac{0.85 - 0.80}{0.028}$$

$$z \approx 1.79$$

15. Polling disclaimer.

a) It is not clear what specific question the pollster asked. Otherwise, they did a great job of identifying the W's.

b) A sample that was stratified by age, sex, region, and education was used.

c) The margin of error was 4%.

d) Since "no more than 1 time in 20 should chance variations in the sample cause the results to vary by more than 4 percentage points", the confidence level is $19/20 = 95\%$.

e) The subgroups had smaller sample sizes than the larger group. The standard errors in these subgroups were larger as a result, and this caused the margins of error to be larger.

f) They cautioned readers about response bias due to wording and order of the questions.

17. Teen deaths 2005.

a) H_0 : The percentage of fatal accidents involving teenage girls is 12.6%, the same as the overall percentage of fatal accidents involving teens . ($p = 0.126$)
H_A : The percentage of fatal accidents involving teenage girls is lower than 12.6%, the overall percentage of fatal accidents involving teens . ($p < 0.126$)

Randomization condition: Assume that the 388 accidents observed are representative of all accidents.

10% condition: The sample of 388 accidents is less than 10% of all accidents.

Success/Failure condition: $np = (388)(0.126) = 48.888$ and $nq = (388)(0.874) = 339.112$ are both greater than 10, so the sample is large enough.

The conditions have been satisfied, so a Normal model can be used to model the sampling distribution of the proportion, with $\mu_{\hat{p}} = p = 0.126$ and

$$\sigma(\hat{p}) = \sqrt{\frac{pq}{n}} = \sqrt{\frac{(0.126)(0.874)}{388}} \approx 0.01685.$$

We can perform a one-proportion z-test. The observed proportion of fatal accidents involving teen girls is $\hat{p} = \dfrac{44}{388} \approx 0.1134$.

Since the *P*-value = 0.227 is high, we fail to reject the null hypothesis. There is little evidence that the proportion of fatal accidents involving teen girls is less than the overall proportion of fatal accidents involving teens.

$$z = \frac{\hat{p} - p_0}{\sqrt{\dfrac{pq}{n}}}$$

$$z = \frac{0.13 - 0.126}{\sqrt{\dfrac{(0.126)(0.874)}{388}}}$$

$$z \approx -0.748$$

$P = 0.227$

$0.1134 \quad 0.126$

$z = -0.748$

b) If the proportion of fatal accidents involving teenage girls is really 12.6%, we expect to see the observed proportion, 11.34%, in about 22.7% of samples of size 388 simply due to sampling variation.

19. Cheating.

a) **Randomization condition:** The 4500 students were selected randomly.

10% condition: 4500 students is less than 10% of all students.

Success/Failure condition: $n\hat{p} = (4500)(0.74) = 3330$ and $n\hat{q} = (4500)(0.26) = 1170$ are both greater than 10, so the sample is large enough.

Since the conditions are met, we can use a one-proportion z-interval to estimate the percentage of students who have cheated at least once.

$$\hat{p} \pm z^* \sqrt{\frac{\hat{p}\hat{q}}{n}} = (0.74) \pm 1.645 \sqrt{\frac{(0.74)(0.26)}{4500}} = (72.9\%, 75.1\%)$$

b) We are 90% confident that between 72.9% and 75.1% of high school students have cheated at least once.

c) About 90% of random samples of size 4500 will produce intervals that contain the true proportion of high school students who have cheated at least once.

d) A 95% confidence interval would be wider. Greater confidence requires a larger margin of error.

21. Cigarettes 2009.

a) H_0: 17.7% of high school students smoke. ($p = 0.177$)
H_A: More than 17.7% of high school students smoke. ($p > 0.177$)

b) **Randomization condition:** The CDC randomly sampled 5080 high school students.
10% condition: The sample of 5080 students is less than 10% of all high school students.
Success/Failure condition: $np = (5080)(0.177) = 899.16$ and $nq = (5080)(0.823) = 4180.84$ are both greater than 10, so the sample is large enough.

The conditions have been satisfied, so a Normal model can be used to model the sampling distribution of the proportion, with $\mu_{\hat{p}} = p = 0.177$ and

$$\sigma(\hat{p}) = \sqrt{\frac{pq}{n}} = \sqrt{\frac{(0.177)(0.823)}{5080}} \approx 0.0054 .$$

c) We can perform a one-proportion z-test. The observed proportion of high school students who smoke is $\hat{p} = 0.195$ This proportion is about 3.36 standard deviations above the hypothesized proportion of smokers.

The P-value of this test is 0.0004.

d) If the proportion of students who smoke is actually 17.7%, the probability that a sample of this size would have a sample proportion of 19.5% or higher is 0.0004.

e) Since the P-value = 0.0004 is low, we reject the null hypothesis. There is strong evidence that greater than 17.7% of all high school students smoked in 2009. The goal is not on track.

f) If the conclusion is incorrect, a Type I error has been made.

23. Teen smoking 2009.

Randomization condition: Assume that the freshman class is representative of all teenagers. This may not be a reasonable assumption. There are many interlocking relationships between smoking, socioeconomic status, and college attendance. This class may not be representative of all teens with regards to smoking simply because they are in college. Be cautious with your conclusions!
10% condition: The freshman class is less than 10% of all teenagers.
Success/Failure condition: $np = (522)(0.195) = 101.79$ and $nq = (522)(0.805) = 420.21$ are both greater than 10.

Therefore, the sampling distribution model for the proportion of 522 students who smoke is Normal, with $\mu_{\hat{p}} = p = 0.195$, and standard deviation

$$\sigma(\hat{p}) = \sqrt{\frac{pq}{n}} = \sqrt{\frac{(0.195)(0.805)}{522}} \approx 0.0173.$$

25% is about 3.2 standard deviations above the expected proportion of smokers. If the true proportion of smokers is 19.5%, the Normal model predicts that the probability that more than 25% of these students smoke is approximately 0.0008. It is very unlikely that more than 25% the freshman class smokes.

25. Errors.

a) Since a treatment (the additive) is imposed, this is an experiment.

b) The company is only interested in a decrease in the percentage of cars needing repairs, so they will perform a one-sided test.

c) The independent laboratory will make a Type I error if they decide that the additive reduces the number of repairs, when it actually makes no difference in the number of repairs.

d) The independent laboratory will make a Type II error if they decide that the additive makes no difference in the number of repairs, when it actually reduces the number of repairs.

e) The additive manufacturer would consider a Type II error more serious. The lab claims that the manufacturer's product doesn't work, and it actually does.

f) Since this was a controlled experiment, the company can conclude that the additive is the reason that the cabs are running better. They should be cautious recommending it for all cars. There is evidence that the additive works well for cabs, which get heavy use. It might not be effective in cars with a different pattern of use than cabs.

27. Fried PCs.

a) H_0: The computer is undamaged.
    H_A: The computer is damaged.

b) The biggest advantage is that all of the damaged computers will be detected, since, historically, damaged computers never pass all the tests. The disadvantage is that only 80% of undamaged computers pass all the tests. The engineers will be classifying 20% of the undamaged computers as damaged.

c) In this example, a Type I error is rejecting an undamaged computer. To allow this to happen only 5% of the time, the engineers would reject any computer that failed 3 or more tests, since 95% of the undamaged computers fail two or fewer tests.

d) The power of the test in part c is 20%, since only 20% of the damaged machines fail 3 or more tests.

e) By declaring computers "damaged" if they fail 2 or more tests, the engineers will be rejecting only 7% of undamaged computers. From 5% to 7% is an increase of 2% in α. Since 90% of the damaged computers fail 2 or more tests, the power of the test is now 90%, a substantial increase.

29. Approval 2008.

H_0 : George W. Bush's April 2008 disapproval rating was 66%. ($p = 0.66$)
H_A : George W. Bush's disapproval rating was higher than 66%. ($p > 0.66$)

Randomization condition: The adults were chosen randomly.
10% condition: 1016 adults are less than 10% of all adults.
Success/Failure condition: $np = (1016)(0.66) = 670.56$ and $nq = (1016)(0.34) = 345.44$ are both greater than 10, so the sample is large enough.

The conditions have been satisfied, so a Normal model can be used to model the sampling distribution of the proportion, with $\mu_{\hat{p}} = p = 0.66$ and

$$\sigma(\hat{p}) = \sqrt{\frac{pq}{n}} = \sqrt{\frac{(0.66)(0.34)}{1016}} \approx 0.015.$$

We can perform a one-proportion z-test. The observed approval rating is $\hat{p} = 0.69$.

The value of z is 2.02. Since the P-value = 0.022 is low, we reject the null hypothesis. There is strong evidence that President George W. Bush's April 2008 disapproval rating was higher than the 66% disapproval rating of President Richard Nixon.

31. Name recognition.

a) The company wants evidence that the athlete's name is recognized more often than 25%.

b) Type I error means that fewer than 25% of people will recognize the athlete's name, yet the company offers the athlete an endorsement contract anyway. In this case, the company is employing an athlete that doesn't fulfill their advertising needs.

Type II error means that more than 25% of people will recognize the athlete's name, but the company doesn't offer the contract to the athlete. In this case, the company is letting go of an athlete that meets their advertising needs.

c) If the company uses a 10% level of significance, the company will hire more athletes that don't have high enough name recognition for their needs. The risk of committing a Type I error is higher.

At the same level of significance, the company is less likely to lose out on athletes with high name recognition. They will commit fewer Type II errors.

33. Dropouts.

Randomization condition: Assume that these subjects are representative of all anorexia nervosa patients.
10% condition: 198 is less than 10% of all patients.
Success/Failure condition: The number of dropouts, 105, and the number of subjects that remained, 93, are both greater than 10, so the samples are both large enough.

Since the conditions have been satisfied, we will find a one-proportion z-interval.

$$\hat{p} \pm z^* \sqrt{\frac{\hat{p}\hat{q}}{n}} = \left(\frac{105}{198}\right) \pm 1.960 \sqrt{\frac{\left(\frac{105}{198}\right)\left(\frac{93}{198}\right)}{198}} = (46\%, 60\%)$$

We are 95% confident that between 46% and 60% of anorexia nervosa patients will drop out of treatment programs. However, this wasn't a random sample of all patients. They were assigned to treatment programs rather than choosing their own. They may have had different experiences if they were not part of an experiment.

35. Speeding.

a) H_0 : The percentage of speeding tickets issued to black drivers is 16%, the same as the percentage of registered drivers who are black. ($p = 0.16$)

H_A : The percentage of speeding tickets issued to black drivers is greater than 16%, the percentage of registered drivers who are black. ($p > 0.16$)

Random condition: Assume that this month is representative of all months with respect to the percentage of tickets issued to black drivers.
10% condition: 324 speeding tickets are less than 10% of all tickets.
Success/Failure condition: $np = (324)(0.16) = 52$ and $nq = (324)(0.84) = 272$ are both greater than 10, so the sample is large enough.

The conditions have been satisfied, so a Normal model can be used to model the sampling distribution of the proportion, with $\mu_{\hat{p}} = p = 0.16$ and

$$\sigma(\hat{p}) = \sqrt{\frac{pq}{n}} = \sqrt{\frac{(0.16)(0.84)}{324}} \approx 0.02037.$$

We can perform a one-proportion z-test.
The observed proportion of tickets issued is $\hat{p} = 0.25$.

$$z = \frac{\hat{p} - p_0}{SD(\hat{p})}$$

Since the *P*-value = 4.96×10^{-6} is very low, we reject the
null hypothesis. There is strong evidence that the
percentage of speeding tickets issued to black drivers is
greater than 16%.

$$z \approx \frac{0.25 - 0.16}{\sqrt{\dfrac{(0.16)(0.84)}{324}}}$$

$$z \approx 4.42$$

b) There is strong evidence of an association between the receipt of a speeding
ticket and race. Black drivers appear to be issued tickets at a higher rate than
expected. However, this does not prove that racial profiling exists. There may
be other factors present.

c) Answers may vary. The primary statistic of interest is the percentage of black
motorists on this section of the New Jersey Turnpike. For example, if 80% of
drivers on this section are black, then 25% of the speeding tickets being issued to
black motorists is not an usually high percentage. In fact, it is probably
unusually low. On the other hand, if only 3% of the motorists on this section of
the turnpike are black, then there is even more evidence that racial profiling may
be occurring.

37. Meals.

H_0: The college student's mean daily food expense is $10. ($\mu = 10$)

H_A: The college student's mean daily food expense is greater than $10. ($\mu > 10$)

Randomization condition: Assume that these days are
representative of all days.
Nearly Normal condition: The histogram of daily expenses is
fairly unimodal and symmetric. It is reasonable to think that
this sample came from a Normal population.

The expenses in the sample had a mean of 11.4243 dollars
and a standard deviation of 8.05794 dollars. Since the
conditions for inference are satisfied, we can model the
sampling distribution of the mean daily expense with a Student's *t* model, with

$14 - 1 = 13$ degrees of freedom, $t_{13}\left(10, \dfrac{8.05794}{\sqrt{14}}\right)$.

We will perform a one-sample *t*-test.

Since the *P*-value = 0.2600 is high, we fail to reject the null hypothesis. There is no evidence that the student's average spending is more than $10 per day.

$$t = \frac{\bar{y} - \mu_0}{SE(\bar{y})}$$

$$t = \frac{11.4243 - 10}{\frac{8.05794}{\sqrt{14}}}$$

$$t \approx 0.66$$

$P = 0.2600$

10   11.42

$t = 0.66$

t_{13}

39. Streams.

Random condition: The researchers randomly selected 172 streams.
10% condition: 172 is less than 10% of all streams.
Success/Failure condition: $n\hat{p} = 69$ and $n\hat{q} = 103$ are both greater than 10, so the sample is large enough.

Since the conditions are met, we can use a one-proportion *z*-interval to estimate the percentage of Adirondack streams with a shale substrate.

$$\hat{p} \pm z^* \sqrt{\frac{\hat{p}\hat{q}}{n}} = \left(\frac{69}{172}\right) \pm 1.960 \sqrt{\frac{\left(\frac{69}{172}\right)\left(\frac{103}{172}\right)}{172}} = (32.8\%, 47.4\%)$$

We are 95% confident that between 32.8% and 47.4% of Adirondack streams have a shale substrate.

41. Bread.

a) Since the histogram shows that the distribution of the number of loaves sold per day is skewed strongly to the right, we can't use the Normal model to estimate the number of loaves sold on the busiest 10% of days.

b) **Randomization condition:** Assume that these days are representative of all days.
Nearly Normal condition: The histogram is skewed strongly to the right. However, since the sample size is large, the Central Limit Theorem guarantees that the distribution of averages will be approximately Normal.

The days in the sample had a mean of 103 loaves sold and a standard deviation of 9 loaves sold. Since the conditions are satisfied, the sampling distribution of the mean can be modeled by a Student's *t*- model, with $100 - 1 = 99$ degrees of freedom. We will use a one-sample *t*-interval with 95% confidence for the mean number of loaves sold. (By hand, use $t_{50}^* \approx 2.403$ from the table.)

c) $$\bar{y} \pm t_{n-1}^* \left(\frac{s}{\sqrt{n}}\right) = 103 \pm t_{99}^* \left(\frac{9}{\sqrt{100}}\right) \approx (101.2, 104.8)$$

We are 95% confident that the mean number of loaves sold per day at the Clarksburg Bakery is between 101.2 and 104.8.

d) We know that in order to cut the margin of error in half, we need to a sample four times as large. If we allow a margin of error that is twice as wide, that would require a sample only one-fourth the size. In this case, our original sample is 100 loaves; so 25 loaves would be a sufficient number to estimate the mean with a margin of error twice as wide.

e) Since the interval is completely above 100 loaves, there is strong evidence that the estimate was incorrect. The evidence suggests that the mean number of loaves sold per day is greater than 100. This difference is statistically significant, but may not be practically significant. It seems like the owners made a pretty good estimate!

43. And it means?

a) The margin of error is $\dfrac{(2391-1644)}{2} = \$373.50$.

b) The insurance agent is 95% confident that the mean loss claimed by clients after home burglaries is between $1644 and $2391.

c) 95% of all random samples of this size will produce intervals that contain the true mean loss claimed.

Chapter 22 – Comparing Groups

Section 22.1

1. Canada.

$$SE\left(\hat{p}_{Can} - \hat{p}_{Am}\right) = \sqrt{\frac{\hat{p}_{Can}\hat{q}_{Can}}{n_{Can}} + \frac{\hat{p}_{Am}\hat{q}_{Am}}{n_{Am}}} = \sqrt{\frac{\left(\frac{192}{960}\right)\left(\frac{768}{960}\right)}{960} + \frac{\left(\frac{170}{1250}\right)\left(\frac{1080}{1250}\right)}{1250}} = 0.0161$$

Section 22.2

3. Canada, deux.

We must assume the data were collected randomly and that the Americans selected are independent of the Canadians selected. Both assumptions should be met. Also, for both groups, we have at least 10 national-born and foreign-born citizens and the sample sizes are less than 10% of the population sizes. All conditions for inference are met.

Section 22.3

5. Canada, trois.

We are 95% confident that, based on these data, the proportion of foreign-born Canadians is between 3.24% and 9.56% more than the proportion of foreign-born Americans.

7. Canada, encore.

If we were to take repeated samples of Canadians and Americans, we would expect 95% of the intervals to contain the true difference in the proportion of foreign-born citizens.

Section 22.4

9. Canada, la fin.

a) $\hat{p}_{Can} - \hat{p}_{Am} = \dfrac{192}{960} - \dfrac{170}{1250} = 0.064$

b) $\hat{p}_{pooled} = \dfrac{192 + 170}{960 + 1250} = \dfrac{362}{2210} \approx 0.1638$

c) $SE_{pooled}\left(\hat{p}_{Can} - \hat{p}_{Am}\right) = \sqrt{\dfrac{\hat{p}_{pooled}\hat{q}_{pooled}}{n_{Can}} + \dfrac{\hat{p}_{pooled}\hat{q}_{pooled}}{n_{Am}}} = \sqrt{\dfrac{\left(\frac{362}{2210}\right)\left(\frac{1848}{2210}\right)}{960} + \dfrac{\left(\frac{362}{2210}\right)\left(\frac{1848}{2210}\right)}{1250}} \approx 0.0159$

d) $z = \dfrac{\hat{p}_{Can} - \hat{p}_{Am}}{SE_{pooled}\left(\hat{p}_{Can} - \hat{p}_{Am}\right)} = \dfrac{0.064}{0.0159} \approx 4.03$

e) Since the *P*-value is < 0.001, which is very low, reject the null hypothesis. There is very strong evidence that the proportion of foreign born citizens is different in Canada than it is in the United States. According to this data, the proportion of foreign born Canadians is likely to be the higher of the two.

Section 22.5

11. Cost of shopping.

We must assume the samples were random or otherwise independent of each other. We also assume that the distributions are roughly normal, so it would be a good idea to check a histogram to make sure there isn't strong skewness or outliers.

13. Cost of shopping, again.

We are 95% confident that the mean purchase amount at Walmart is between $1.85 and $14.15 less than the mean purchase amount at Target.

Section 22.6

15. Cost of shopping, three.

The difference is –$8 with an SE of 3.115, so the *t*-stat is –2.569. With 162.75 (or 163) df, the P-value is 0.011 which is less than 0.05. Reject the null hypothesis that the means are equal. There is evidence that the mean purchase amounts at the two stores are not the same. These data suggest that the mean purchase amount at Target is lower than the mean purchase price at Walmart.

Section 20.7

17. Cost of shopping, once more.

The *t*-statistic is –2.561 using the pooled estimate of the standard deviation, 20.06. There are 163 df so the P-value is still 0.011. We reach the same conclusion as before. Because the sample standard deviations are nearly the same and the sample sizes are large, the pooled test is essentially the same as the two-sample *t*-test.

19. Cost of shopping, last look.

No. The two-sample test is almost always the safer choice. In this case, the variances are likely to be quite different. The purchase prices of Italian sports cars are much higher and may be more variable than the domestic prices. They should use the two-sample t-test.

21. Online social networking.

It is very unlikely that samples would show an observed difference this large if, in fact, there was no real difference between the proportion of American adults who visited Facebook on a daily basis in 2013 and the proportion of American adults who visited Facebook on a daily basis 2010.

Chapter Exercises.

23. Revealing information.

This test is not appropriate for these data, since the responses are not from independent groups, but are from the same individuals. The independent samples condition has been violated.

25. Gender gap.

a) This is a stratified random sample, stratified by gender.

b) We would expect the difference in proportions in the sample to be the same as the difference in proportions in the population, with the percentage of respondents with a favorable impression of the candidate 6 percentage points higher among males.

c) The standard deviation of the difference in proportions is:

$$SD(\hat{p}_M - \hat{p}_F) = \sqrt{\frac{\hat{p}_M \hat{q}_M}{n_M} + \frac{\hat{p}_F \hat{q}_F}{n_F}} = \sqrt{\frac{(0.59)(0.41)}{300} + \frac{(0.53)(0.47)}{300}} \approx 4\%$$

d)

Difference in proportion with favorable impression (Male - Female)

e) The campaign could certainly be misled by the poll. According to the model, a poll showing little difference could occur relatively frequently. That result is only 1.5 standard deviations below the expected difference in proportions.

27. Arthritis.

a) **Randomization condition:** Americans age 65 and older were selected randomly.
10% condition: 1012 men and 1062 women are less than 10% of all men and women.
Independent groups assumption: The sample of men and the sample of women were drawn independently of each other.
Success/Failure condition: $n\hat{p}$ (men) = 411, $n\hat{q}$ (men) = 601, $n\hat{p}$ (women) = 535, and $n\hat{q}$ (women) = 527 are all greater than 10, so the samples are both large enough.

Since the conditions have been satisfied, we will find a two-proportion z-interval.

b)

$$\left(\hat{p}_F - \hat{p}_M\right) \pm z^* \sqrt{\frac{\hat{p}_F \hat{q}_F}{n_F} + \frac{\hat{p}_M \hat{q}_M}{n_M}} = \left(\tfrac{535}{1062} - \tfrac{411}{1012}\right) \pm 1.960 \sqrt{\frac{\left(\tfrac{535}{1062}\right)\left(\tfrac{527}{1062}\right)}{1062} + \frac{\left(\tfrac{411}{1012}\right)\left(\tfrac{601}{1012}\right)}{1012}}$$

$$= \left(0.055,\ 0.140\right)$$

c) We are 95% confident that the proportion of American women age 65 and older who suffer from arthritis is between 5.5% and 14.0% higher than the proportion of American men the same age who suffer from arthritis.

d) Since the interval for the difference in proportions of arthritis sufferers does not contain 0, there is strong evidence that arthritis is more likely to afflict women than men.

29. Pets.

a) $SE\left(\hat{p}_{Herb} - \hat{p}_{None}\right) = \sqrt{\dfrac{\hat{p}_{Herb} \hat{q}_{Herb}}{n_{Herb}} + \dfrac{\hat{p}_{None} \hat{q}_{None}}{n_{None}}} = \sqrt{\dfrac{\left(\tfrac{473}{827}\right)\left(\tfrac{354}{827}\right)}{827} + \dfrac{\left(\tfrac{19}{130}\right)\left(\tfrac{111}{130}\right)}{130}} = 0.035$

b) Randomization condition: Assume that the dogs studied were representative of all dogs.

10% condition: 827 dogs from homes with herbicide used regularly and 130 dogs from homes with no herbicide used are less than 10% of all dogs.

Independent groups assumption: The samples were drawn independently of each other.

Success/Failure condition: $n\hat{p}$ (herb) = 473, $n\hat{q}$ (herb) = 354, $n\hat{p}$ (none) = 19, and $n\hat{q}$ (none) = 111 are all greater than 10, so the samples are both large enough.

Since the conditions have been satisfied, we will find a two-proportion z-interval.

$$\left(\hat{p}_{Herb} - \hat{p}_{None}\right) \pm z^* \sqrt{\frac{\hat{p}_{Herb} \hat{q}_{Herb}}{n_{Herb}} + \frac{\hat{p}_{None} \hat{q}_{None}}{n_{None}}}$$

$$= \left(\tfrac{473}{827} - \tfrac{19}{130}\right) \pm 1.960 \sqrt{\frac{\left(\tfrac{473}{827}\right)\left(\tfrac{354}{827}\right)}{827} + \frac{\left(\tfrac{19}{130}\right)\left(\tfrac{111}{130}\right)}{130}} = \left(0.356,\ 0.495\right)$$

c) We are 95% confident that the proportion of pets with a malignant lymphoma in homes where herbicides are used is between 35.6 and 49.5 percentage points higher than the proportion with lymphoma in homes where no pesticides are used.

31. Prostate cancer.

a) This is an experiment. Men were randomly assigned to have surgery or not.

b) **Randomization condition:** Men were randomly assigned to treatment groups.
Independent groups assumption: The survival rates of the two groups are not related.
Success/Failure condition: $n\hat{p}$ (no surgery) = 31 , $n\hat{q}$ (no surgery) = 317, $n\hat{p}$ (surgery) = 16, and $n\hat{q}$ (surgery) = 331 are all greater than 10, so the samples are both large enough.

Since the conditions have been satisfied, we will find a two-proportion z-interval.

$$\left(\hat{p}_{None} - \hat{p}_{Surg}\right) \pm z^* \sqrt{\frac{\hat{p}_{None}\hat{q}_{None}}{n_{None}} + \frac{\hat{p}_{Surg}\hat{q}_{Surg}}{n_{Surg}}}$$

$$= \left(\tfrac{31}{348} - \tfrac{16}{347}\right) \pm 1.960 \sqrt{\frac{\left(\tfrac{31}{348}\right)\left(\tfrac{317}{348}\right)}{348} + \frac{\left(\tfrac{16}{347}\right)\left(\tfrac{331}{347}\right)}{347}} = \left(0.006,\ 0.080\right)$$

c) We are 95% confident that the survival rate of patients who have surgery for prostate cancer is between 0.6 and 8.0 percentage points higher than the survival rate of patients who do not have surgery. Since the interval is completely above zero, there is evidence that surgery may be effective in preventing death from prostate cancer.

33. Ear infections.

a) **Randomization condition:** The babies were randomly assigned to the two treatment groups.
Independent groups assumption: The groups were assigned randomly, so the groups are not related.
Success/Failure condition: $n\hat{p}$ (vaccine) = 333, $n\hat{q}$ (vaccine) = 2122, $n\hat{p}$ (none) = 499, and $n\hat{q}$ (none) = 1953 are all greater than 10, so the samples are both large enough.

Since the conditions have been satisfied, we will find a two-proportion z-interval.

b) $\left(\hat{p}_{None} - \hat{p}_{Vacc}\right) \pm z^* \sqrt{\dfrac{\hat{p}_{None}\hat{q}_{None}}{n_{None}} + \dfrac{\hat{p}_{Vacc}\hat{q}_{Vacc}}{n_{Vacc}}}$

$$= \left(\tfrac{499}{2452} - \tfrac{333}{2455}\right) \pm 1.960 \sqrt{\frac{\left(\tfrac{499}{2452}\right)\left(\tfrac{1953}{2452}\right)}{2452} + \frac{\left(\tfrac{333}{2455}\right)\left(\tfrac{2122}{2455}\right)}{2455}} = \left(0.047,\ 0.089\right)$$

c) We are 95% confident that the proportion of unvaccinated babies who develop ear infections is between 4.7 and 8.9 percentage points higher than the proportion of vaccinated babies who develop ear infections. The vaccinations appear to be effective, especially considering the 20% infection rate among the unvaccinated. A reduction of 5% to 9% is meaningful.

35. Another ear infection.

a) H_0: The proportion of vaccinated babies who get ear infections is the same as the proportion of unvaccinated babies who get ear infections.
$$\left(p_{Vacc} = p_{None} \quad \text{or} \quad p_{Vacc} - p_{None} = 0\right)$$

H_A: The proportion of vaccinated babies who get ear infections is the lower than the proportion of unvaccinated babies who get ear infections.
$$\left(p_{Vacc} < p_{None} \quad \text{or} \quad p_{Vacc} - p_{None} < 0\right)$$

b) Since 0 is not in the confidence interval, reject the null hypothesis. There is evidence that the vaccine reduces the rate of ear infections.

c) If we think that the vaccine really reduces the rate of ear infections and it really does not reduce the rate of ear infections, then we have committed a Type I error.

d) Babies would be given ineffective vaccines.

37. Teen smoking.

a) This is a prospective observational study.

b) H_0: The proportion of teen smokers among the group whose parents disapprove of smoking is the same as the proportion of teen smokers among the group whose parents are lenient about smoking.
$$\left(p_{Dis} = p_{Len} \quad \text{or} \quad p_{Dis} - p_{Len} = 0\right)$$

H_A: The proportion of teen smokers among the group whose parents disapprove of smoking is lower than the proportion of teen smokers among the group whose parents are lenient about smoking.
$$\left(p_{Dis} < p_{Len} \quad \text{or} \quad p_{Dis} - p_{Len} < 0\right)$$

c) **Randomization condition:** Assume that the teens surveyed are representative of all teens.
10% condition: 284 and 41 are both less than 10% of all teens.
Independent groups assumption: The groups were surveyed independently.
Success/Failure condition: $n\hat{p}$ (disapprove) = 54, $n\hat{q}$ (disapprove) = 230, $n\hat{p}$ (lenient) = 11, and $n\hat{q}$ (lenient) = 30 are all greater than 10, so the samples are both large enough.

Since the conditions have been satisfied, we will model the sampling distribution of the difference in proportion with a Normal model with mean 0 and standard deviation estimated by

$$SE_{pooled}\left(\hat{p}_{Dis} - \hat{p}_{Len}\right) = \sqrt{\frac{\hat{p}_{pooled}\,\hat{q}_{pooled}}{n_{Dis}} + \frac{\hat{p}_{pooled}\,\hat{q}_{pooled}}{n_{Len}}} = \sqrt{\frac{\left(\frac{65}{325}\right)\left(\frac{260}{325}\right)}{284} + \frac{\left(\frac{65}{325}\right)\left(\frac{260}{325}\right)}{41}} = 0.0668\,.$$

d) The observed difference between the proportions is $0.190 - 0.268 = -0.078$.

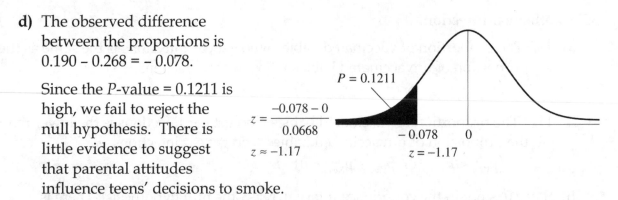

Since the *P*-value = 0.1211 is high, we fail to reject the null hypothesis. There is little evidence to suggest that parental attitudes influence teens' decisions to smoke.

$$z = \frac{-0.078 - 0}{0.0668}$$

$$z \approx -1.17$$

$P = 0.1211$

$-0.078 \qquad 0$

$z = -1.17$

e) If there is no difference in the proportions, there is about a 12% chance of seeing the observed difference or larger by natural sampling variation.

f) If teens' decisions about smoking *are* influenced, we have committed a Type II error.

g) Since the conditions have already been satisfied in a previous exercise, we will find a two-proportion *z*-interval.

$$\left(\hat{p}_{Dis} - \hat{p}_{Len}\right) \pm z^* \sqrt{\frac{\hat{p}_{Dis}\hat{q}_{Dis}}{n_{Dis}} + \frac{\hat{p}_{Len}\hat{q}_{Len}}{n_{Len}}}$$

$$= \left(\tfrac{54}{284} - \tfrac{11}{41}\right) \pm 1.960 \sqrt{\frac{\left(\tfrac{54}{284}\right)\left(\tfrac{230}{284}\right)}{284} + \frac{\left(\tfrac{11}{41}\right)\left(\tfrac{30}{41}\right)}{41}} = \left(-0.221,\ 0.065\right)$$

h) We are 95% confident that the proportion of teens whose parents disapprove of smoking who will eventually smoke is between 22.1 percentage points lower and 6.5 percentage points higher than for teens with parents who are lenient about smoking.

i) We expect 95% of random samples of this size to produce intervals that contain the true difference between the proportions.

39. Birthweight.

a) H_0: The proportion of low birthweight is the same. $\left(p_{Exp} = p_{Not} \text{ or } p_{Exp} - p_{Not} = 0\right)$

H_A: The proportion of low birthweight is higher for women exposed to soot and ash. $\left(p_{Exp} > p_{Not} \text{ or } p_{Exp} - p_{Not} > 0\right)$

Randomization condition: Assume the women are representative of all women.
10% condition: 182 and 2300 are both less than 10% of all women.
Independent groups assumption: The groups don't appear to be associated, with respect to soot and ash exposure, but all of the women were in New York. There may be a confounding variable explaining any relationship between exposure and birthweight.

Success/Failure condition: $n\hat{p}$ (Exposed) = 15, $n\hat{q}$ (Exposed) = 167, $n\hat{p}$ (Not) = 92, and $n\hat{q}$ (Not) = 2208 are all greater than 10. All of the samples are large enough.

Since the conditions have been satisfied, we will model the sampling distribution of the difference in proportion with a Normal model with mean 0 and standard deviation estimated by

$$SE_{pooled}\left(\hat{p}_{Exp} - \hat{p}_{Not}\right) = \sqrt{\frac{\hat{p}_P \hat{q}_P}{n_{Exp}} + \frac{\hat{p}_P \hat{q}_P}{n_{Not}}} = \sqrt{\frac{\left(\frac{107}{2482}\right)\left(\frac{2375}{2482}\right)}{182} + \frac{\left(\frac{107}{2482}\right)\left(\frac{2375}{2482}\right)}{2300}} \approx 0.0156.$$

The observed difference between the proportions is: 0.08 – 0.04 = 0.04.

Since the *P*-value = 0.005 is low, we reject the null hypothesis. There is strong evidence that the proportion of low birthweight babies is higher in the women exposed to soot and ash after the World Trade Center attacks.

$$z = \frac{0.08 - 0.04}{0.0156}$$

$$z \approx 2.56$$

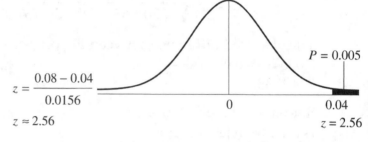

b) $\left(\hat{p}_{Exp} - \hat{p}_{Not}\right) \pm z^* \sqrt{\dfrac{\hat{p}_{Exp} \hat{q}_{Exp}}{n_{Exp}} + \dfrac{\hat{p}_{Not} \hat{q}_{Not}}{n_{Not}}}$

$$= \left(\frac{15}{182} - \frac{92}{2300}\right) \pm 1.960 \sqrt{\frac{\left(\frac{15}{182}\right)\left(\frac{167}{182}\right)}{182} + \frac{\left(\frac{92}{2300}\right)\left(\frac{2208}{2300}\right)}{2300}} = (0.002, 0.083)$$

We are 95% confident that the proportion of low birthweight babies is between 0.2 and 8.3 percentage points higher for mothers exposed to soot and ash after the World Trade Center attacks, than the proportion of low birthweight babies for mothers not exposed.

41. Mammograms.

a) H_0: The proportion of deaths from breast cancer is the same for women who never had a mammogram as for women who had mammograms.
$\left(p_N = p_M \text{ or } p_N - p_M = 0\right)$

H_A: The proportion of deaths from breast cancer is greater for women who never had a mammogram than for women who had mammograms.
$\left(p_N > p_M \text{ or } p_N - p_M > 0\right)$

Randomization condition: Assume the women are representative of all women.
10% condition: 30,565 and 30,131 are both less than 10% of all women.
Independent groups assumption: We must assume that these groups are independent.

Success/Failure condition: $n\hat{p}$ (never) = 196, $n\hat{q}$ (never) = 30,369, $n\hat{p}$ (mammogram) = 153, and $n\hat{q}$ (mammogram) = 29,978 are all greater than 10, so both samples are large enough.

Since the conditions have been satisfied, we will model the sampling distribution of the difference in proportion with a Normal model with mean 0 and standard deviation estimated by:

$$SE_{pooled}\left(\hat{p}_N - \hat{p}_M\right) = \sqrt{\frac{\hat{p}_{pooled}\hat{q}_{pooled}}{n_N} + \frac{\hat{p}_{pooled}\hat{q}_{pooled}}{n_M}}$$

$$= \sqrt{\frac{\left(\frac{349}{60,696}\right)\left(\frac{60,347}{60,696}\right)}{30,565} + \frac{\left(\frac{349}{60,696}\right)\left(\frac{60,347}{60,696}\right)}{30,131}} \approx 0.000614$$

The observed difference between the proportions is:
0.006413 – 0.005078 =
0.001335.

Since the *P*-value = 0.0148 is low, we reject the null hypothesis. There is strong evidence that the breast cancer mortality rate is higher among

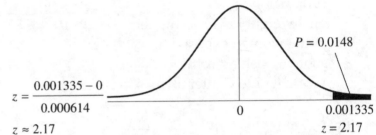

$$z = \frac{0.001335 - 0}{0.000614}$$

$$z \approx 2.17$$

women that have never had a mammogram. However, the large sample sizes involved may have yielded a result that is statistically significant, but not practically significant. These data suggest a difference in mortality rate of only about 0.1 percentage points.

b) If there is actually no difference in the mortality rates, we have committed a Type I error.

43. Pain.

a) **Randomization condition:** The patients were randomly selected AND randomly assigned to treatment groups. If that's not random enough for you, I don't know what is!
10% condition: 112 and 108 are both less than 10% of all people with joint pain.
Success/Failure condition: $n\hat{p}$ (A) = 84, $n\hat{q}$ (A) = 28, $n\hat{p}$ (B) = 66, and $n\hat{q}$ (B) = 42 are all greater than 10, so both samples are large enough.

Since the conditions are met, we can use a one-proportion z-interval to estimate the percentage of patients who may get relief from medication A.

$$\hat{p} \pm z^* \sqrt{\frac{\hat{p}\hat{q}}{n}} = \left(\frac{84}{112}\right) \pm 1.960 \sqrt{\frac{\left(\frac{84}{112}\right)\left(\frac{28}{112}\right)}{112}} = (67.0\%, 83.0\%)$$

We are 95% confident that between 67.0% and 83.0% of patients with joint pain will find medication A to be effective.

b) Since the conditions were met in part a, we can use a one-proportion z-interval to estimate the percentage of patients who may get relief from medication B.

$$\hat{p} \pm z^* \sqrt{\frac{\hat{p}\hat{q}}{n}} = \left(\frac{66}{108}\right) \pm 1.960 \sqrt{\frac{\left(\frac{66}{108}\right)\left(\frac{42}{108}\right)}{108}} = (51.9\%, 70.3\%)$$

We are 95% confident that between 51.9% and 70.3% of patients with joint pain will find medication B to be effective.

c) The 95% confidence intervals overlap, which might lead one to believe that there is no evidence of a difference in the proportions of people who find each medication effective. However, if one was lead to believe that, one should proceed to part…

d) Most of the conditions were checked in part a. We only have one more to check: **Independent groups assumption:** The groups were assigned randomly, so there is no reason to believe there is a relationship between them.

Since the conditions have been satisfied, we will find a two-proportion z-interval.

$$\left(\hat{p}_A - \hat{p}_B\right) \pm z^* \sqrt{\frac{\hat{p}_A \hat{q}_A}{n_A} + \frac{\hat{p}_B \hat{q}_B}{n_B}}$$

$$= \left(\frac{84}{112} - \frac{66}{108}\right) \pm 1.960 \sqrt{\frac{\left(\frac{84}{112}\right)\left(\frac{28}{112}\right)}{112} + \frac{\left(\frac{66}{108}\right)\left(\frac{42}{108}\right)}{112}} = (0.017, 0.261)$$

We are 95% confident that the proportion of patients with joint pain who will find medication A effective is between 1.70 and 26.1 percentage points higher than the proportion of patients who will find medication B effective.

e) The interval does not contain zero. There is evidence that medication A is more effective than medication B.

f) The two-proportion method is the proper method. By attempting to use two, separate, confidence intervals, you are adding standard deviations when looking for a difference in proportions. We know from our previous studies that *variances* add when finding the standard deviation of a difference. The two-proportion method does this.

45. Convention bounce.

> **Randomization condition:** The polls were conducted randomly.
> **10% condition:** 1500 and 1500 are both less than 10% of all voters.
> **Independent groups assumption:** The groups were chosen independently.
> **Success/Failure condition:** $n\hat{p}$ (after) = 705, $n\hat{q}$ (after) = 795, $n\hat{p}$ (before) = 735, and $n\hat{q}$ (women) = 765 are all greater than 10, so both samples are large enough.

Since the conditions have been satisfied, we will find a two-proportion z-interval.

$$\left(\hat{p}_{After} - \hat{p}_{Before}\right) \pm z^* \sqrt{\frac{\hat{p}_{After}\hat{q}_{After}}{n_{After}} + \frac{\hat{p}_{Before}\hat{q}_{Before}}{n_{Before}}}$$

$$= (0.49 - 0.47) \pm 1.960 \sqrt{\frac{(0.49)(0.51)}{1500} + \frac{(0.47)(0.53)}{1500}} = (-0.016, 0.056)$$

We can be 95% confident that the proportion of likely voters who favored John Kerry was between 1.6 percentage points lower and 5.6 percentage points higher after the convention, compared to before the convention. Since zero is contained in the interval, it is plausible that there was no difference in Kerry support. The poll showed no evidence of a convention bounce.

47. Sensitive men.

H_0 : The proportion of 18-24-year-old men who are comfortable talking about their problems is the same as the proportion of 25-34-year old men.

$$\left(p_{Young} = p_{Old} \text{ or } p_{Young} - p_{Old} = 0\right)$$

H_A : The proportion of 18-24-year-old men who are comfortable talking about their problems is higher than the proportion of 25-34-year old men.

$$\left(p_{Young} > p_{Old} \text{ or } p_{Young} - p_{Old} > 0\right)$$

Randomization condition: We assume the respondents were chosen randomly.
10% condition: 129 and 184 are both less than 10% of all people.
Independent groups assumption: The groups were chosen independently.
Success/Failure condition: $n\hat{p}$ (young) = 80, $n\hat{q}$ (young) = 49, $n\hat{p}$ (old) = 98, and $n\hat{q}$ (old) = 86 are all greater than 10, so both samples are large enough.

Since the conditions have been satisfied, we will model the sampling distribution of the difference in proportion with a Normal model with mean 0 and standard deviation estimated
by:.

$$SE_{pooled}\left(\hat{p}_{Young} - \hat{p}_{Old}\right) = \sqrt{\frac{\hat{p}_{pooled}\hat{q}_{pooled}}{n_Y} + \frac{\hat{p}_{pooled}\hat{q}_{pooled}}{n_O}} = \sqrt{\frac{\left(\frac{178}{313}\right)\left(\frac{135}{313}\right)}{129} + \frac{\left(\frac{178}{313}\right)\left(\frac{135}{313}\right)}{184}} \approx 0.05687$$

The observed difference between the proportions is:
0.620 – 0.533 = 0.087.

Since the P-value = 0.0619 is high, we fail to reject the null hypothesis. There is little evidence that the proportion of 18-24-year-old men who are comfortable talking about their problems is higher than the proportion of 25-34-year-old men who are comfortable. *Time* magazine's interpretation is questionable.

$$z = \frac{0.087 - 0}{0.05687}$$

$$z \approx 1.54$$

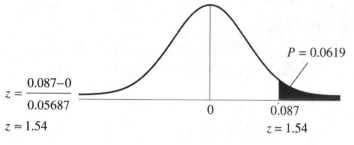

$P = 0.0619$

0 0.087

$z = 1.54$

49. Food preference.

H_0: The proportion of people who agree with the statement is the same in rural and urban areas. $\left(p_{Urban} = p_{Rural}\ \text{ or }\ p_{Urban} - p_{Rural} = 0\right)$

H_A: The proportion of people who agree with the statement differs in rural and urban areas. $\left(p_{Urban} \neq p_{Rural}\ \text{ or }\ p_{Urban} - p_{Rural} \neq 0\right)$

Randomization condition: The respondents were chosen randomly.
10% condition: 646 and 154 are both less than 10% of all urban and rural people, respectively.
Independent groups assumption: The groups were chosen independently.
Success/Failure condition: $n\hat{p}$ (urban) = 417, $n\hat{q}$ (urban) = 229, $n\hat{p}$ (rural) = 78, and $n\hat{q}$ (rural) = 76 are all greater than 10, so both samples are large enough.

Since the conditions have been satisfied, we will model the sampling distribution of the difference in proportion with a Normal model with mean 0 and standard deviation estimated by:

$$SE_{pooled}\left(\hat{p}_{Urban} - \hat{p}_{Rural}\right) = \sqrt{\frac{\hat{p}_{pooled}\hat{q}_{pooled}}{n_{Urban}} + \frac{\hat{p}_{pooled}\hat{q}_{pooled}}{n_{Rural}}} = \sqrt{\frac{\left(\frac{495}{800}\right)\left(\frac{305}{800}\right)}{646} + \frac{\left(\frac{495}{800}\right)\left(\frac{305}{800}\right)}{154}} \approx 0.0436$$

The observed difference between the proportions is: $\frac{417}{646} - \frac{78}{154} \approx 0.139$

Since the P-value = 0.0014 is low, reject the null hypothesis. There is evidence that the proportion of people who agree with the statement is not the same in urban and rural areas. These data suggest that the proportion is higher in urban areas than in rural areas.

$$z = \frac{0.139 - 0}{0.0436}$$

$$z \approx 3.19$$

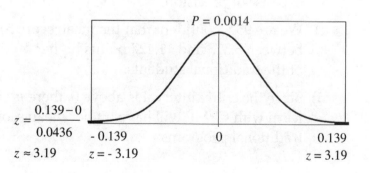

$P = 0.0014$

-0.139 0 0.139

$z = -3.19$ $z = 3.19$

51. Hot dogs.

Yes, the 95% confidence interval would contain 0. The high P-value means that we lack evidence of a difference, so 0 is a possible value for $\mu_{Meat} - \mu_{Beef}$.

53. Hot dogs, second helping.

a) Plausible values for $\mu_{Meat} - \mu_{Beef}$ are all negative, so the mean fat content is probably higher for beef hot dogs.

b) The fact that the confidence interval does not contain 0 indicates that the difference is significant.

c) The corresponding alpha level is 10%.

55. Hot dogs, last one.

a) False. The confidence interval is about means, not about individual hot dogs.

b) False. The confidence interval is about means, not about individual hot dogs.

c) True.

d) False. Confidence intervals based on other samples will also try to estimate the true difference in population means. There's not reason to expect other samples to conform to this result.

e) True.

57. Learning math.

a) The margin of error of this confidence interval is (11.427 – 5.573)/2 = 2.927 points.

b) The margin of error for a 98% confidence interval would have been larger. The critical value of t^* is larger for higher confidence levels. We need a wider interval to increase the likelihood that we catch the true mean difference in test scores within our interval. In other words, greater confidence comes at the expense of precision.

c) We are 95% confident that the mean score for the CPMP math students will be between 5.573 and 11.427 points higher on this assessment than the mean score of the traditional students.

d) Since the entire interval is above 0, there is strong evidence that students who learn with CPMP will have higher mean scores is algebra than those in traditional programs.

59. CPMP, again.

a) H$_0$: The mean score of CPMP students is the same as the mean score of traditional students. $\left(\mu_C = \mu_T \text{ or } \mu_C - \mu_T = 0\right)$

H$_A$: The mean score of CPMP students is different from the mean score of traditional students. $\left(\mu_C \neq \mu_T \text{ or } \mu_C - \mu_T \neq 0\right)$

b) **Independent groups assumption:** Scores of students from different classes should be independent.
Randomization condition: Although not specifically stated, classes in this experiment were probably randomly assigned to either CPMP or traditional curricula.
Nearly Normal condition: We don't have the actual data, so we can't check the distribution of the sample. However, the samples are large. The Central Limit Theorem allows us to proceed.

Since the conditions are satisfied, we can use a two-sample *t*-test with 583 degrees of freedom (from the computer).

c) If the mean scores for the CPMP and traditional students are really equal, there is less than a 1 in 10,000 chance of seeing a difference as large or larger than the observed difference just from natural sampling variation.

d) Since the *P*-value < 0.0001, reject the null hypothesis. There is strong evidence that the CPMP students have a different mean score than the traditional students. The evidence suggests that the CPMP students have a higher mean score.

61. Commuting.

a) **Independent groups assumption:** Since the choice of route was determined at random, the commuting times for Route A are independent of the commuting times for Route B.
Randomization condition: The man randomly determined which route he would travel on each day.
Nearly Normal condition: The histograms of travel times for the routes are roughly unimodal and symmetric. (Given)

Since the conditions are satisfied, it is appropriate to model the sampling distribution of the difference in means with a Student's *t*-model, with 33.1 degrees of freedom (from the approximation formula). We will construct a two-sample *t*-interval, with 95% confidence.

$$\left(\overline{y}_B - \overline{y}_A\right) \pm t_{df}^{*} \sqrt{\frac{s_B^2}{n_B} + \frac{s_A^2}{n_A}} = (43 - 40) \pm t_{33.1}^{*} \sqrt{\frac{2^2}{20} + \frac{3^2}{20}} \approx (1.36, 4.64)$$

We are 95% confident that Route B has a mean commuting time between 1.36 and 4.64 minutes longer than the mean commuting time of Route A.

b) Since 5 minutes is beyond the high end of the interval, there is no evidence that the Route B is an average of 5 minutes longer than Route A. It appears that the old-timer may be exaggerating the average difference in commuting time.

63. View of the water.

Independent groups assumption: Since the 170 properties were randomly selected, the groups should be independent.
Randomization condition: The 170 properties were selected randomly.
Nearly Normal condition: The boxplots of sale prices are roughly symmetric. The plots show several outliers, but the sample sizes are large. The Central Limit Theorem allows us to proceed.

Since the conditions are satisfied, it is appropriate to model the sampling distribution of the difference in means with a Student's t-model, with 105.48 degrees of freedom (from the approximation formula). We will construct a two-sample t-interval, with 95% confidence.

$$(\bar{y}_W - \bar{y}_N) \pm t^*_{df}\sqrt{\frac{s_W^2}{n_W} + \frac{s_N^2}{n_N}}$$

$$= (319,906.40 - 219,896.60) \pm t^*_{105.48}\sqrt{\frac{153,303.80^2}{70} + \frac{94,627.15^2}{100}}$$

$$\approx (\$59121, \$140898)$$

We are 95% confident that waterfront property has a mean selling price that is between \$59,121 and \$140,898 higher, on average, than non-waterfront property.

65. Cereal.

Independent groups assumption: The percentage of sugar in the children's cereals is unrelated to the percentage of sugar in adult's cereals.
Randomization condition: It is reasonable to assume that the cereals are representative of all children's cereals and adult cereals, in regard to sugar content.
Nearly Normal condition: The histogram of adult cereal sugar content is skewed to the right, but the sample sizes are of reasonable size. The Central Limit Theorem allows us to proceed.

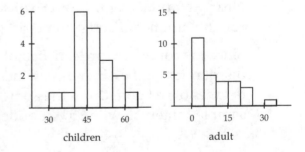

Since the conditions are satisfied, it is appropriate to model the sampling distribution of the difference in means with a Student's *t*-model, with 42 degrees of freedom (from the approximation formula). We will construct a two-sample *t*-interval, with 95% confidence.

$$(\bar{y}_C - \bar{y}_A) \pm t^*_{df}\sqrt{\frac{s_C^2}{n_C} + \frac{s_A^2}{n_A}} = (46.8 - 10.1536) \pm t^*_{42}\sqrt{\frac{6.41838^2}{19} + \frac{7.61239^2}{28}} \approx (32.49,\ 40.80)$$

We are 95% confident that children's cereals have a mean sugar content that is between 32.49% and 40.80% higher than the mean sugar content of adult cereals.

67. Reading.

H_0: The mean reading comprehension score of students who learn by the new method is the same as the mean score of students who learn by traditional methods. $\left(\mu_N = \mu_T \text{ or } \mu_N - \mu_T = 0\right)$

H_A: The mean reading comprehension score of students who learn by the new method is greater than the mean score of students who learn by traditional methods. $\left(\mu_N > \mu_T \text{ or } \mu_N - \mu_T > 0\right)$

Independent groups assumption: Student scores in one group should not have an impact on the scores of students in the other group.
Randomization condition: Students were randomly assigned to classes.
Nearly Normal condition: The stem-and-leaf plots show distributions of scores that are unimodal and symmetric.

Since the conditions are satisfied, it is appropriate to model the sampling distribution of the difference in means with a Student's *t*-model, with 33.4 degrees of freedom (from the approximation formula). We will perform a two-sample *t*-test. We know:

$$\bar{y}_N = 51.7222 \qquad \bar{y}_T = 41.8182$$
$$s_N = 11.7062 \qquad s_T = 16.5979$$
$$n_N = 18 \qquad n_T = 22$$

The sampling distribution model has mean 0, with standard error:

$$SE(\bar{y}_N - \bar{y}_T) = \sqrt{\frac{11.7062^2}{18} + \frac{16.5979^2}{22}} \approx 4.487 .$$

The observed difference between the mean scores is 51.7222 – 41.8182 ≈ 9.904.

Since the *P*-value = 0.0168 is low, we reject the null hypothesis. There is evidence that the students taught using the new

$$t = \frac{(\bar{y}_N - \bar{y}_T) - (0)}{SE(\bar{y}_N - \bar{y}_T)}$$

$$t \approx \frac{9.904}{4.487}$$

$$t \approx 2.207$$

activities have a higher mean score on the reading comprehension test than the students taught using traditional methods.

The control group contains both the highest and lowest scores, so Tukey's test is not appropriate. The Rank-sum test gives a *P*-value of 0.0045, so we reject the null hypothesis. There is evidence that the students taught using the new activities have a higher mean score on the reading comprehension test than the students taught using traditional methods.

69. Hurricanes 2013.

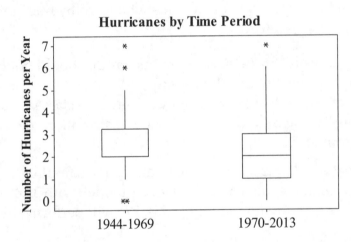

There are several concerns here. First, we don't have a random sample. We have to assume that the actual number of hurricanes in a given year is a random sample of the hurricanes that might occur under similar weather conditions. Also, the data for 1944–1969 are not symmetric and have four outliers. The outliers will tend to make the average for the period 1944–1969 higher. These data are not appropriate for inference. The boxplots provide little evidence of a change in the mean number of hurricanes in the two periods.

71. Home runs 2013.

a) The boxplots of the average number of home runs hit at the ballparks in the two leagues are at the right. Both distributions appear at least roughly symmetric, with roughly the same center, around 2 home runs. The distribution of average number of home runs hit is more spread out for the National League. There are outliers in the American League's distribution. (There may be only a high outlier, depending on how quartiles are calculated.)

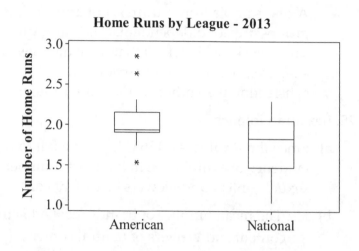

Home Runs by League - 2013

b) $\bar{y} \pm t^*_{n-1}\left(\dfrac{s}{\sqrt{n}}\right) = 2.0693 \pm t^*_{14}\left(\dfrac{0.3324}{\sqrt{15}}\right) \approx (1.89,\ 2.25)$

We are 95% confident that the mean number of home runs hit per game in American League stadiums is between 1.89 and 2.25.

c) The average of 1.96 home runs hit per game in Coors Field is not unusual. It isn't even the highest average in the National League.

d) If you attempt to use two confidence intervals to assess a difference in means, you are actually adding standard deviations. But it's the variances that add, not the standard deviations. The two-sample difference of means procedure takes this into account.

e)

$$(\bar{y}_A - \bar{y}_N) \pm t^*_{df}\sqrt{\dfrac{s_A^2}{n_A} + \dfrac{s_N^2}{n_N}}$$

$$= (2.0693 - 1.7653) \pm t^*_{26.77}\sqrt{\dfrac{0.3324^2}{15} + \dfrac{0.3833^2}{15}} \approx (0.035,\ 0.573)$$

f) We are 95% confident that the mean number of home runs in American League stadiums is between 0.035 and 0.573 home runs higher than the mean number of home runs in National League stadiums.

g) Since the interval does not contain 0, there is evidence of a difference in the mean number of home runs hit per game in the stadiums of the two leagues. American League stadiums have a significantly higher average number of home runs than National League stadiums.

73. Job satisfaction.

A two-sample *t*-procedure is not appropriate for these data, because the two groups are not independent. They are before and after satisfaction scores for the same workers. Workers that have high levels of job satisfaction before the exercise program is implemented may tend to have higher levels of job satisfaction than other workers after the program as well.

75. Sex and violence.

a) Since the *P*-value = 0.136 is high, we fail to reject the null hypothesis. There is no evidence of a difference in the mean number of brands recalled by viewers of sexual content and viewers of violent content.

b) H_0: The mean number of brands recalled is the same for viewers of sexual content and viewers of neutral content. $\left(\mu_S = \mu_N \text{ or } \mu_S - \mu_N = 0\right)$

H_A: The mean number of brands recalled is different for viewers of sexual content and viewers of neutral content. $\left(\mu_S \neq \mu_N \text{ or } \mu_S - \mu_N \neq 0\right)$

Independent groups assumption: Recall of one group should not affect recall of another.
Randomization condition: Subjects were randomly assigned to groups.
Nearly Normal condition: The samples are large.

Since the conditions are satisfied, it is appropriate to model the sampling distribution of the difference in means with a Student's *t*-model, with 214 degrees of freedom (from the approximation formula). We will perform a two-sample *t*-test.

The sampling distribution model has mean 0, with standard error:

$$SE(\bar{y}_S - \bar{y}_N) = \sqrt{\frac{1.76^2}{108} + \frac{1.77^2}{108}} \approx 0.24 \, .$$

The observed difference between the mean scores is 1.71 – 3.17 = – 1.46.

Since the *P*-value = 5.5×10^{-9} is low, we reject the null hypothesis. There is strong evidence that the mean number of brand names recalled is different for viewers of sexual content and viewers of neutral content. The evidence suggests that viewers of neutral ads remember more brand names on average than viewers of sexual content.

$$t = \frac{(\bar{y}_S - \bar{y}_N) - (0)}{SE(\bar{y}_S - \bar{y}_N)}$$

$$t \approx \frac{-1.46}{0.24}$$

$$t \approx -6.08$$

77. Hungry?

H_0: The mean number of ounces of ice cream people scoop is the same for large and small bowls. $\left(\mu_{big} = \mu_{small} \text{ or } \mu_{big} - \mu_{small} = 0\right)$

H_A: The mean number of ounces of ice cream people scoop is the different for large and small bowls. $\left(\mu_{big} \neq \mu_{small} \text{ or } \mu_{big} - \mu_{small} \neq 0\right)$

Independent groups assumption: The amount of ice cream scooped by individuals should be independent.
Randomization condition: Subjects were randomly assigned to groups.
Nearly Normal condition: Assume that this condition is met.

Since the conditions are satisfied, it is appropriate to model the sampling distribution of the difference in means with a Student's *t*-model, with 34 degrees of freedom (from the approximation formula). Perform a two-sample *t*-test.

The sampling distribution model has mean 0, with standard error:

$$SE(\bar{y}_{big} - \bar{y}_{small}) = \sqrt{\frac{2.91^2}{22} + \frac{1.84^2}{26}} \approx 0.7177 \text{ oz.}$$

The observed difference between the mean amounts is 6.58 – 5.07 = 1.51 oz.

$$t = \frac{(\bar{y}_{big} - \bar{y}_{small}) - (0)}{SE(\bar{y}_{big} - \bar{y}_{small})}$$

$$t \approx \frac{1.51}{0.7177}$$

$$t \approx 2.104$$

Since the *P*-value of 0.0428 is low, we reject the null hypothesis. There is strong evidence that the that the mean amount of ice cream people put into a bowl is related to the size of the bowl. People tend to put more ice cream into the large bowl, on average, than the small bowl.

79. Swim the Lake 2013 revisited.

a)

$$(\bar{y}_M - \bar{y}_W) \pm t^*_{df} \sqrt{\frac{s_M^2}{n_M} + \frac{s_W^2}{n_W}}$$

$$= (1226.04 - 1257.09) \pm t^*_{35.7} \sqrt{\frac{382.85^2}{23} + \frac{261.10^2}{34}} \approx (-216.74, 154.64)$$

If the assumptions and conditions are met, we can be 95% confident that the interval –216.74 to 154.64 minutes contains the true difference in mean crossing times between men and women. Because the interval includes zero, we cannot be confident that there is any difference at all.

b) **Independent groups assumption:** The times from the two groups are likely to be independent of one another, provided that these were all individual swims.
Randomization condition: The times are not a random sample from any identifiable population, but it is likely that the times are representative of times from swimmers who might attempt a challenge such as this. Hopefully, these times were recorded from different swimmers.
Nearly Normal condition: The boxplots show two high outliers for the men and some skewness for both. Removing the outliers may make the difference in times larger, but there is no justification for doing so. The histograms are unimodal; but somewhat skewed to the right.

We are reluctant to draw any conclusions about the difference in mean times it takes men or women to swim the lake. The sample is not random, we have no way of knowing if it is representative, and the data are skewed with some outliers.

81. Running heats London.

H_0: The mean time to finish is the same for heats 2 and 5. $\left(\mu_2 = \mu_5 \text{ or } \mu_2 - \mu_5 = 0\right)$

H_A: The mean time is not the same for heats 2 and 5. $\left(\mu_2 \neq \mu_5 \text{ or } \mu_2 - \mu_5 \neq 0\right)$

Independent groups assumption: The two heats were independent.
Randomization condition: Runners were randomly assigned.
Nearly Normal condition: The boxplots show an outlier in the distribution of times in heat 2. We will perform the test twice, with and without the outlier.

Since the conditions are satisfied, it is appropriate to model the sampling distribution of the difference in means with a Student's *t*-model, with 8.39 degrees of freedom (from the approximation formula). We will perform a two-sample *t*-test.

The sampling distribution model has mean 0, with standard error:

$$SE(\bar{y}_2 - \bar{y}_5) = \sqrt{\frac{3.3618249^2}{6} + \frac{2.0996325^2}{6}} \approx 1.6181 .$$

The observed difference between mean times is
53.2267 – 53.1483 = 0.0784.

$$t = \frac{(\bar{y}_2 - \bar{y}_5) - (0)}{SE(\bar{y}_2 - \bar{y}_5)}$$

Since the *P*-value = 0.9625 is high, we fail to reject the null hypothesis. There is no evidence that the mean time to finish differs between the two heats.

$$t \approx \frac{0.0784}{1.6181}$$

$$t \approx 0.048$$

Without the outlier, it is appropriate to model the sampling distribution of the difference in means with a Student's *t*-model, with 8.79 degrees of freedom (from the approximation formula). We will perform a two-sample *t*-test.

The sampling distribution model has mean 0, with standard error:

$$SE(\bar{y}_2 - \bar{y}_5) = \sqrt{\frac{1.4601267^2}{5} + \frac{2.0996325^2}{6}} \approx 1.0776 .$$

The observed difference between mean times is
51.962 – 53.1483 = –1.1863.

$$t = \frac{(\bar{y}_2 - \bar{y}_5) - (0)}{SE(\bar{y}_2 - \bar{y}_5)}$$

Since the *P*-value = 0.3001 is high, we fail to reject the null hypothesis. There is no evidence that the mean time to finish differs between the two heats.

$$t \approx \frac{-1.1863}{1.0776}$$

$$t \approx -1.10$$

There are only two values that exceed the maximum and minimum in the other set, so Tukey's test does no find a significant difference in mean 400 meter times between the two heats.

The Rank-sum test has a *P*-value of 0.70, so there is no evidence of a difference in mean 400 meter times here, either.

83. Tees.

H$_0$: The mean ball velocity is the same for regular and Stinger tees.
$$(\mu_S = \mu_R \text{ or } \mu_S - \mu_R = 0)$$

H$_A$: The mean ball velocity is higher for the Stinger tees. $(\mu_S > \mu_R \text{ or } \mu_S - \mu_R > 0)$

Assuming the conditions are satisfied, it is appropriate to model the sampling distribution of the difference in means with a Student's *t*-model, with 7.03 degrees of freedom (from the approximation formula). We will perform a two-sample *t*-test.

The sampling distribution model has mean 0, with standard error:

$$SE(\bar{y}_S - \bar{y}_R) = \sqrt{\frac{0.41^2}{6} + \frac{0.89^2}{6}} \approx 0.4000 .$$

The observed difference between the mean velocities is 128.83 – 127 = 1.83.

$$t = \frac{(\bar{y}_S - \bar{y}_R) - (0)}{SE(\bar{y}_S - \bar{y}_R)}$$

Since the P-value = 0.0013, we reject the null hypothesis. There is strong evidence that the mean ball velocity for stinger tees is higher than the mean velocity for regular tees.

$$t \approx \frac{1.83}{0.4000}$$

$$t \approx 4.57$$

85. Music and memory.

a) H_0: The mean memory test score is the same for those who listen to Mozart as it is for those who listen to rap music. $(\mu_M = \mu_R$ or $\mu_M - \mu_R = 0)$

H_A: The mean memory test score is greater for those who listen to Mozart than it is for those who listen to rap music. $(\mu_M > \mu_R$ or $\mu_M - \mu_R > 0)$

Independent groups assumption: The groups' memory scores are not related.
Randomization condition: Subjects were randomly assigned to groups.
Nearly Normal condition: We don't have the actual data. We will assume that the distributions of the populations of memory test scores are Normal.

Since the conditions are satisfied, it is appropriate to model the sampling distribution of the difference in means with a Student's t-model, with 45.88 df (from the approx. formula). We will perform a two-sample t-test.

The sampling distribution model has mean 0, with standard error:

$$SE(\bar{y}_M - \bar{y}_R) = \sqrt{\frac{3.19^2}{20} + \frac{3.99^2}{29}} \approx 1.0285.$$

The observed difference between the mean number of objects remembered is 10.0 – 10.72 = – 0.72.

$$t = \frac{(\bar{y}_M - \bar{y}_R) - (0)}{SE(\bar{y}_M - \bar{y}_R)}$$

$$t \approx \frac{-0.72}{1.0285}$$

$$t \approx -0.70$$

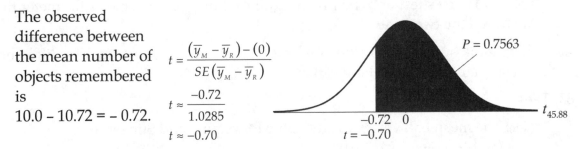

$P = 0.7563$

-0.72 0

$t = -0.70$

$t_{45.88}$

Since the P-value = 0.7563 is high, we fail to reject the null hypothesis. There is no evidence that the mean number of objects remembered by those who listen to Mozart is higher than the mean number of objects remembered by those who listen to rap music.

b) $(\bar{y}_M - \bar{y}_N) \pm t^*_{df} \sqrt{\frac{s_M^2}{n_M} + \frac{s_N^2}{n_N}} = (10.0 - 12.77) \pm t^*_{19.09} \sqrt{\frac{3.19^2}{20} + \frac{4.73^2}{13}} \approx (-5.351, -0.189)$

We are 90% confident that the mean number of objects remembered by those who listen to Mozart is between 0.189 and 5.352 objects lower than the mean of those who listened to no music.

Chapter 23 – Paired Samples and Blocks

Section 23.1

1. Which method?

a) Paired. Each individual has two scores, so they are certainly associated.

b) Not paired. The scores of males and females are independent.

c) Paired. Each student was surveyed twice, so there responses are associated.

Section 23.2

3. Cars and trucks.

a) No. The vehicles have no natural pairing.

b) Possibly. The data are quantitative and paired by vehicle.

c) The sample size is large, but there is at least one extreme outlier that should be investigated before applying these methods.

Section 23.3

5. Cars and trucks again.

$$\bar{d} \pm t^*_{n-1}\left(\frac{s_d}{\sqrt{n}}\right) = 7.37 \pm t^*_{631}\left(\frac{2.52}{\sqrt{632}}\right) \approx (7.17, 7.57)$$

We are 95% confident that the interval 7.17 mpg to 7.57 mpg captures the true improvements in highway gas mileage compared to city gas mileage.

Section 23.4

7. Blocking cars and trucks.

The difference between fuel efficiency of cars and that of trucks can be large, but isn't relevant to the question asked about highway vs. city driving. Pairing places each vehicle in its own block to remove that variation from consideration.

Chapter Exercises

9. More eggs?

a) Randomly assign 50 hens to each of the two kinds of feed. Compare the mean egg production of the two groups at the end of one month.

b) Randomly divide the 100 hens into two groups of 50 hens each. Feed the hens in the first group the regular feed for two weeks, then switch to the additive for 2 weeks. Feed the hens in the second group the additive for two weeks, and then switch to the regular feed for two weeks. Subtract each hen's "regular" egg production from her "additive" egg production, and analyze the mean difference in egg production.

$$
\text{100 hens} \underset{\text{Random}}{<}
\begin{array}{l}
\text{Group 1} \;-\; \begin{array}{l}\text{additive first 2 weeks}\\ \text{regular second 2 weeks}\end{array} \\[1em]
\text{Group 2} \;-\; \begin{array}{l}\text{regular first 2 weeks}\\ \text{additive second 2 weeks}\end{array}
\end{array}
\;>\; \begin{array}{l}\text{Analyze differences in}\\ \text{egg production}\end{array}
$$

c) The matched pairs design in part b is the stronger design. Hens vary in their egg production regardless of feed. This design controls for that variability by matching the hens with themselves.

11. Sex sells?

a) Randomly assign half of the volunteers to watch ads with sexual images, and assign the other half to watch ads without the sexual images. Record the number of items remembered. Then have each group watch the other type of ad. Record the number of items recalled. Examine the difference in the number of items remembered for each person.

$$
\text{Volunteers} \underset{\text{Random}}{<}
\begin{array}{l}
\text{Group 1} \;-\; \text{Ads with sexual images first} \\[1em]
\text{Group 2} \;-\; \text{Ads with no sexual images first}
\end{array}
\;>\; \begin{array}{l}\text{Analyze mean}\\ \text{difference in number}\\ \text{of items remembered}\end{array}
$$

b) Randomly assign half of the volunteers to watch ads with sexual images, and assign the other half to watch ads without the sexual images. Record the number of items remembered. Compare the mean number of products remembered by each group.

$$
\text{Volunteers} \underset{\text{Random}}{<}
\begin{array}{l}
\text{Group 1} \;-\; \text{Ads with sexual images} \\[1em]
\text{Group 2} \;-\; \text{Ads with no sexual images}
\end{array}
\;>\; \begin{array}{l}\text{Compare mean}\\ \text{number of items}\\ \text{remembered}\end{array}
$$

13. Women.

a) The paired *t*-test is appropriate. The labor force participation rate for two different years was paired by city.

b) Since the *P*-value = 0.0244, there is evidence of a difference in the average labor force participation rate for women between 1968 and 1972. The evidence suggests an increase in the participation rate for women.

15. Friday the 13th, traffic.

a) The paired *t*-test is appropriate, since we have pairs of Fridays in 5 different months. Data from adjacent Fridays within a month may be more similar than randomly chosen Fridays.

b) Since the *P*-value = 0.0212, there is evidence that the mean number of cars on the M25 motorway on Friday the 13th is less than the mean number of cars on the previous Friday.

c) We don't know if these Friday pairs were selected at random. Obviously, if these are the Fridays with the largest differences, this will affect our conclusion. The Nearly Normal condition appears to be met by the differences, but the sample size of five pairs is small.

17. Online insurance I.

Adding variances requires that the variables be independent. These price quotes are for the same cars, so they are paired. Drivers quoted high insurance premiums by the local company will be likely to get a high rate from the online company, too.

19. Online insurance II.

a) The histogram would help you decide whether the online company offers cheaper insurance. We are concerned with the difference in price, not the distribution of each set of prices.

b) Insurance cost is based on risk, so drivers are likely to see similar quotes from each company, making the differences relatively smaller.

c) The price quotes are paired. They were for a random sample of the agent's customers and the histogram of differences looks approximately Normal.

21. Online insurance III.

H_0: The mean difference between online and local insurance rates is zero. $\left(\mu_{Local-Online} = 0 \right)$

H_A: The mean difference is greater than zero. $\left(\mu_{Local-Online} > 0 \right)$

Since the conditions are satisfied (in a previous exercise), the sampling distribution of the difference can be modeled with a Student's *t*-model with 10 – 1 = 9 degrees of freedom, $t_9 \left(0, \dfrac{175.663}{\sqrt{10}} \right)$.

We will use a paired *t*-test, with $\bar{d} = 45.9$.

Since the *P*-value = 0.215 is high, we fail to reject the null hypothesis. There is no evidence that online insurance premiums are lower on average.

$$t = \frac{\bar{d} - 0}{\frac{s}{\sqrt{n}}}$$

$$t = \frac{45.9 - 0}{\frac{175.663}{\sqrt{10}}}$$

$$t \approx 0.83$$

$P = 0.215$

$0 \quad 45.9$

$t = 0.83$

t_9

23. City Temperatures.

a) **Paired data assumption:** The data are paired by city.
Randomization condition: These cities might not be representative of all cities, so be cautious in generalizing the results.
Normal population assumption: The histogram of differences between January and July mean temperature is roughly unimodal and symmetric, but

July - January Temperatures

shows a low outlier. This is Auckland, New Zealand, a city in the southern hemisphere. Seasons here would be the opposite of the rest of the cities, which are in the northern hemisphere. It should be set aside.

b) With Auckland set aside, the conditions are satisfied. The sampling distribution of the difference can be modeled with a Student's *t*-model with 11 – 1 = 10 degrees of freedom. We will find a paired *t*-interval, with 95% confidence.

$$\bar{d} \pm t^*_{n-1}\left(\frac{s_d}{\sqrt{n}}\right) = 33.7429 \pm t^*_{34}\left(\frac{14.9968}{\sqrt{34}}\right) \approx (30.94, 39.53)$$

We are 95% confident that the average high temperature in northern hemisphere cities in July is an average of between 30.94° to 39.53° higher than in January.

25. Push-ups.

Independent groups assumption: The group of boys is independent of the group of girls.
Randomization condition: Assume that students are assigned to gym classes at random.
Nearly Normal condition: The histograms of the number of push-ups from each group are roughly unimodal and symmetric.

Since the conditions are satisfied, it is appropriate to model the sampling distribution of the difference in means with a Student's *t*-model, with 21 degrees of freedom (from the approximation formula). We will construct a two-sample *t*-interval, with 90% confidence.

$$(\bar{y}_B - \bar{y}_G) \pm t^*_{df} \sqrt{\frac{s^2_B}{n_B} + \frac{s^2_G}{n_G}} = (23.8333 - 16.5000) \pm t^*_{21} \sqrt{\frac{7.20900^2}{12} + \frac{8.93919^2}{12}} \approx (1.6,\, 13.0)$$

We are 90% confident that, at Gossett High, the mean number of push-ups that boys can do is between 1.6 and 13.0 more than the mean for the girls.

27. Job satisfaction.

a) Use a paired *t*-test.

Paired data assumption: The data are before and after job satisfaction rating for the same workers.
Randomization condition: The workers were randomly selected to participate.
Nearly Normal conditon: The histogram of differences between before and after job satisfaction ratings is roughly unimodal and symmetric.

After-Before

b) H_0: The mean difference in before and after job satisfaction scores is zero, and the exercise program is not effective at improving job satisfaction. $(\mu_d = 0)$

H_A: The mean difference is greater than zero, and the exercise program is effective at improving job satisfaction. $(\mu_d > 0)$

Since the conditions are satisfied, the sampling distribution of the difference can be modeled with a Student's *t*-model with $10 - 1 = 9$ degrees of freedom, $t_9\left(0, \dfrac{7.47217}{\sqrt{10}}\right)$. We will use a paired *t*-test, with $\bar{d} = 8.5$.

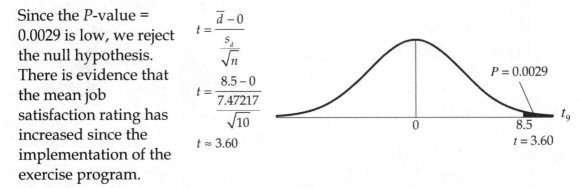

Since the *P*-value = 0.0029 is low, we reject the null hypothesis. There is evidence that the mean job satisfaction rating has increased since the implementation of the exercise program.

$$t = \frac{\bar{d} - 0}{\dfrac{s_d}{\sqrt{n}}}$$

$$t = \frac{8.5 - 0}{\dfrac{7.47217}{\sqrt{10}}}$$

$$t \approx 3.60$$

$P = 0.0029$

$t = 3.60$

c) We concluded that there was an increase job satisfaction rating. If we are wrong, and there actually was no increase, we have committed a Type I error.

d) The *P*-value of the sign test is 0.1094, which would lead us to fail to reject the null hypothesis that the median difference was 0. This is a different conclusion than the paired *t*-test. However, since the conditions for the paired *t*-test are met, we should use those results. The *t*-test has more power.

29. Yogurt.

H_0: The mean difference in calories between servings of strawberry and vanilla yogurt is zero. $(\mu_d = 0)$

H_A: The mean difference in calories between servings of strawberry and vanilla yogurt is different from zero. $(\mu_d \neq 0)$

Paired data assumption: The yogurt is paired by brand.
Randomization condition: Assume that these brands are representative of all brands.
Normal population assumption: The histogram of differences in calorie content between strawberry and vanilla shows an outlier, Great Value. When the outlier is eliminated, the histogram of differences is roughly unimodal and symmetric.

With outlier   Without outlier

When Great Value yogurt is removed, the conditions are satisfied. The sampling distribution of the difference can be modeled with a Student's *t*-model with

11 – 1 = 10 degrees of freedom, $t_{10}\left(0, \dfrac{18.0907}{\sqrt{11}}\right)$.

We will use a paired *t*-test, with $\bar{d} \approx 4.54545$.

Since the *P*-value = 0.4241 is high, we fail to reject the null hypothesis. There is no evidence of a mean difference in calorie content between strawberry yogurt and vanilla yogurt.

$$t = \frac{\bar{d} - 0}{\dfrac{s_d}{\sqrt{n}}}$$

$$t = \frac{4.54545 - 0}{\dfrac{18.0907}{\sqrt{11}}}$$

$$t \approx 0.833$$

31. Braking test.

a) **Randomization Condition:** These cars are not a random sample, but are probably representative of all cars in terms of stopping distance.
Nearly Normal Condition: A histogram of the stopping distances is skewed to the right, but this may just be sampling variation from a Normal population. The "skew" is only a couple of stopping distances. We will proceed cautiously.

The cars in the sample had a mean stopping distance of 138.7 feet and a standard deviation of 9.66149 feet. Since the conditions have been satisfied, construct a one-sample t-interval, with $10 - 1 = 9$ degrees of freedom, at 95% confidence.

$$\bar{y} \pm t^*_{n-1}\left(\frac{s}{\sqrt{n}}\right) = 138.7 \pm t^*_9\left(\frac{9.66149}{\sqrt{10}}\right) \approx (131.8,\ 145.6)$$

We are 95% confident that the mean dry pavement stopping distance for cars with this type of tires is between 131.8 and 145.6 feet. This estimate is based on an assumption that these cars are representative of all cars and that the population of stopping distances is Normal.

b) **Paired data assumption:** The data are paired by car.
Randomization condition: Assume that the cars are representative of all cars.
Normal population assumption: The difference in stopping distance for car #4 is an outlier, at only 12 feet. After excluding this difference, the histogram of differences is unimodal and symmetric.

Wet-Dry

Since the conditions are satisfied, the sampling distribution of the difference can be modeled with a Student's t-model with $9 - 1 = 8$ degrees of freedom. We will find a paired t-interval, with 95% confidence.

$$\bar{d} \pm t^*_{n-1}\left(\frac{s_d}{\sqrt{n}}\right) = 55 \pm t^*_8\left(\frac{10.2103}{\sqrt{9}}\right) \approx (47.2,\ 62.8)$$

With car #4 removed, we are 95% confident that the mean increase in stopping distance on wet pavement is between 47.2 and 62.8 feet. (If you leave the outlier in, the interval is 38.8 to 62.6 feet, but you should remove it! This procedure is sensitive to outliers!)

33. Tuition 2013.

a) **Paired data assumption:** The data are
paired by college.
Randomization condition: The
colleges were selected randomly.
Normal population assumption: The
tuition difference for UC Irvine, at
$22,878, is an outlier, as is the tuition
difference for New College of Florida
at $23,029. Once these have been set
aside, the histogram of the differences

is roughly unimodal and slightly skewed to the right. This should be fine for
inference in a sample of 17 colleges.

Since the conditions are satisfied, the sampling distribution of the difference can
be modeled with a Student's t-model with $17 - 1 = 16$ degrees of freedom. We
will find a paired t-interval, with 90% confidence.

$$\bar{d} \pm t^*_{n-1}\left(\frac{s_d}{\sqrt{n}}\right) = 7963.65 \pm t^*_{16}\left(\frac{3560.53}{\sqrt{17}}\right) \approx (6456, 9471)$$

b) With outliers removed, we are 90% confident that the mean increase in tuition
for nonresidents versus residents is between about $6456 and $9471. (If you left
the outliers in your data, the interval is about $7235 to $11,848, but you should
set them aside! This procedure is sensitive to the presence of outliers!)

c) There is no evidence to suggest that the magazine made a false claim. An
increase of $7000 for nonresidents is contained within our 90% confidence
interval.

35. Strikes.

a) Since 60% of 50 pitches is 30 pitches, the Little Leaguers would have to throw an
average of more than 30 strikes in order to give support to the claim made by the
advertisements.

H_0: The mean number of strikes thrown by Little Leaguers who have completed
the training is 30. $\left(\mu_A = 30\right)$

H_A: The mean number of strikes thrown by Little Leaguers who have completed
the training is greater than 30. $\left(\mu_A > 30\right)$

Randomization Condition: Assume that these players are representative of all Little League pitchers.
Nearly Normal Condition: The histogram of the number of strikes thrown after the training is roughly unimodal and symmetric.

After

The pitchers in the sample threw a mean of 33.15 strikes, with a standard deviation of 2.32322 strikes. Since the conditions for inference are satisfied, we can model the sampling distribution of the mean number of strikes thrown with a Student's *t* model, with 20 – 1 = 19 degrees of freedom, $t_{19}\left(30, \dfrac{2.32322}{\sqrt{20}}\right)$.

We will perform a one-sample *t*-test.

Since the *P*-value = 3.92×10^{-6} is very low, we reject the null hypothesis. There is strong evidence that the mean number of strikes that Little Leaguers can throw after the training is more than 30. (This test says nothing about the effectiveness of the training; just that Little Leaguers can throw more than 60% strikes on average after completing the training. This might not be an improvement.)

$$t = \frac{\bar{y}_A - \mu_0}{\dfrac{s_A}{\sqrt{n_A}}}$$

$$t = \frac{33.15 - 30}{\dfrac{2.32322}{\sqrt{20}}}$$

$$t = 6.06$$

b) H_0: The mean difference in number of strikes thrown before and after the training is zero. $\left(\mu_d = 0\right)$

H_A: The mean difference in number of strikes thrown before and after the training is greater than zero. $\left(\mu_d > 0\right)$

Paired data assumption: The data are paired by pitcher.
Randomization condition: Assume that these players are representative of all Little League pitchers.
Normal population assumption: The histogram of differences is roughly unimodal and symmetric.

After-Before

Since the conditions are satisfied, the sampling distribution of the difference can be modeled with a Student's *t*-model with 20 – 1 = 19 degrees of freedom, $t_{19}\left(0, \dfrac{3.32297}{\sqrt{19}}\right)$.

We will use a paired *t*-test, with $\bar{d} = 0.1$.

Since the *P*-value = 0.4472 is high, we fail to reject the null hypothesis. There is no evidence of a mean difference in number of strikes thrown before and after the training. The training does not appear to be effective.

$$t = \frac{\bar{d} - 0}{\frac{s_d}{\sqrt{n}}}$$

$$t = \frac{0.1 - 0}{\frac{3.32297}{\sqrt{20}}}$$

$$t \approx 0.135$$

$P = 0.4472$

$0 \quad 0.10$

$t = 0.135$

t_{19}

c) The sign test has a *P*-value of 0.7597, which would lead us to fail to reject the hypothesis that the median difference was 0. This conclusion is consistent with the paired *t*-test.

37. Wheelchair marathon 2013.

a) The data are certainly paired. Even if the individual times show a trend of improving speed over time, the differences may well be independent of each other. They are subject to random year-to-year fluctuations, and we may believe that these data are representative of similar races. We don't have any information with which to check the Nearly Normal condition.

b) $\bar{d} \pm t_{n-1}^{*}\left(\dfrac{s_d}{\sqrt{n}}\right) = -4.88 \pm t_{36}^{*}\left(\dfrac{35.227}{\sqrt{37}}\right) \approx (-16.63,\ 6.87)$

We are 95% confident that the interval –16.63 to 6.87 minutes contains the true mean time difference for women's wheelchair times and men's running times.

c) The interval contains zero, so we would not reject a null hypothesis of no mean difference at a significance level of 0.05. We are unable to discern a difference between the female wheelchair times and the male running times.

39. BST.

a) **Paired data assumption:** We are testing the same cows, before and after injections of BST.
Randomization condition: These cows are likely to be representative of all Ayrshires.
Normal population assumption: We don't have the list of individual differences, so we can't look at a histogram. The sample is large, so we may proceed.

Since the conditions are satisfied, the sampling distribution of the difference can be modeled with a Student's *t*-model with 60 – 1 = 59 degrees of freedom. We will find a paired *t*-interval, with 95% confidence.

b) $\bar{d} \pm t_{n-1}^{*}\left(\dfrac{s_d}{\sqrt{n}}\right) = 14 \pm t_{59}^{*}\left(\dfrac{5.2}{\sqrt{60}}\right) \approx (12.66,\ 15.34)$

c) We are 95% confident that the mean increase in daily milk production for Ayshire cows after BST injection is between 12.66 and 15.34 pounds.

d) 25% of 47 pounds is 11.75 pounds. According to the interval generated in part b, the average increase in milk production is more than this, so the farmer can justify the extra expense for BST.

Chapter 24 – Comparing Counts

Section 24.1

1. **Human births.**

   a) If there were no "seasonal effect" we would expects 25% of births to occur in each season. The expected number of births is 0.25(120) = 30 births per season.

   b) $\chi^2 = \dfrac{(25-30)^2}{30} + \dfrac{(35-30)^2}{30} + \dfrac{(32-30)^2}{30} + \dfrac{(28-30)^2}{30} \approx 1.933$

   c) There are 4 seasons, so there are 4 – 1 = 3 degrees of freedom.

3. **Human births, again.**

   a) The mean of the χ^2 distribution is the number of degrees of freedom, so we would expect the χ^2 statistic to be 3 if there were no seasonal effect.

   b) Since 1.933 is less than the mean of 3, it does not seem large in comparison.

   c) We should fail to reject the null hypothesis. These data do not provide evidence of a seasonal effect on human births.

   d) The critical value of χ^2 with 3 degrees of freedom and $\alpha = 0.05$ is 7.815.

   e) Since $\chi^2 = 1.933$ is less than the critical value, 7.815, we fail to reject the null hypothesis. There is no evidence of a seasonal effect on human births.

5. **Customer ages.**

   a) The null hypothesis is that the age distributions of the customers are the same at the two branches.

   b) There are three age groups and one variable, location of the branch. This is a χ^2 test of homogeneity.

   c) Expected Counts

	Age			
	Less than 30	30-55	56 or older	Total
In-Town Branch	25	45	30	100
Mall Branch	25	45	30	100
Total	50	90	60	200

   d) $\chi^2 = \dfrac{(20-25)^2}{25} + \dfrac{(40-45)^2}{45} + \dfrac{(40-30)^2}{30} + \dfrac{(30-25)^2}{25} + \dfrac{(50-45)^2}{45} + \dfrac{(20-30)^2}{30} \approx 9.778$

   e) There are 2 rows and 3 columns, so there are (2 – 1)(3 – 1) = 2 degrees of freedom.

f) The probability of having a χ^2 value over 9.778 with df = 2 is 0.0075.

g) Since the P-value is so low, reject the null hypothesis. There is strong evidence that the distribution of ages is not the same at the two branches. The mall branch had more customers under 30, and fewer customers 56 or older, than expected. The in-town branch had more customers 56 and older, and fewer customers under 30, than expected.

Chi-Square, df=2

Section 24.2

7. **Human births, last time.**

a) The standardized residuals are:

$$c_1 = \frac{(25-30)}{\sqrt{30}} \approx -0.913 \qquad c_2 = \frac{(35-30)}{\sqrt{30}} \approx 0.913$$

$$c_3 = \frac{(32-30)}{\sqrt{30}} \approx 0.365 \qquad c_4 = \frac{(28-30)}{\sqrt{30}} \approx -0.365$$

b) None of the standardized residuals are large. Since they are z-scores, they are actually quite small.

c) We did not reject the null hypothesis, so we should expect the standardized residuals to be relatively small.

Section 24.3

9. **Internet use poll.**

a) The null hypothesis is that the response to the question is independent of race.

Expected Counts	Did you use the Internet yesterday?		
	Yes	No	Total
White	2506.38	895.62	3402
Black	338.90	121.10	460
Hispanic/Other	445.73	159.27	605
Total	3291	1176	4467

b)

$$\chi^2 = \frac{(2546 - 2506.38)^2}{2506.38} + \frac{(856 - 895.62)^2}{895.62}$$
$$+ \frac{(314 - 338.90)^2}{338.90} + \frac{(146 - 121.10)^2}{121.10}$$
$$+ \frac{(431 - 445.73)^2}{445.73} + \frac{(174 - 159.27)^2}{159.27} \approx 11.176$$

c) There are 3 rows and 2 columns, so there are $(3 - 1)(2 - 1) = 2$ degrees of freedom.

d) The probability of having a χ^2 value over 11.176 with df = 2 is 0.0037.

e) Since the P-value is low, reject the null hypothesis. There is evidence of an association between the response to the question and race. Black respondents indicated that they had not used the Internet yesterday more frequently than we expected.

Chapter Exercises.

11. Which test?

a) Chi-square test of Independence. We have one sample and two variables. We want to see if the variable *account type* is independent of the variable *trade type*.

b) Some other statistics test. The variable *account size* is quantitative, not counts.

c) Chi-square test of Homogeneity. We have two samples (residential and non-residential students), and one variable, *courses*. We want to see if the distribution of *courses* is the same for the two groups.

13. Dice.

a) If the die were fair, you'd expect each face to show 10 times.

b) Use a chi-square test for goodness-of-fit. We are comparing the distribution of a single variable (outcome of a die roll) to an expected distribution.

c) H_0: The die is fair. (All faces have the same probability of coming up.)

H_A: The die is not fair. (Some faces are more or less likely to come up than others.)

d) **Counted data condition:** We are counting the number of times each face comes up.
Randomization condition: Die rolls are random and independent of each other.
Expected cell frequency condition: We expect each face to come up 10 times, and 10 is greater than 5.

e) Under these conditions, the sampling distribution of the test statistic is χ^2 on 6 – 1 = 5 degrees of freedom. We will use a chi-square goodness-of-fit test.

f)

Face	Observed	Expected	Residual = $(Obs - Exp)$	$(Obs - Exp)^2$	Component = $\dfrac{(Obs - Exp)^2}{Exp}$
1	11	10	1	1	0.1
2	7	10	– 3	9	0.9
3	9	10	– 1	1	0.1
4	15	10	5	25	2.5
5	12	10	2	4	0.4
6	6	10	– 4	16	1.6
					$\sum$ = 5.6

g) Since the *P*-value = 0.3471 is high, we fail to reject the null hypothesis. There is no evidence that the die is unfair.

15. Nuts.

a) The weights of the nuts are quantitative. Chi-square goodness-of-fit requires counts.

b) In order to use a chi-square test, you could count the number of each type of nut. However, it's not clear whether the company's claim was a percentage by number or a percentage by weight.

17. NYPD and race.

H_0: The distribution of ethnicities in the police department represents the distribution of ethnicities of the youth of New York City.

H_A: The distribution of ethnicities in the police department does not represent the distribution of ethnicities of the youth of New York City.

Counted data condition: The percentages reported must be converted to counts.
Randomization condition: Assume that the current NYPD is representative of recent departments with respect to ethnicity.
Expected cell frequency condition: The expected counts are all much greater than 5.

(Note: The observed counts should be whole numbers. They are actual policemen. The expected counts may be decimals, since they behave like averages.)

Ethnicity	Observed	Expected
White	16965	7644.852
Black	3796	7383.042
Latino	5001	8247.015
Asian	367	2382.471
Other	52	523.620

Under these conditions, the sampling distribution of the test statistic is χ^2 on 5 – 1 = 4 degrees of freedom. We will use a chi-square goodness-of-fit test.

$$\chi^2 = \sum_{all\,cells} \frac{(Obs - Exp)^2}{Exp} = \frac{(16965 - 7644.852)^2}{7644.852} + \frac{(3796 - 7383.042)^2}{7383.042} + \frac{(5001 - 8247.015)^2}{8247.015}$$

$$+ \frac{(367 - 2382.471)^2}{2382.471} + \frac{(52 - 523.620)^2}{523.620} \approx 16,500$$

With χ^2 of over 16,500, on 4 degrees of freedom, the *P*-value is essentially 0, so we reject the null hypothesis. There is strong evidence that the distribution of ethnicities of NYPD officers does not represent the distribution of ethnicities of the youth of New York City. Specifically, the proportion of white officers is much higher than the proportion of white youth in the community. As one might expect, there are also lower proportions of officers who are black, Latino, Asian, and other ethnicities than we see in the youth in the community.

19. Fruit flies.

a) H$_0$: The ratio of traits in this type of fruit fly is 9:3:3:1, as genetic theory predicts.

H$_A$: The ratio of traits in this type of fruit fly is not 9:3:3:1.

Counted data condition: The data are counts.
Randomization condition: Assume that these flies are representative of all fruit flies of this type.
Expected cell frequency condition: The expected counts are all greater than 5.

trait	Observed	Expected
YN	59	56.25
YS	20	18.75
EN	11	18.75
ES	10	6.25

Under these conditions, the sampling distribution of the test statistic is χ^2 on 4 – 1 = 3 degrees of freedom. We will use a chi-square goodness-of-fit test.

$$\chi^2 = \sum_{all\,cells} \frac{(Obs - Exp)^2}{Exp} = \frac{(59 - 56.25)^2}{56.25} + \frac{(20 - 18.75)^2}{18.75} + \frac{(11 - 18.75)^2}{18.75} + \frac{(10 - 6.25)^2}{6.25} \approx 5.671$$

With $\chi^2 \approx 5.671$, on 3 degrees of freedom, the *P*-value = 0.1288 is high, so we fail to reject the null hypothesis. There is no evidence that the ratio of traits is different than the theoretical ratio predicted by the genetic model. The observed results are consistent with the genetic model.

b) With $\chi^2 \approx 11.342$, on 3 degrees of freedom, the P-value = 0.0100 is low, so we reject the null hypothesis. There is strong evidence that the ratio of traits is different than the theoretical ratio predicted by the genetic model. Specifically, there is evidence that the normal wing length occurs less frequently than expected and the short wing length occurs more frequently than expected.

trait	Observed	Expected
YN	118	112.5
YS	40	37.5
EN	22	37.5
ES	20	12.5

c) At first, this seems like a contradiction. We have two samples with exactly the same ratio of traits. The smaller of the two provides no evidence of a difference, yet the larger one provides strong evidence of a difference. This is explained by the sample size. In general, large samples decrease the proportion of variation from the true ratio. Because of the relatively small sample in the first test, we are unwilling to say that there is a difference. There just isn't enough evidence. But the larger sample allows us to be more certain about the difference.

21. Hurricane frequencies.

a) We would expect 96/16 = 6 hurricanes per time period.

b) We are comparing the distribution of the number of hurricanes, a single variable, to a theoretical distribution. A Chi-square test for goodness-of-fit is appropriate.

c) H_0: The number of large hurricanes remains constant over decades.

H_A: The number of large hurricanes has changed.

d) There are 16 time periods, so there are 16 – 1 = 15 degrees of freedom.

e) $P(\chi^2_{df=15} > 12.67) \approx 0.63$

f) The very high P-value means that these data offer no evidence that the number of hurricanes large hurricanes has changed.

23. Childbirth, part 1.

a) There are two variables, breastfeeding and having an epidural, from a single group of births. We will perform a Chi-square test for Independence.

b) H_0: Breastfeeding success is independent of having an epidural.

H_A: There is an association between breastfeeding success and having an epidural.

25. Childbirth, part 2.

a) The table has 2 rows and 2 columns, so there are $(2-1) \times (2-1) = 1$ degree of freedom.

b) We expect $\dfrac{474}{1178} \approx 40.2\%$ of all babies to not be breastfeeding after 6 months, so we expect that 40.2% of the 396 epidural babies, or 159.34, to not be breastfeeding after 6 months.

c) Breastfeeding behavior should be independent for these babies. They are fewer than 10% of all babies, and we assume they are representative of all babies. We have counts, and all the expected cells are at least 5.

27. Childbirth, part 3.

a) $\dfrac{(Obs - Exp)^2}{Exp} = \dfrac{(190 - 159.34)^2}{159.34} = 5.90$ **b)** $P(\chi^2_{df=1} > 14.87) < 0.005$

c) The *P*-value is very low, so reject the null hypothesis. There's strong evidence of an association between having an epidural and subsequent success in breastfeeding.

29. Childbirth, part 4.

a) $c = \dfrac{Obs - Exp}{\sqrt{Exp}} = \dfrac{190 - 159.34}{\sqrt{159.34}} = 2.43$

b) It appears that babies whose mothers had epidurals during childbirth are much more likely to be breastfeeding 6 months later.

31. Childbirth, part 5.

These factors would not have been mutually exclusive. There would be yes or no responses for every baby for each.

33. Titanic.

a) $P(\text{crew}) = \dfrac{885}{2201} \approx 0.402$ **b)** $P(\text{third and alive}) = \dfrac{178}{2201} \approx 0.081$

c) $P(\text{alive} \mid \text{first}) = \dfrac{P(\text{alive and first})}{P(\text{first})} = \dfrac{202/2201}{325/2201} = \dfrac{202}{325} \approx 0.622$

d) The overall chance of survival is $\dfrac{710}{2201} \approx 0.323$, so we would expect about 32.3% of the crew, or about 285.48 members of the crew, to survive.

e) H_0: Survival was independent of status on the ship.

H_A: Survival depended on status on the ship.

f) The table has 2 rows and 4 columns, so there are $(2-1) \times (4-1) = 3$ degrees of freedom.

g) With $\chi^2 \approx 187.8$, on 3 degrees of freedom, the *P*-value is essentially 0, so we reject the null hypothesis. There is strong evidence survival depended on status. First-class passengers were more likely to survive than any other class or crew.

35. Titanic again.

First class passengers were most likely to survive, while third class passengers and crew were under-represented among the survivors.

37. Cranberry juice.

a) This is an experiment. Volunteers were assigned to drink a different beverage.

b) We are concerned with the proportion of urinary tract infections among three different groups. We will use a chi-square test for homogeneity.

c) H$_0$: The proportion of urinary tract infection is the same for each group.

H$_A$: The proportion of urinary tract infection is different among the groups.

d) **Counted data condition:** The data are counts.
Randomization condition: Although not specifically stated, we will assume that the women were randomly assigned to treatments.
Expected cell frequency condition: The expected counts are all greater than 5.

	Cranberry (Obs/Exp)	Lactobacillus (Obs/Exp)	Control (Obs/Exp)
Infection	8 / 15.333	20 / 15.333	18 / 15.333
No infection	42 / 34.667	30 / 34.667	32 / 34.667

e) The table has 2 rows and 3 columns, so there are $(2-1)\times(3-1)=2$ degrees of freedom.

f) $\chi^2 = \sum_{all\,cells} \frac{(Obs - Exp)^2}{Exp} \approx 7.776$

P-value ≈ 0.020.

g) Since the *P*-value is low, we reject the null hypothesis. There is strong evidence of difference in the proportion of urinary tract infections for cranberry juice drinkers, lactobacillus drinkers, and women that drink neither of the two beverages.

h) A table of the standardized residuals is below, calculated by using $c = \dfrac{Obs - Exp}{\sqrt{Exp}}$.

	Cranberry	Lactobacillus	Control
Infection	–1.87276	1.191759	0.681005
No infection	1.245505	–0.79259	–0.45291

There is evidence that women who drink cranberry juice are less likely to develop urinary tract infections, and women who drank lactobacillus are more likely to develop urinary tract infections.

39. Montana.

a) We have one group, categorized according to two variables, political party and being male or female, so we will perform a chi-square test for independence.

b) H_0: Political party is independent of being male or female in Montana.

H_A: There is an association between political party and being male or female in Montana.

c) **Counted data condition:** The data are counts.
Randomization condition: Although not specifically stated, we will assume that the poll was conducted randomly.
Expected cell frequency condition: The expected counts are all greater than 5.

	Democrat (Obs / Exp)	Republican (Obs / Exp)	Independent (Obs / Exp)
Male	36 / 43.663	45 / 40.545	24 / 20.792
Female	48 / 40.337	33 / 37.455	16 / 19.208

d) Under these conditions, the sampling distribution of the test statistic is χ^2 on 2 degrees of freedom. We will use a chi-square test for independence.

$$\chi^2 = \sum_{all\,cells} \frac{(Obs - Exp)^2}{Exp} \approx 4.851$$

The *P*-value ≈ 0.0884

e) Since the *P*-value ≈ 0.0884 is fairly high, we fail to reject the null hypothesis. There is little evidence of an association between being male or female and political party in Montana.

41. Montana revisited.

H_0: Political party is independent of region in Montana.

H_A: There is an association between political party and region in Montana.

Counted data condition: The data are counts.
Randomization condition: Although not specifically stated, we will assume that the poll was conducted randomly.

Expected cell frequency condition: All expected counts are greater than 5.

	Democrat (Obs / Exp)	Republican (Obs / Exp)	Independent (Obs / Exp)
West	39 / 28.277	17 / 26.257	12 / 13.465
Northeast	15 / 23.703	30 / 22.01	12 / 11.287
Southeast	30 /32.02	31 / 29.733	16 / 15.248

Under these conditions, the sampling distribution of the test statistic is χ^2 on 4 degrees of freedom. We will use a chi-square test for independence.

$$\chi^2 = \sum_{all\,cells} \frac{(Obs - Exp)^2}{Exp} \approx 13.849,$$

and the *P*-value ≈ 0.0078

Since the *P*-value ≈ 0.0078 is low, reject the null hypothesis. There is strong evidence of an association between region and political party in Montana. Residents in the West are more likely to be Democrats than Republicans, and residents in the Northeast are more likely to be Republicans than Democrats.

43. Grades.

a) We have two groups, students of Professor Alpha and students of Professor Beta, and we are concerned with the distribution of one variable, grade. We will perform a chi-square test for homogeneity.

b) H$_0$: The distribution of grades is the same for the two professors.

   H$_A$: The distribution of grades is different for the two professors.

c) The expected counts are organized in the table below:

	Prof. Alpha	Prof. Beta
A	6.667	5.333
B	12.778	10.222
C	12.222	9.778
D	6.111	4.889
F	2.222	1.778

Since three cells have expected counts less than 5, the chi-square procedures are not appropriate. Cells would have to be combined in order to proceed. (We will do this in another exercise.)

45. Grades again.

a) **Counted data condition:** The data are counts.

Randomization condition: Assume that these students are representative of all students that have ever taken courses from the professors.

Expected cell frequency condition: The expected counts are all greater than 5.

	Prof. Alpha (Obs / Exp)	Prof. Beta (Obs / Exp)
A	3 / 6.667	9 / 5.333
B	11 / 12.778	12 / 10.222
C	14 / 12.222	8 / 9.778
Below C	12 / 8.333	3 / 6.667

b) Under these conditions, the sampling distribution of the test statistic is χ^2 on 3 degrees of freedom, instead of 4 degrees of freedom before the change in the table. We will use a chi-square test for homogeneity.

c) With $\chi^2 = \sum_{all\,cells} \dfrac{(Obs - Exp)^2}{Exp} \approx 9.306$,

the P-value ≈ 0.0255.

$P = 0.0255$, $df = 3$, χ^2, 0, 9.306

Since the P-value = 0.0255 is low, we reject the null hypothesis. There is evidence that the grade distributions for the two professors are different. Professor Alpha gives fewer As and more grades below C than Professor Beta.

47. Racial steering.

H_0: There is no association between race and section of the complex in which people live.

H_A: There is an association between race and section of the complex in which people live.

Counted data condition: The data are counts.

Randomization condition: Assume that the recently rented apartments are representative of all apartments in the complex.

	White (Obs / Exp)	Black (Obs / Exp)
Section A	87 / 76.179	8 / 18.821
Section B	83 / 93.821	34 / 23.179

Expected cell frequency condition: The expected counts are all greater than 5.

Under these conditions, the sampling distribution of the test statistic is χ^2 on 1 degree of freedom. We will use a chi-square test for independence.

$$\chi^2 = \sum_{all\,cells} \dfrac{(Obs - Exp)^2}{Exp} \approx \dfrac{(87 - 76.179)^2}{76.179} + \dfrac{(8 - 18.821)^2}{18.821} + \dfrac{(83 - 93.821)^2}{93.821} + \dfrac{(34 - 23.179)^2}{23.179}$$

$$\approx 1.5371 + 6.2215 + 1.2481 + 5.0517$$

$$\approx 14.058$$

With $\chi^2 \approx 14.058$, on 1 degree of freedom, the *P*-value ≈ 0.0002.

Since the *P*-value ≈ 0.0002 is low, we reject the null hypothesis. There is strong evidence of an association between race and the section of the apartment complex in which people live. An examination of the components shows us that whites are much more likely to rent in Section A (component = 6.2215), and blacks are much more likely to rent in Section B (component = 5.0517).

49. Steering revisited.

a) H₀: The proportion of whites who live in Section A is the same as the proportion of blacks who live in Section A. $\left(p_W = p_B \;\; \text{or} \;\; p_W - p_B = 0 \right)$

H_A: The proportion of whites who live in Section A is different than the proportion of blacks who live in Section A. $\left(p_W \neq p_B \;\; \text{or} \;\; p_W - p_B \neq 0 \right)$

Independence assumption: Assume that people rent independent of the section.
Independent samples condition: The groups are not associated.
Success/Failure condition: $n\hat{p}$ (white) = 87, $n\hat{q}$ (white) = 83, $n\hat{p}$ (black) = 8, and $n\hat{q}$ (black) = 34. These are not all greater than 10, since the number of black renters in Section A is only 8, but it is close to 10, and the others are large. It should be safe to proceed.

Since the conditions have been satisfied, we will model the sampling distribution of the difference in proportion with a Normal model with mean 0 and standard deviation estimated by

$$SE_{pooled}\left(\hat{p}_W - \hat{p}_B \right) = \sqrt{\frac{\hat{p}_{pooled}\hat{q}_{pooled}}{n_W} + \frac{\hat{p}_{pooled}\hat{q}_{pooled}}{n_B}} = \sqrt{\frac{\left(\frac{95}{212}\right)\left(\frac{117}{212}\right)}{170} + \frac{\left(\frac{95}{212}\right)\left(\frac{117}{212}\right)}{42}} \approx 0.0856915.$$

The observed difference between the proportions is:
$0.5117647 - 0.1904762 = 0.3212885$.

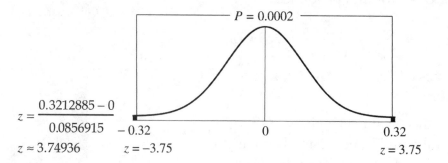

$z = \dfrac{0.3212885 - 0}{0.0856915}$

$z \approx 3.74936$

(You have to use a *ridiculous* number of decimal places to get this to come out "right". This is to done to illustrate the point of the question. DO NOT DO THIS! Use technology.)

Since the P-value = 0.0002 is low, we reject the null hypothesis. There is strong evidence of a difference in the proportion of whites and blacks living in Section A. The evidence suggests that the proportion of all whites living in Section A is much higher than the proportion of all black residents living in Section A.

The value of z for this test was approximately 3.74936. $z^2 \approx (3.74936)^2 \approx 14.058$, the same as the value for χ^2 in the previous exercise.

b) The resulting P-values were both approximately 0.0002. The two tests are equivalent.

51. Pregnancies.

H_0: Pregnancy outcome is independent of age.

H_A: There is an association between pregnancy outcome and age.

Counted data condition: The data are counts.
Randomization condition: Assume that these women are representative of all pregnant women.
Expected cell frequency condition: The expected counts are all greater than 5.

	Early Preterm (Obs/Exp)	Late Preterm (Obs/Exp)
Under 20	129 / 116.55	270 / 282.45
20 – 29	243 / 249.75	612 / 605.25
30 – 39	165 / 172.05	424 / 416.95
40 and over	18 / 16.65	39 / 40.35

Under these conditions, the sampling distribution of the test statistic is χ^2 on 3 degrees of freedom. We will use a chi-square test for independence.

With $\chi^2 = \sum_{all\,cells} \dfrac{(Obs - Exp)^2}{Exp} \approx 2.699$, on 3 degrees of freedom, the P-value $\approx$ 0.440.

Since the P-value is high, we fail to reject the null hypothesis. There is no evidence of an association between pregnancy outcome and age.

Chi-Square, df=3

Chapter 25 – Inferences for Regression

Section 25.2

1. Graduation rates 2013.

The scatterplot shows a linear relationship with equal spread about a regression line throughout the range of *Acceptance Rates*. The residual plot has no structure, and there are not any striking outliers. The histogram of residuals is symmetric and bell-shaped. All conditions are satisfied to proceed with the regression analysis.

Section 25.3

3. Graduation rates 2013, part II.

The error standard deviation is $s = 00254$ or 2.54 percentage points. Since the range of *Graduation Rates* is only about 17 percentage points, this number is fairly small and helps us to understand the amount of spread about the regression model. The standard deviation of the residuals is only about 2.54 percentage points.

5. Graduation rates 2013, part III.

The standard error for the slope tells us how much the slope of the regression equation would vary from sample to sample. If we took many samples, and found a regression equation for each, we would expect most of these slopes to be within a couple of standard deviations of the true slope of the association between graduation rate and acceptance rate.

Section 25.4

7. Graduation 2013, part IV.

Since the *P*-value is so low, we reject the null hypothesis. The administrators can conclude that there is evidence of a relationship between *Graduation Rates* and *Admission Rates* for these prestigious schools. It seems the lower the *Admission Rate*, the higher the *Graduation Rate*.

9. Graduation 2013, part V.

The administrators can conclude, with 95% confidence, that schools with *Admission Rates* that are lower by 1 percentage point will have, on average, *Graduation Rates* that are higher by between 0.238 and 0.332 percentage points.

Section 25.5

11. Graduation 2013, part VI.

The standard error for the mean will be smaller. We can be much more precise about the prediction for the mean graduation rate of all colleges compared to the prediction for just one.

Section 25.6

13. Graduation 2013, part VII.

The administrators can expect, with 95% confidence, that schools that admit 33% of applicants will have average graduation rates between 87.35% and 88.88%.

15. Graduation 2013, part VIII.

The administrators can be 95% confident that if their *Admission Rate* is 33%, their *Graduation Rate* would be between 82.96% and 93.28%.

Chapter Exercises.

17. Graduation 2013 party.

The sentence in the exercise is implying a cause-and-effect relationship. It seems to suggest that if they reduce the admission rate, then the graduation rate must increase. These data don't support that strong a conclusion. We have no evidence of what effects a change in admission rate might have.

19. Tracking hurricanes 2012.

a) The equation of the line of best fit for these data points is

$24\widehat{Error} = 133.0 - 2.06(Year)$, where *Year* is measured in years since 1970.

According to the linear model, the error made in predicting a hurricane's path was about 133 nautical miles, on average, in 1970. It has been declining at rate of about 2.06 nautical miles per year.

b) H_0: There has been no change in prediction accuracy. $(\beta_1 = 0)$

H_A: There has been a change in prediction accuracy. $(\beta_1 \neq 0)$

c) Assuming the conditions have been met, the sampling distribution of the regression slope can be modeled by a Student's *t*-model with 43 – 2 = 41 degrees of freedom. We will use a regression slope *t*-test.

The value of $t = -10.4$. The *P*-value ≤ 0.0001 means that the association we see in the data is unlikely to occur by chance. We reject the null hypothesis, and conclude that there is strong evidence that the prediction accuracies have in fact been changing during the time period.

d) 72.5% of the variation in the prediction accuracy is accounted for by the linear model based on year.

21. Movie budgets.

a) $\widehat{Budget} = -63.9981 + 1.02648(RunTime)$. The model suggests that each additional minute of run time for a movie costs about $1,026,000.

b) A negative intercept makes no sense, but the *P*-value of 0.07 indicates that we can't discern a difference between our estimated value and zero. It makes sense that a movie of zero length should cost $0.

c) Amounts by which movie costs differ from predictions made by this model vary, with a standard deviation of about $33 million.

d) The standard error of the slope is 0.1541 million dollars per minute.

e) If we constructed other models based on different samples of movies, we'd expect the slopes of the regression lines to vary, with a standard deviation of about $154,000 per minute.

23. Movie budgets, the sequel.

a) **Straight enough condition:** The scatterplot is straight enough, and the residuals plot looks unpatterned.
Independence assumption: The residuals plot shows no evidence of dependence.
Does the plot thicken? condition: The residuals plot shows no obvious trends in the spread.
Nearly Normal condition, Outlier condition: The Normal probability plot is reasonably straight.

b) Since conditions have been satisfied, the sampling distribution of the regression slope can be modeled by a Student's *t*-model with (120 – 2) = 118 degrees of freedom.

$$b_1 \pm t^*_{n-2} \times SE(b_1) = 1.02648 \pm (t^*_{118}) \times 0.1541 \approx (0.72, \, 1.33)$$

We are 95% confident that the cost of making longer movies increases at a rate of between 0.72 and 1.33 million dollars per minute.

25. Hot dogs.

a) H_0: There's no association between calories and sodium content of all-beef hot dogs. $(\beta_1 = 0)$

H_A: There is an association between calories and sodium content. $(\beta_1 \neq 0)$

b) Assuming the conditions have been met, the sampling distribution of the regression slope can be modeled by a Student's *t*-model with (13 – 2) = 11 degrees of freedom. We will use a regression slope *t*-test. The equation of the line of best fit for these data points is: $\widehat{Sodium} = 90.9783 + 2.29959(Calories)$

The value of $t = 4.10$. The P-value of 0.0018 means that the association we see in the data is very unlikely to occur by chance alone. We reject the null hypothesis, and conclude that there is evidence of a linear association between the number of calories in all-beef hotdogs and their sodium content. Because of the positive slope, there is evidence that hot dogs with more calories generally have higher sodium contents.

27. Second frank.

a) Among all-beef hot dogs with the same number of calories, the sodium content varies, with a standard deviation of about 60 mg.

b) The standard error of the slope of the regression line is 0.5607 milligrams of sodium per calorie.

c) If we tested many other samples of all-beef hot dogs, the slopes of the resulting regression lines would be expected to vary, with a standard deviation of about 0.56 mg of sodium per calorie.

29. Last dog.

$$b_1 \pm t_{n-2}^* \times SE(b_1) = 2.29959 \pm (2.201) \times 0.5607 \approx (1.06, 3.54)$$

We are 95% confident that for every additional calorie, all-beef hot dogs have, on average, between 1.06 and 3.54 mg more sodium.

31. Marriage age 2011.

a) H_0: The difference in age between men and women at first marriage has not been decreasing since 1975. $(\beta_1 = 0)$

H_A: The difference in age between men and women at first marriage has been decreasing since 1975. $(\beta_1 < 0)$

b) **Straight enough condition:** The scatterplot is straight enough.
Independence assumption: We are examining a relationship over time, so there is reason to be cautious, but the residuals plot shows no evidence of dependence.
Does the plot thicken? condition: The residuals plot shows no obvious trends in the spread.
Nearly Normal condition, Outlier condition: The histogram is reasoanbly unimodal and symmetric, and shows no obvious skewness or outliers.

c) Since conditions have been satisfied, the sampling distribution of the regression slope can be modeled by a Student's t-model with $37 - 2 = 35$ degrees of freedom. We will use a regression slope t-test. The equation of the line of best fit for these data points is: $(\overline{Men - Women}) = 42.566 - 0.0203(Year)$

The value of $t = -6.14$. The P-value of less than 0.0001 (even though this is the value for a two-tailed test, it is still very small) means that the association we see in the data is unlikely to occur by chance. We reject the null hypothesis, and conclude that there is strong evidence of a negative linear relationship between difference in age at first marriage and year. The difference in marriage age between men and women appears to be decreasing over time.

33. Marriage age 2011, again.

$$b_1 \pm t^*_{n-2} \times SE(b_1) = -0.020 \pm (2.030) \times 0.0033 \approx (-0.027, -0.013)$$

We are 95% confident that the mean difference in age between men and women at first marriage decreases by between 0.013 and 0.027 years in age for each year that passes.

35. Fuel economy.

a) H_0: There is no linear relationship between the weight of a car and its mileage.
   $$(\beta_1 = 0)$$

   H_A: There is a linear relationship between the weight of a car and its mileage.
   $$(\beta_1 \neq 0)$$

b) **Straight enough condition:** The scatterplot is straight enough to try a linear model.
   Independence assumption: The residuals plot is scattered.
   Does the plot thicken? condition: The residuals plot indicates some possible "thickening" as the predicted values increases, but it's probably not enough to worry about.
   Nearly Normal condition, Outlier condition: The histogram of residuals is unimodal and symmetric, with one possible outlier. With the large sample size, it is okay to proceed.

   Since conditions have been satisfied, the sampling distribution of the regression slope can be modeled by a Student's t-model with $50 - 2 = 48$ degrees of freedom. We will use a regression slope t-test. The equation of the line of best fit for these data points is: $\widehat{MPG} = 48.7393 - 8.2136(Weight)$, where $Weight$ is measured in thousands of pounds.

   The value of $t = -12.2$. The P-value of less than 0.0001 means that the association we see in the data is unlikely to occur by chance. We reject the null hypothesis, and conclude that there is strong evidence of a linear relationship between weight of a car and its mileage. Cars that weigh more tend to have lower gas mileage.

37. Fuel economy, part II.

a) Since conditions have been satisfied in Exercise 35, the sampling distribution of the regression slope can be modeled by a Student's t-model with $(50 - 2) = 48$ degrees of freedom. (Use $t^*_{45} = 2.014$ from the table.) We will use a regression slope t-interval, with 95% confidence.

$$b_1 \pm t^*_{n-2} \times SE(b_1) = -8.2136 \pm (2.014) \times 0.674 \approx (-9.57, -6.86)$$

b) We are 95% confident that the mean mileage of cars decreases by between 6.86 and 9.57 miles per gallon for each additional 1000 pounds of weight.

39. Fuel economy, part III.

a) The regression equation predicts that cars that weigh 2500 pounds will have a mean fuel efficiency of $48.7393 - 8.2136(2.5) = 28.2053$ miles per gallon.

$$\hat{y}_v \pm t^*_{n-2} \sqrt{SE^2(b_1) \cdot (x_v - \bar{x})^2 + \frac{s_e^2}{n}}$$

$$= 28.2053 \pm (2.014) \sqrt{0.674^2 \cdot (2.5 - 2.8878)^2 + \frac{2.413^2}{50}} \approx (27.35, 28.20)$$

We are 95% confident that cars weighing 2500 pounds will have mean fuel efficiency between 27.35 and 28.20 miles per gallon.

b) The regression equation predicts that cars that weigh 3450 pounds will have a mean fuel efficiency of $48.7393 - 8.2136(3.45) = 20.40238$ miles per gallon.

$$\hat{y}_v \pm t^*_{n-2} \sqrt{SE^2(b_1) \cdot (x_v - \bar{x})^2 + \frac{s_e^2}{n} + s_e^2}$$

$$= 20.40238 \pm (2.014) \sqrt{0.6738^2 \cdot (3.45 - 2.8878)^2 + \frac{2.413^2}{50} + 2.413^2} \approx (15.44, 25.37)$$

We are 95% confident that a car weighing 3450 pounds will have fuel efficiency between 15.44 and 25.37 miles per gallon.

41. Cereals.

a) H_0: There is no linear relationship between the number of calories and the sodium content of cereals. $(\beta_1 = 0)$

H_A: There is a linear relationship between the number of calories and the sodium content of cereals. $(\beta_1 \neq 0)$

Since these data were judged acceptable for inference, the sampling distribution of the regression slope can be modeled by a Student's t-model with $77 - 2 = 75$ degrees of freedom. We will use a regression slope t-test. The equation of the line of best fit for these data points is: $\widehat{Sodium} = 21.4143 + 1.29357(Calories)$.

The value of $t = 2.73$. The P-value of 0.0079 means that the association we see in the data is unlikely to occur by chance. We reject the null hypothesis, and conclude that there is strong evidence of a linear relationship between the number of calories and sodium content of

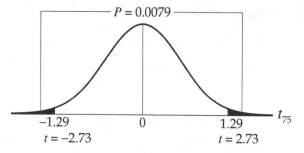

cereals. Cereals with higher numbers of calories tend to have higher sodium contents.

b) Only 9% of the variability in sodium content can be explained by the number of calories. The residual standard deviation is 80.49 mg, which is pretty large when you consider that the range of sodium content is only 320 mg. Although there is strong evidence of a linear association, it is too weak to be of much use. Predictions would tend to be very imprecise.

43. Cereals, part 2.

Straight enough condition: The scatterplot is not straight.
Independence assumption: The residuals plot shows a curved pattern.
Does the plot thicken? condition: The spread of the residuals is not consistent. The residuals plot "thickens" as the predicted values increase.
Nearly Normal condition, Outlier condition: The histogram of residuals is skewed to the right, with an outlier.

These data are not appropriate for inference.

45. Streams.

a) H_0: There is no linear relationship between BCI and pH. $(\beta_1 = 0)$
   H_A: There is a linear relationship between BCI and pH. $(\beta_1 \neq 0)$

b) Assuming the conditions for inference are satisfied, the sampling distribution of the regression slope can be modeled by a Student's t-model with $163 - 2 = 161$ degrees of freedom. We will use a regression slope t-test. The equation of the line of best fit for these data points is: $\widehat{BCI} = 2733.37 - 197.694(pH)$.

c) The value of $t \approx -7.73$. The P-value (two-sided!) of essentially 0 means that the association we see in the data is unlikely to occur by chance. We reject the null hypothesis, and conclude that there is strong evidence of a linear relationship between BCI and pH. Streams with higher pH tend to have lower BCI.

$$t = \frac{b_1 - \beta_1}{SE(b_1)}$$

$$t = \frac{-197.694 - 0}{25.57}$$

$$t \approx -7.73$$

47. Ozone and population.

a) H_0: There is no linear relationship between population and ozone level. $(\beta_1 = 0)$

H_A: There is a positive linear relationship between population and ozone level. $(\beta_1 > 0)$

Assuming the conditions for inference are satisfied, the sampling distribution of the regression slope can be modeled by a Student's t-model with $16 - 2 = 14$ degrees of freedom. We will use a regression slope t-test. The equation of the line of best fit for these data points is: $\widehat{Ozone} = 18.892 + 6.650(Population)$, where ozone level is measured in parts per million and population is measured in millions.

The value of $t \approx 3.48$.
The P-value of 0.0018 means that the association we see in the data is unlikely to occur by chance. We

$$t = \frac{b_1 - \beta_1}{SE(b_1)}$$

$$t = \frac{6.650 - 0}{1.910}$$

$$t \approx 3.48$$

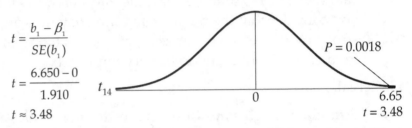

reject the null hypothesis, and conclude that there is strong evidence of a positive linear relationship between ozone level and population. Cities with larger populations tend to have higher ozone levels.

b) City population is a good predictor of ozone level. Population explains 84% of the variability in ozone level and s is just over 5 parts per million.

49. Ozone, again

a) $b_1 \pm t_{n-2}^* \times SE(b_1) = 6.65 \pm (1.761) \times 1.910 \approx (3.29,\ 10.01)$

We are 90% confident that each additional million people will increase mean ozone levels by between 3.29 and 10.01 parts per million.

b) The regression equation predicts that cities with a population of 600,000 people will have ozone levels of $18.892 + 6.650(0.6) = 22.882$ parts per million.

$$\hat{y}_v \pm t_{n-2}^* \sqrt{SE^2(b_1) \cdot (x_v - \bar{x})^2 + \frac{s_e^2}{n}}$$

$$= 22.882 \pm (1.761) \sqrt{1.91^2 \cdot (0.6 - 1.7)^2 + \frac{5.454^2}{16}} \approx (18.47,\ 27.29)$$

We are 90% confident that the mean ozone level for cities with populations of 600,000 will be between 18.47 and 27.29 parts per million.

51. Tablet computers 2014.

a) Since there are 34 – 2 = 32 degrees of freedom, there were 34 tablet computers tested.

b) **Straight enough condition:** The scatterplot is roughly straight, but scattered.
Independence assumption: The residuals plot shows no pattern.
Does the plot thicken? condition: The spread of the residuals is consistent.
Nearly Normal condition: The Normal probability plot of residuals is reasonably straight.

c) H_0: There is no linear relationship between maximum brightness and battery life. $(\beta_1 = 0)$

   H_A: There is a positive linear relationship between maximum brightness and battery life. $(\beta_1 > 0)$

   Since the conditions for inference are satisfied, the sampling distribution of the regression slope can be modeled by a Student's t-model with 34 – 2 = 32 degrees of freedom. We will use a regression slope t-test. The equation of the line of best fit for these data points is: $\widehat{Hours} = 5.38719 + 0.00904(ScreenBrightness)$, battery life measure in hours and screen brightness measured in cd/m².

   The value of $t \approx 2.02$. The P-value of 0.0522 means that the association we see in the data is not unlikely to occur by chance. We fail to reject the null hypothesis, and conclude that there is little evidence of a positive linear relationship between battery life and screen brightness.

d) Since $R^2 = 11.3\%$, only 11.3% of the variability in battery life can be accounted for by screen brightness. The residual standard deviation is 2.128 hours. That's pretty large, considering the range of battery life is only about 9 hours. Even if we concluded that there was some evidence of a linear association, it is too weak to be of much use. Predictions would tend to be very imprecise.

e) The equation of the line of best fit for these data points is:
   $\widehat{Hours} = 5.38719 + 0.00904(ScreenBrightness)$, battery life measure in hours and screen brightness measured in cd/m².

f) There are 32 degrees of freedom, so use $t^*_{32} = 1.694$.

   $b_1 \pm t^*_{n-2} \times SE(b_1) = 0.00904 \pm (1.694) \times 0.0045 \approx (0.00142, 0.0167)$

g) We are 90% confident that the mean battery life increases by between 0.00142 and 0.0167 hours for each additional cd/m² of screen brightness.

53. Body fat.

a) H_0: There is no linear relationship between waist size and percent body fat. $(\beta_1 = 0)$

H_A: There is a linear relationship between waist size and percent body fat. $(\beta_1 \neq 0)$

Straight enough condition: The scatterplot is straight enough.

Independence assumption: The residuals plot shows no pattern.

Does the plot thicken? condition: The spread of the residuals is consistent.

Nearly Normal condition, Outlier condition: The Normal probability plot of residuals is straight, and the histogram of the residuals is unimodal and symmetric with no outliers.

Since the conditions for inference are inference are satisfied, the sampling distribution of the regression slope can be modeled by a Student's t-model with $(20 - 2) = 18$ degrees of freedom. We will use a regression slope t-test.

Dependent variable is: **Body Fat %**
No Selector
R squared = 78.7%   R squared (adjusted) = 77.5%
s = 4.540 with 20 - 2 = 18 degrees of freedom

The equation of the line of best fit for these data points is:

$$\overline{\%BodyFat} = -62.5573 + 2.22152(Waist).$$

Source	Sum of Squares	df	Mean Square	F-ratio
Regression	1366.79	1	1366.79	66.3
Residual	370.960	18	20.6089	

Variable	Coefficient	s.e. of Coeff	t-ratio	prob
Constant	-62.5573	10.16	-6.16	≤ 0.0001
Waist (in)	2.22152	0.2728	8.14	≤ 0.0001

The value of $t \approx 8.14$. The P-value of essentially 0 means that the association we see in the data is unlikely to occur by chance. We reject the null hypothesis, and conclude that there is strong evidence of a linear relationship between waist size and percent body fat. People with larger waists tend to have a higher percentage of body fat.

b) The regression equation predicts that people with 40-inch waists will have
$-62.5573 + 2.22152(40) = 26.3035\%$ body fat. The average waist size of the people
sampled was approximately 37.05 inches.

$$\hat{y}_v \pm t^*_{n-2} \sqrt{SE^2(b_1) \cdot (x_v - \bar{x})^2 + \frac{s_e^2}{n}}$$

$$= 26.3035 \pm (2.101)\sqrt{0.2728^2 \cdot (40 - 37.05)^2 + \frac{4.54^2}{20}}$$

$$\approx (23.58, \ 29.03)$$

We are 95% confident that the mean percent body fat for people with 40-inch
waists is between 23.58% and 29.03%.

55. Grades.

a) The regression output is to the right.
The model is:

$$\widehat{Midterm2} = 12.005 + 0.721(Midterm1)$$

Dependent variable is: **Midterm 2**

No Selector

R squared = 19.9%   R squared (adjusted) = 18.6%

s = 16.78 with 64 - 2 = 62 degrees of freedom

Source	Sum of Squares	df	Mean Square	F-ratio
Regression	4337.14	1	4337.14	15.4
Residual	17459.5	62	281.604	

Variable	Coefficient	s.e. of Coeff	t-ratio	prob
Constant	12.0054	15.96	0.752	0.4546
Midterm 1	0.720990	0.1837	3.92	0.0002

b) **Straight enough condition:** The scatterplot shows a weak, positive relationship
between Midterm 2 score and Midterm 1 score. There are several outliers, but
removing them only makes the relationship slightly stronger. The relationship is
straight enough to try linear regression.
Independence assumption: The residuals plot shows no pattern..
Does the plot thicken? condition: The spread of the residuals is consistent.
Nearly Normal condition, Outlier condition: The histogram of the residuals is
unimodal, slightly skewed with several possible outliers. The Normal
probability plot shows some slight curvature.

Since we had some difficulty with the conditions for inference, we should be cautious in making conclusions from these data. The small P-value of 0.0002 for the slope would indicate that the slope is statistically distinguishable from zero, but the R^2 value of 0.199 suggests that the relationship is weak. Midterm 1 isn't a useful predictor of Midterm 2.

c) The student's reasoning is not valid. The R^2 value is only 0.199 and the value of s is 16.8 points. Although correlation between Midterm 1 and Midterm 2 may be statistically significant, it isn't of much practical use in predicting Midterm 2 scores. It's too weak.

57. Strike two.

H_0: Effectiveness is independent of the player's initial ability. $(\beta_1 = 0)$

H_A: Effectiveness of the video depends on the player's initial ability. $(\beta_1 \neq 0)$

Straight enough condition: The scatterplot is straight enough.
Independence assumption: The residuals plot shows no pattern.
Does the plot thicken? condition: The spread of the residuals is consistent.
Nearly Normal condition: The Normal probability plot is straight.

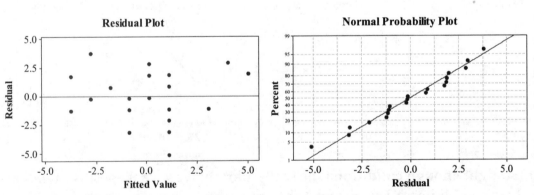

Since the conditions for inference are inference are satisfied, the sampling distribution of the regression slope can be modeled by a Student's t-model with $20 - 2 = 18$ degrees of freedom. We will use a regression slope t-test.

The equation of the line of best fit for these data points is:

$$\widehat{(After - Before)} = 32.316 - 0.9748(Before),$$ where we are counting the number of strikes thrown before and after the training program.

The value of $t \approx -4.34$. Since the P-value is 0.004, reject the null hypothesis, and conclude that there is evidence of a linear relationship between the player's intial ability and the effectiveness of the program. The negative slope indicates that the method is more effective for those whose initial performance was poorest and less effective for those whose initial performance was better. This appears to be a case of regression to the mean. Those who were above average initially tended to do worse after training. Those who were below average initially tended to improve.

59. Education and mortality.

a) **Straight enough condition:** The scatterplot is straight enough.
Independence assumption: The residuals plot shows no pattern. If these cities are representative of other cities, we can generalize our results.
Does the plot thicken? condition: The spread of the residuals is consistent.
Nearly Normal condition, Outlier condition: The histogram of the residuals is unimodal and symmetric with no outliers.

b) H_0: There is no linear relationship between education and mortality. $(\beta_1 = 0)$

H_A: There is a linear relationship between education and mortality. $(\beta_1 \neq 0)$

Since the conditions for inference are inference are satisfied, the sampling distribution of the regression slope can be modeled by a Student's t-model with $58 - 2 = 56$ degrees of freedom. We will use a regression slope t-test. The equation of the line of best fit for these data points is:

$\widehat{Mortality} = 1493.26 - 49.9202(Education)$.

The value of $t \approx -6.24$. The P-value of essentially 0 means that the association we see in the data is unlikely to occur by chance. We reject the null hypothesis, and conclude that there is strong evidence of a linear relationship between the level of education in a city and its mortality rate. Cities with lower education levels tend to have higher mortality rates.

$$t = \frac{b_1 - \beta_1}{SE(b_1)}$$

$$t = \frac{-49.9202 - 0}{8.000}$$

$$t \approx -6.24$$

c) We cannot conclude that getting more education is likely to prolong your life. Association does not imply causation. There may be lurking variables involved.

d) For 95% confidence, $t_{56}^* \approx 2.00327$.

$$b_1 \pm t_{n-2}^* \times SE(b_1) = -49.9202 \pm (2.003) \times 8.000 \approx (-65.95, -33.89)$$

e) We are 95% confident that the mean number of deaths per 100,000 people decreases by between 33.89 and 65.95 deaths for an increase of one year in average education level.

f) The regression equation predicts that cities with an adult population with an average of 12 years of school will have a mortality rate of $1493.26 - 49.9202(12) = 894.2176$ deaths per 100,000. The average education level was 11.0328 years.

$$\hat{y}_v \pm t^*_{n-2}\sqrt{SE^2(b_1)\cdot(x_v - \bar{x})^2 + \frac{s_e^2}{n}}$$

$$= 894.2176 \pm (2.003)\sqrt{8.00^2 \cdot (12 - 11.0328)^2 + \frac{47.92^2}{58}} \approx (874.239,\ 914.196)$$

We are 95% confident that the mean mortality rate for cities with an average of 12 years of schooling is between 874.239 and 914.196 deaths per 100,000 residents.

61. Right-to-work laws.

a) $\overline{\text{Logit}\left(Right-to-work\right)} = 6.19951 - 0.106155\,publ - 0.222957\,pvt$

b) Yes, logistic regression seems appropriate here. The response variable is dichotomous and there are no outliers in the predictor variables.

Review of Part VI – Accessing Associations Between Variables

1. Herbal cancer.

H_0: The cancer rate for those taking the herb is the same as the cancer rate for those not taking the herb. $\left(p_{Herb} = p_{Not} \text{ or } p_{Herb} - p_{Not} = 0\right)$

H_A: The cancer rate for those taking the herb is higher than the cancer rate for those not taking the herb. $\left(p_{Herb} > p_{Not} \text{ or } p_{Herb} - p_{Not} > 0\right)$

3. Surgery and germs.

a) Lister imposed a treatment, the use of carbolic acid as a disinfectant. This is an experiment.

b) H_0: The survival rate when carbolic acid is used is the same as the survival rate when carbolic acid is not used. $\left(p_C = p_N \text{ or } p_C - p_N = 0\right)$

H_A: The survival rate when carbolic acid is used is greater than the survival rate when carbolic acid is not used. $\left(p_C > p_N \text{ or } p_C - p_N > 0\right)$

Randomization condition: There is no mention of random assignment. Assume that the two groups of patients were similar, and amputations took place under similar conditions, with the use of carbolic acid being the only variable.
10% condition: 40 and 35 are both less than 10% of all possible amputations.
Independent samples condition: It is reasonable to think that the groups were not related in any way.
Success/Failure condition: $n\hat{p}$ (carbolic acid) = 34, $n\hat{q}$ (carbolic acid) = 6, $n\hat{p}$ (none) = 19, and $n\hat{q}$ (none) = 16. The number of patients who died in the carbolic acid group is only 6, but the expected number of deaths using the pooled proportion, $n\hat{q}_{pooled} = (40)(\frac{22}{75}) = 11.7$, so the samples are both large enough.

Since the conditions have been satisfied, we will perform a two-proportion z-test. We will model the sampling distribution of the difference in proportion with a Normal model with mean 0 and standard deviation estimated by:

$$SE_{pooled}\left(\hat{p}_C - \hat{p}_N\right) = \sqrt{\frac{\hat{p}_{pooled}\hat{q}_{pooled}}{n_C} + \frac{\hat{p}_{pooled}\hat{q}_{pooled}}{n_N}} = \sqrt{\frac{\left(\frac{53}{75}\right)\left(\frac{22}{75}\right)}{40} + \frac{\left(\frac{53}{75}\right)\left(\frac{22}{75}\right)}{35}} \approx 0.1054.$$

The observed difference between the proportions is:
0.85 – 0.5429 = 0.3071.

Since the *P*-value = 0.0018 is low, we reject the null hypothesis. There is strong evidence that the survival rate is higher when carbolic acid is used to disinfect the operating room than when carbolic acid is not used.

$$z = \frac{0.3071 - 0}{0.1054}$$

$$z \approx 2.91$$

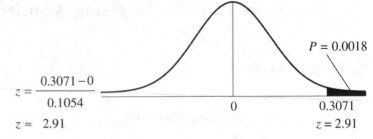

c) We don't know whether or not patients were randomly assigned to treatments, and we don't know whether or not blinding was used.

5. Twins.

H_0: The proportion of preterm twin births in 2000 is the same as the proportion of preterm twin births in 2010. $\left(p_{2000} = p_{2010} \ \text{or} \ p_{2000} - p_{2010} = 0 \right)$

H_A: The proportion of preterm twin births in 1990 is the less than the proportion of preterm twin births in 2000. $\left(p_{2000} < p_{2010} \ \text{or} \ p_{2000} - p_{2010} < 0 \right)$

Randomization condition: Assume that these births are representative of all twin births.
10% condition: 43 and 48 are both less than 10% of all twin births.
Independent samples condition: The samples are from different years, so they are unlikely to be related.
Success/Failure condition: $n\hat{p}\,(2000) = 20$, $n\hat{q}\,(2000) = 23$, $n\hat{p}\,(2010) = 26$, and $n\hat{q}\,(2010) = 22$ are all greater than 10, so both samples are large enough.

Since the conditions have been satisfied, we will perform a two-proportion *z*-test. We will model the sampling distribution of the difference in proportion with a Normal model with mean 0 and standard deviation estimated by:

$$SE_{pooled}\left(\hat{p}_{2000} - \hat{p}_{2010} \right) = \sqrt{\frac{\hat{p}_{pooled}\hat{q}_{pooled}}{n_{2000}} + \frac{\hat{p}_{pooled}\hat{q}_{pooled}}{n_{2010}}} = \sqrt{\frac{\left(\frac{46}{91}\right)\left(\frac{45}{91}\right)}{43} + \frac{\left(\frac{46}{91}\right)\left(\frac{45}{91}\right)}{48}} \approx 0.1050$$

The observed difference between the proportions is: 0.4651 – 0.5417 = – 0.0766

Since the *P*-value = 0.2329 is high, we fail to reject the null hypothesis. There is no evidence of an increase in the proportion of preterm twin births from 2000 to 2010, at least not at this large city hospital.

$$z = \frac{-0.0766 - 0}{0.1050}$$

$$z \approx -0.73$$

7. **Eclampsia deaths.**

a) H_0: The proportion of pregnant women who die after developing eclampsia is the same for women taking magnesium sulfide as it is for women not taking magnesium sulfide. $(p_{MS} = p_N \text{ or } p_{MS} - p_N = 0)$

H_A: The proportion of pregnant women who die after developing eclampsia is lower for women taking magnesium sulfide than for women not taking magnesium sulfide. $(p_{MS} < p_N \text{ or } p_{MS} - p_N < 0)$

b) **Randomization condition:** Although not specifically stated, these results are from a large-scale experiment, which was undoubtedly properly randomized. **10% condition:** 40 and 96 are less than 10% of all pregnant women. **Independent samples condition:** Subjects were randomly assigned to the treatments. **Success/Failure condition:** $n\hat{p}$ (mag. sulf.) = 11, $n\hat{q}$ (mag. sulf.) = 29, $n\hat{p}$ (placebo) = 20, and $n\hat{q}$ (placebo) = 76 are all greater than 10, so both samples are large enough.

Since the conditions have been satisfied, we will perform a two-proportion z-test. We will model the sampling distribution of the difference in proportion with a Normal model with mean 0 and standard deviation estimated by:

$$SE_{pooled}\left(\hat{p}_{MS} - \hat{p}_N\right) = \sqrt{\frac{\hat{p}_{pooled}\hat{q}_{pooled}}{n_{MS}} + \frac{\hat{p}_{pooled}\hat{q}_{pooled}}{n_N}} = \sqrt{\frac{\left(\frac{31}{136}\right)\left(\frac{105}{136}\right)}{40} + \frac{\left(\frac{31}{136}\right)\left(\frac{105}{136}\right)}{96}} \approx 0.07895.$$

c) The observed difference between the proportions is: 0.275 – 0.2083 = 0.0667

Since the *P*-value = 0.8008 is high, we fail to reject the null hypothesis. There is no evidence that the proportion of women who may die after developing eclampsia is lower for women taking magnesium sulfide than for women who are not taking the drug.

$z = \dfrac{0.0667 - 0}{0.07895}$

$z \approx 0.84$

d) There is not sufficient evidence to conclude that magnesium sulfide is effective in preventing death when eclampsia develops.

e) If magnesium sulfide is effective in preventing death when eclampsia develops, then we have made a Type II error.

f) To increase the power of the test to detect a decrease in death rate due to magnesium sulfide, we could increase the sample size or increase the level of significance.

g) Increasing the sample size lowers variation in the sampling distribution, but may be costly. The sample size is already quite large. Increasing the level of significance increases power by increasing the likelihood of rejecting the null hypothesis, but increases the chance of making a Type I error, namely thinking that magnesium sulfide is effective when it is not.

9. **More errors.**

a) Since a treatment (the additive) is imposed, this is an experiment.

b) The company is only interested in an increase in fuel economy, so they will perform a one-sided test.

c) The company will make a Type I error if they decide that the additive increases the fuel economy, when it actually makes no difference in the fuel economy.

d) The company will make a Type II error if they decide that the additive does not increase the fuel economy, when it actually increases fuel economy.

e) The additive manufacturer would consider a Type II error more serious. If the test caused the company to conclude that the manufacturer's product didn't work, and it actually did, the manufacturer would lose sales, and the reputation of their product would suffer.

f) Since this was a controlled experiment, the company can conclude that the additive is the reason that the fuel economy has increased. They should be cautious recommending it for all cars. There is evidence that the additive works well for fleet vehicles, which get heavy use. It might not be effective in cars with a different pattern of use than fleet vehicles.

11. **Crawling.**

a) H_0: The mean age at which babies begin to crawl is the same whether the babies were born in January or July. $\left(\mu_{Jan} = \mu_{July} \text{ or } \mu_{Jan} - \mu_{July} = 0 \right)$

H_A: There is a difference in the mean age at which babies begin to crawl, depending on whether the babies were born in January or July.
$\left(\mu_{Jan} \neq \mu_{July} \text{ or } \mu_{Jan} - \mu_{July} \neq 0 \right)$

Independent groups assumption: The groups of January and July babies are independent.
Randomization condition: Although not specifically stated, we will assume that the babies are representative of all babies.
Nearly Normal condition: We don't have the actual data, so we can't check the distribution of the sample. However, the samples are fairly large. The Central Limit Theorem allows us to proceed.

Since the conditions are satisfied, it is appropriate to model the sampling distribution of the difference in means with a Student's *t*-model, with 43.68 degrees of freedom (from the approximation formula).

We will perform a two-sample *t*-test. The sampling distribution model has mean 0, with standard error: $SE(\bar{y}_{Jan} - \bar{y}_{July}) = \sqrt{\dfrac{7.08^2}{32} + \dfrac{6.91^2}{21}} \approx 1.9596$.

The observed difference between the mean ages is 29.84 – 33.64 = – 3.8 weeks.

Since the *P*-value = 0.0590 is fairly low, we reject the null hypothesis. There is some evidence that mean age at which babies crawl is different for January and July babies. July babies appear to

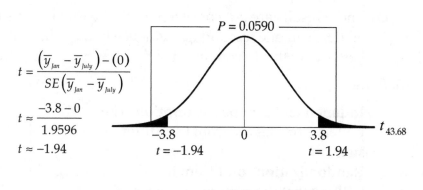

$$t = \frac{\left(\bar{y}_{Jan} - \bar{y}_{July}\right) - (0)}{SE\left(\bar{y}_{Jan} - \bar{y}_{July}\right)}$$

$$t \approx \frac{-3.8 - 0}{1.9596}$$

$$t \approx -1.94$$

crawl a bit earlier than July babies, on average. Since the evidence is not strong, we might want to do some more research into this claim.

b) H₀: The mean age at which babies begin to crawl is the same whether the babies were born in April or October. $\left(\mu_{Apr} = \mu_{Oct} \text{ or } \mu_{Apr} - \mu_{Oct} = 0\right)$

H_A: There is a difference in the mean age at which babies begin to crawl, depending on whether the babies were born in April or October. $\left(\mu_{Apr} \neq \mu_{Oct} \text{ or } \mu_{Apr} - \mu_{Oct} \neq 0\right)$

The conditions (with minor variations) were checked in part a.

Since the conditions are satisfied, it is appropriate to model the sampling distribution of the difference in means with a Student's *t*-model, with 59.40 degrees of freedom (from the approximation formula).

We will perform a two-sample *t*-test. The sampling distribution model has mean 0, with standard error: $SE(\bar{y}_{Apr} - \bar{y}_{Oct}) = \sqrt{\dfrac{6.21^2}{26} + \dfrac{7.29^2}{44}} \approx 1.6404$.

The observed difference between the mean ages is 31.84 – 33.35 = – 1.51 weeks.

Since the *P*-value = 0.3610 is high, we fail to reject the null hypothesis. There is no evidence that mean age at which babies crawl is different for April and October babies.

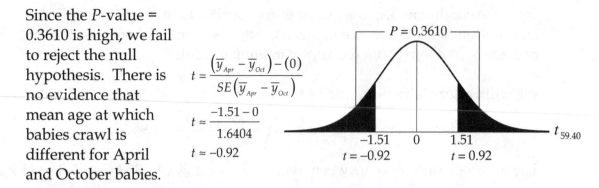

$$t = \frac{(\bar{y}_{Apr} - \bar{y}_{Oct}) - (0)}{SE(\bar{y}_{Apr} - \bar{y}_{Oct})}$$

$$t \approx \frac{-1.51 - 0}{1.6404}$$

$$t \approx -0.92$$

c) These results are not consistent with the researcher's claim. We have slight evidence in one test and no evidence in the other. The researcher will have to do better than this to convince us!

13. Pottery.

Independent groups assumption: The pottery samples are from two different sites.
Randomization condition: It is reasonable to think that the pottery samples are representative of all pottery at that site with respect to aluminum oxide content.
Nearly Normal condition: The histograms of aluminum oxide content are roughly unimodal and symmetric.

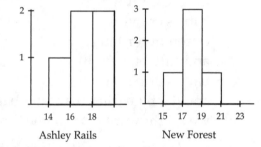

Since the conditions are satisfied, it is appropriate to model the sampling distribution of the difference in means with a Student's *t*-model, with 7 degrees of freedom (from the approximation formula). We will construct a two-sample *t*-interval, with 95% confidence.

$$(\bar{y}_{AR} - \bar{y}_{NF}) \pm t_{df}^{*} \sqrt{\frac{s_{AR}^{2}}{n_{AR}} + \frac{s_{NF}^{2}}{n_{NF}}}$$

$$= (17.32 - 18.18) \pm t_{7}^{*} \sqrt{\frac{1.65892^{2}}{5} + \frac{1.77539^{2}}{5}} \approx (-3.37, 1.65)$$

We are 95% confident that the difference in the mean percentage of aluminum oxide content of the pottery at the two sites is between –3.37% and 1.65%. Since 0 is in the interval, there is no evidence that the aluminum oxide content at the two sites is different. It would be reasonable for the archaeologists to think that the same ancient people inhabited the sites.

15. Feeding fish.

a) If there is no difference in the average fish sizes, the chance of observing a difference this large, or larger, just by natural sampling variation is 0.1%.

b) There is evidence that largemouth bass that are fed a natural diet are larger. The researchers would advise people who raise largemouth bass to feed them a natural diet.

c) If the advice is incorrect, the researchers have committed a Type I error.

17. Age.

a) **Independent groups assumption:** The group of patients with and without cardiac disease are not related in any way.
Randomization condition: Assume that these patients are representative of all people.
Normal population assumption: We don't have the actual data, so we will assume that the population of ages of patients is Normal.

Since the conditions are satisfied, it is appropriate to model the sampling distribution of the difference in means with a Student's t-model, with 670 degrees of freedom (from the approximation formula). We will construct a two-sample t-interval, with 95% confidence.

$$\left(\bar{y}_{Card} - \bar{y}_{None}\right) \pm t^*_{df}\sqrt{\frac{s^2_{Card}}{n_{Card}} + \frac{s^2_{None}}{n_{None}}} = (74.0 - 69.8) \pm t^*_{670}\sqrt{\frac{7.9^2}{450} + \frac{8.7^2}{2397}} \approx (3.39,\ 5.01)$$

We are 95% confident that the mean age of patients with cardiac disease is between 3.39 and 5.01 years higher than the mean age of patients without cardiac disease.

b) Older patients are at greater risk for a variety of health problems. If an older patient does not survive a heart attack, the researchers will not know to what extent depression was involved, because there will be a variety of other possible variables influencing the death rate. Additionally, older patients may be more (or less) likely to be depressed than younger ones.

19. Eating disorders.

a) **Randomization condition:** Hopefully, the students were selected randomly.
10% condition: 150 and 200 are less than 10% of all students.
Independent samples condition: The groups are independent.
Success/Failure condition: $n\hat{p}$ (Muslim) = 46, $n\hat{q}$ (Muslim) = 104, $n\hat{p}$ (Christian) = 34, and $n\hat{q}$ (Christian) = 166 are all greater than 10, so the samples are both large enough.

Since the conditions have been satisfied, we will find a two-proportion z-interval.

$$\left(\hat{p}_M - \hat{p}_C\right) \pm z^* \sqrt{\frac{\hat{p}_M \hat{q}_M}{n_M} + \frac{\hat{p}_C \hat{q}_C}{n_C}}$$

$$= \left(\tfrac{46}{150} - \tfrac{34}{200}\right) \pm 1.960 \sqrt{\frac{\left(\tfrac{46}{150}\right)\left(\tfrac{104}{150}\right)}{150} + \frac{\left(\tfrac{34}{200}\right)\left(\tfrac{166}{200}\right)}{200}} = \left(0.046,\ 0.227\right)$$

We are 95% confident that the percentage of Muslim students who have an eating disorder is between 4.6 and 22.7 percentage points higher than the percentage of Christian students who have an eating disorder.

b) Although caution in generalizing must be used since the study was restricted to the Spanish city of Ceuta, it appears there is a true difference in the prevalence of eating disorders. We can conclude this because the entire interval is above 0.

21. Teach for America.

H_0: The mean score of students with certified teachers is the same as the mean score of students with uncertified teachers. $\left(\mu_C = \mu_U \text{ or } \mu_C - \mu_U = 0\right)$

H_A: The mean score of students with certified teachers is greater than as the mean score of students with uncertified teachers. $\left(\mu_C > \mu_U \text{ or } \mu_C - \mu_U > 0\right)$

Independent groups assumption: The certified and uncertified teachers are independent groups.
Randomization condition: Assume the students studied were representative of all students.
Nearly Normal condition: We don't have the actual data, so we can't look at the graphical displays, but the sample sizes are large, so we can proceed.

Since the conditions are satisfied, it is appropriate to model the sampling distribution of the difference in means with a Student's t-model, with 86 degrees of freedom (from the approximation formula).

We will perform a two-sample t-test. The sampling distribution model has mean 0, with standard error: $SE(\bar{y}_C - \bar{y}_U) = \sqrt{\dfrac{9.31^2}{44} + \dfrac{9.43^2}{44}} \approx 1.9977$.

The observed difference between the mean scores is 35.62 – 32.48 = 3.14.

Since the *P*-value = 0.0598 is fairly low, we reject the null hypothesis. There is some evidence that students with certified teachers had mean scores higher than students with uncertified teachers. However, since the *P*-value is not extremely low, further investigation is recommended.

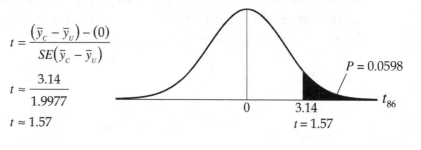

$$t = \frac{(\bar{y}_c - \bar{y}_U) - (0)}{SE(\bar{y}_c - \bar{y}_U)}$$

$$t \approx \frac{3.14}{1.9977}$$

$$t \approx 1.57$$

P = 0.0598

t_{86}

3.14

$t = 1.57$

23. Teach for America, Part II.

H_0: The mean score of students with certified teachers is the same as the mean score of students with uncertified teachers. $\left(\mu_C = \mu_U \ \text{ or } \ \mu_C - \mu_U = 0\right)$

H_A: The mean score of students with certified teachers is different than the mean score of students with uncertified teachers. $\left(\mu_C \neq \mu_U \ \text{ or } \ \mu_C - \mu_U \neq 0\right)$

Mathematics: Since the *P*-value = 0.002 is low, we reject the null hypothesis. There is strong evidence that students with certified teachers have different mean math scores than students with uncertified teachers. Students with certified teachers do better.

Language: Since the *P*-value = 0.045 is fairly low, we reject the null hypothesis. There is evidence that students with certified teachers have different mean language scores than students with uncertified teachers. Students with certified teachers do better. However, since the *P*-value is not extremely low, further investigation is recommended.

25. Insulin and diet.

a) H_0: People with high dairy consumption have IRS at the same rate as those with low dairy consumption. $\left(p_{High} = p_{Low} \ \text{ or } \ p_{High} - p_{Low} = 0\right)$

H_A: People with high dairy consumption have IRS at a different rate than those with low dairy consumption. $\left(p_{High} \neq p_{Low} \ \text{ or } \ p_{High} - p_{Low} \neq 0\right)$

Random condition: Assume the people studied are representative of all people.
10% condition: 102 and 190 are both less than 10% of all people.
Independent samples condition: The two groups are not related.
Success/Failure condition: $n\hat{p}$ (high) = 24, $n\hat{q}$ (high) = 78, $n\hat{p}$ (low) = 85, and $n\hat{q}$ (low) = 105 are all greater than 10, so the samples are both large enough.

Since the conditions have been satisfied, we will model the sampling distribution of the difference in proportion with a Normal model with mean 0 and standard deviation estimated by

$$SE_{pooled}\left(\hat{p}_{High} - \hat{p}_{Low}\right) = \sqrt{\frac{\hat{p}_{pooled}\hat{q}_{pooled}}{n_{High}} + \frac{\hat{p}_{pooled}\hat{q}_{pooled}}{n_{Low}}}$$

$$= \sqrt{\frac{(0.373)(0.627)}{102} + \frac{(0.373)(0.627)}{190}} \approx 0.05936$$

The observed difference between the proportions is 0.2352 – 0.4474 = – 0.2122.

Since the *P*-value = 0.0004 is very low, we reject the null hypothesis. There is strong evidence that the proportion of people with IRS is different for those

$$z = \frac{-0.2122 - 0}{0.05936}$$

$$z \approx -3.57$$

P = 0.0004

−0.2122   0   0.2122
z = −3.57   *z* = 3.57

who with high dairy consumption compared to those with low dairy consumption. People who consume dairy products more than 35 times per week appear less likely to have IRS than those who consume dairy products fewer than 10 times per week.

b) There is evidence of an association between the low consumption of dairy products and IRS, but that does not prove that dairy consumption influences the development of IRS. This is an observational study, and a controlled experiment is required to prove cause and effect.

27. Genetics.

H_0: The proportions of traits are as specified by the ratio 1:3:3:9.
H_A: The proportions of traits are not as specified.

Counted data condition: The data are counts.
Randomization condition: Assume that these students are representative of all people.

Expected cell frequency condition: The expected counts (shown in the table) are all greater than 5.

Under these conditions, the sampling distribution of the test statistic is χ^2 on 4 – 1 = 3 degrees of freedom. We will use a chi-square goodness-of-fit test.

Trait	Observed	Expected	Residual = $(Obs - Exp)$	$(Obs - Exp)^2$	Component = $\dfrac{(Obs - Exp)^2}{Exp}$
Attached, noncurling	10	7.625	2.375	5.6406	0.73975
Attached, curling	22	22.875	– 0.875	0.7656	0.03347
Free, noncurling	31	22.875	8.125	66.0156	2.8859
Free, curling	59	68.625	– 9.625	92.6406	1.35
					$\sum \approx 5.01$

$\chi^2 = 5.01$. Since the P-value = 0.1711 is high, we fail to reject the null hypothesis.

There is no evidence that the proportions of traits are anything other than 1:3:3:9.

29. Hard water.

a) H_0: There is no linear relationship between calcium concentration in water and mortality rates for males. $(\beta_1 = 0)$

H_A: There is a linear relationship between calcium concentration in water and mortality rates for males. $(\beta_1 \neq 0)$

b) Assuming the conditions have been satisfied, the sampling distribution of the regression slope can be modeled by a Student's t-model with $(61 - 2) = 59$ degrees of freedom. We will use a regression slope t-test. The equation of the line of best fit for these data points is: $\widehat{Mortality} = 1676 - 3.23(Calcium)$, where mortality is measured in deaths per 100,000, and calcium concentration is measured in parts per million.

$t = \dfrac{b_1 - \beta_1}{SE(b_1)}$

$t = \dfrac{-3.23 - 0}{0.48}$

$t \approx -6.73$

The value of $t = -6.73$. The P-value of less than 0.0001 means that the association we see in the data is unlikely to occur by chance. We reject the null hypothesis, and conclude that there is strong evidence of a linear relationship between calcium concentration and mortality. Towns with higher calcium concentrations tend to have lower mortality rates.

c) For 95% confidence, use $t^*_{59} \approx 2.001$, or estimate from the table $t^*_{50} \approx 2.009$.

$b_1 \pm t^*_{n-2} \times SE(b_1) = -3.23 \pm (2.001) \times 0.48 \approx (-4.19, -2.27)$

d) We are 95% confident that the average mortality rate decreases by between 2.27 and 4.19 deaths per 100,000 for each additional part per million of calcium in drinking water.

31. Wild horses.

a) Since there are 36 degrees of freedom, 38 herds of wild horses were studied.

b) **Straight enough condition:** The scatterplot is straight enough to try linear regression.
Independence assumption: The residuals plot shows no pattern.
Does the plot thicken? condition: The spread of the residuals is consistent.
Nearly Normal condition: The histogram of residuals is unimodal and symmetric.

c) Since the conditions for inference are satisfied, the sampling distribution of the regression slope can be modeled by a Student's t-model with $(38 - 2) = 36$ degrees of freedom. We will use a regression slope t-interval, with 95% confidence. Use $t_{35}^* \approx 2.030$ as an estimate.

$$b_1 \pm t_{n-2}^* \times SE(b_1) = 0.153969 \pm (2.030) \times 0.0114 \approx (0.131, 0.177)$$

d) We are 95% confident that the mean number of foals in a herd increases by between 0.131 and 0.177 foals for each additional adult horse.

e) The regression equation predicts that herds with 80 adults will have $-1.57835 + 0.153969(80) = 10.73917$ foals. The average size of the herds sampled is 110.237 adult horses. Use $t_{36}^* \approx 1.6883$, or use an estimate of $t_{35}^* \approx 1.690$, from the table.

$$\hat{y}_v \pm t_{n-2}^* \sqrt{SE^2(b_1) \cdot (x_v - \bar{x})^2 + \frac{s_e^2}{n} + s_e^2}$$

$$= 10.73917 \pm (1.6883)\sqrt{0.0114^2 \cdot (80 - 110.237)^2 + \frac{4.941^2}{38} + 4.941^2}$$

$$\approx (2.26, 19.21)$$

We are 90% confident that number of foals in a herd of 80 adult horses will be between 2.26 and 19.21. This prediction interval is too wide to be of much use.

33. AP Statistics scores 2010.

a) H_0: The distribution of AP Statistics scores at Ithaca High School is the same as it is nationally.
H_A: The distribution of AP Statistics scores at Ithaca High School is different than it is nationally.

Counted data condition: The data are counts.
Randomization condition: Assume that this group of students is representative of all years at Ithaca High School.
Expected cell frequency condition: The expected counts (shown in the table) are all greater than 5.

Under these conditions, the sampling distribution of the test statistic is χ^2 on $5 - 1 = 4$ degrees of freedom. We will use a chi-square goodness-of-fit test.

Score	Observed	Expected	Residual = $(Obs - Exp)$	Standardized Residual = $\dfrac{(Obs - Exp)}{\sqrt{Exp}}$	Component = $\dfrac{(Obs - Exp)^2}{Exp}$
5	26	12.416	4.445	3.8551	14.862
4	36	21.728	2.792	3.0618	9.3745
3	19	22.795	– 3.795	– 0.7949	0.63181
2	10	17.654	– 7.654	– 1.822	3.3184
1	6	22.407	– 16.41	– 3.466	12.014
					$\sum \approx 40.20$

$\chi^2 \approx 40.20$. Since the *P*-value is essentially 0, we reject the null hypothesis.

There is strong evidence that the distribution of scores at Ithaca High School is different than the national distribution. Students at IHS get fewer scores of 2 and 1 than expected, and more scores of 4 and 5 than expected.

b) H_0: Gender and AP Statistics score are independent at Ithaca High School.

H_A: There is an association between gender and AP Statistics score at Ithaca High School.

Counted data condition: The data are counts.
Randomization condition: Assume this year's students are representative of all years.
Expected cell frequency condition: After combining the cells for scores of 2 and 1, the expected counts are all greater than 5.

	Boys (*Obs / Exp*)	Girls (*Obs / Exp*)
5	13 / 13.67	13 / 12.33
4	21 / 18.928	15 / 17.072
3	6 / 9.9897	13 / 9.0103
2 or 1	11 / 8.4124	5 / 7.5876

Under these conditions, the sampling distribution of the test statistic is χ^2 on 3 degrees of freedom. We will use a chi-square test for independence. (This is a test for independence, since we have one group that has been classified according to two variables, gender and score. However, if you said it was a test for homogeneity, since you were comparing two groups, no one would get terribly upset!)

With $\chi^2 = \sum_{all\,cells} \dfrac{(Obs - Exp)^2}{Exp} \approx 5.59$,

the P-value ≈ 0.1336.

Since P-value ≈ 0.1336 is high, we fail to reject the null hypothesis. There is no evidence of an association between gender and score at Ithaca High School. The boys seem to do just as well as the girls.

35. Twins, again.

H_0: The distributions of pregnancy durations are the same for the three years.

H_A: The distributions of pregnancy durations are different for the three years.

Counted data condition: The data are counts.
Independence assumption: Assume that the durations of the pregnancies are mutually independent.
Expected cell frequency condition: The expected counts are all greater than 5.

	1990 (Obs / Exp)	1995 (Obs / Exp)	2000 (Obs / Exp)
Preterm (induced or Cesarean)	11 / 12.676	13 / 13.173	19 / 17.150
Preterm (without procedures)	13 / 13.266	14 / 13.786	18 / 17.948
Term or postterm	27 / 25.058	26 / 26.04	32 / 33.902

Under these conditions, the sampling distribution of the test statistic is χ^2 on 4 degrees of freedom. We will use a chi-square test for homogeneity.

$\chi^2 = \sum_{all\,cells} \dfrac{(Obs - Exp)^2}{Exp} \approx 0.69$,

and the P-value ≈ 0.9526.

Since the P-value ≈ 0.9526 is high, we fail to reject the null hypothesis. There is no evidence that the distributions of the durations of pregnancies are different for the three years. It does not appear that the way the hospital deals with twin pregnancies has changed.

37. Age and party.

a) There is one sample, classified according to two different variables, so we will perform a chi-square test for independence.

b) H$_0$: There is no association between age and political party for white voters.

H$_A$: There is an association between age and political party for white voters.

Counted data condition: The data are counts.
Randomization condition: These data are from a representative phone survey.
Expected cell frequency condition: The expected counts are all greater than 5.

	Leaning Republican (Obs / Exp)	Leaning Democrat (Obs / Exp)	Other (Obs / Exp)
18 – 29	274 / 271.97	216 / 205.84	36 / 48.197
30 – 49	888 / 835.03	581 / 631.99	146 / 147.98
50 – 64	1173 / 1213	962 / 918.05	211 / 214.96
65 +	1062 / 1077	812 / 815.13	209 / 190.86

Under these conditions, the sampling distribution of the test statistic is χ^2 on 6 degrees of freedom. We will use a chi-square test for independence.

$$\chi^2 = \sum_{all\,cells} \frac{(Obs - Exp)^2}{Exp} \approx 16.54, \text{ and the } P\text{-value} \approx 0.0111.$$

c) Since the *P*-value ≈ 0.0111 is low, we reject the null hypothesis. There is evidence of an association between age and political party for white voters.

The table of standardized residuals is useful for the analysis of the differences. Looking for the largest standardized residuals, we can see there are more Republicans and fewer Democrats among those 30

	Standardized Residuals		
	Leaning Republican	Leaning Democrat	Other
18 – 29	0.12	0.71	-1.76
30 – 49	1.83	-2.03	-0.16
50 – 64	-1.15	1.45	-0.27
65 +	-0.46	-0.11	1.31

– 49 than we expect. There are also fewer voters classified as "Other" than we would expect.

39. Depression and the Internet.

a) H$_0$: There is no linear relationship between depression and Internet usage. $(\beta_1 = 0)$

H$_A$: There is a linear relationship. $(\beta_1 \neq 0)$

Since the conditions for inference are satisfied (given), the sampling distribution of the regression slope can be modeled by a Student's t-model with $(162 - 2) = 160$ degrees of freedom. We will use a regression slope t-test. The equation of the line of best fit for these data points is:

$\widehat{DepressionAfter} = 0.565485 + 0.019948(InternetUsage)$.

The value of $t \approx 2.76$. The P-value of 0.0064 means that the association we see in the data is unlikely to occur by chance. We reject the null hypothesis, and conclude that there is strong evidence of a linear relationship between depression and Internet usage. Those with high levels of Internet usage tend to

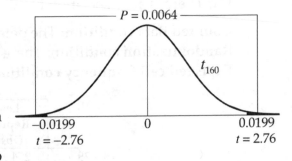

have high levels of depression. It should be noted, however, that although the evidence is strong, the association is quite weak, with $R^2 = 4.6\%$. The regression analysis only explains 4.6% of the variation in depression level.

b) The study says nothing about causality, merely association. Furthermore, there are almost certainly other factors involved. In fact, if 4.6% of the variation in depression level is related to Internet usage, the other 95.4% of the variation must be related to something else!

c) H_0: The mean difference in depression before and after the experiment is zero. $(\mu_d = 0)$

H_A: The mean difference in depression before and after the experiment is different than zero. $(\mu_d \neq 0)$

Since the conditions are satisfied (given), the sampling distribution of the difference can be modeled with a Student's t-model with $162 - 1 = 161$ degrees of freedom, $t_{161}\left(0, \dfrac{0.552417}{\sqrt{162}}\right)$.

We will use a paired t-test, with $\bar{d} = -0.118457$.

Since the P-value = 0.0070 is very low, we reject the null hypothesis. There is strong evidence that the mean depression level changed over the course of the experiment. These data suggest that depression levels actually decreased.

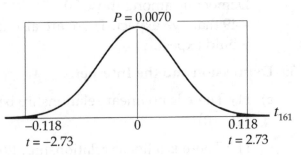

41. Family planning.

H₀: Unplanned pregnancies and education level are independent.

Hₐ: There is an association between unplanned pregnancies and education level.

Counted data condition: The percentages must be converted to counts.
Randomization condition: Assume that these women are representative of all women.
Expected cell frequency condition: The expected counts are all greater than 5.

Under these conditions, the sampling distribution of the test statistic is χ^2 on 1 degree of freedom. We will use a chi-square test for independence.

$$\chi^2 = \sum_{all\,cells} \frac{(Obs - Exp)^2}{Exp} \approx 40.71,$$ and the P-value is essentially 0.

Since the P-value is essentially 0, we reject the null hypothesis. There is strong evidence of an association between unplanned pregnancies and education level. More educated women tend to have fewer unplanned pregnancies.

43. Togetherness.

a) H₀: There is no linear relationship number of meals eaten as a family and grades. $(\beta_1 = 0)$

Hₐ: There is a linear relationship. $(\beta_1 \neq 0)$

Since the conditions for inference are satisfied (given), the sampling distribution of the regression slope can be modeled by a Student's t-model with (142 – 2) = 140 degrees of freedom. We will use a regression slope t-test. The equation of the line of best fit for these data points is: $\widehat{GPA} = 2.7288 + 0.1093(Meals / Week)$.

$$t = \frac{b_1 - \beta_1}{SE(b_1)}$$
$$t = \frac{0.1093 - 0}{0.0263}$$
$$t \approx 4.16$$

The value of $t \approx 4.16$. The P-value of less than 0.0001 means that the association we see in the data is unlikely to occur by chance. We reject the null hypothesis, and conclude that there is strong evidence of a linear relationship between grades and the number of meals eaten as a family. Students whose families eat together relatively frequently tend to have higher grades than those whose families don't eat together as frequently.

b) This relationship would not be particularly useful for predicting a student's grade point average. $R^2 = 11.0\%$, which means that only 11% of the variation in GPA can be explained by the number of meals eaten together per week.

c) These conclusions are not contradictory. There is strong evidence that the slope is not zero, and that means strong evidence of a linear relationship. This does not mean that the relationship itself is strong, or useful for predictions.

45. Juvenile offenders.

a) **Randomization condition:** We will assume that the youths studied are representative of other youths that might receive this therapy.
10% condition: 125 and 125 are less than 10% of all such youths.
Independent samples condition: The groups are independent.
Success/Failure condition: $n\hat{p}$ (Ind) = 19, $n\hat{q}$ (Ind) = 106, $n\hat{p}$ (MST) = 5, and $n\hat{q}$ (MST) = 120 are all not greater than 10, but with only one equal to 5, the test two-proportion interval should be reliable.

Since the conditions have been satisfied, we will find a two-proportion z-interval.

$$\left(\hat{p}_I - \hat{p}_M\right) \pm z^* \sqrt{\frac{\hat{p}_I \hat{q}_I}{n_I} + \frac{\hat{p}_M \hat{q}_M}{n_M}}$$

$$= \left(\tfrac{19}{125} - \tfrac{5}{125}\right) \pm 2.576 \sqrt{\frac{\left(\tfrac{19}{125}\right)\left(\tfrac{106}{125}\right)}{125} + \frac{\left(\tfrac{5}{125}\right)\left(\tfrac{120}{125}\right)}{125}} = \left(0.0178,\, 0.206\right)$$

We are 99% confident that the percentage of violent felony arrest among juveniles who receive individual therapy is between 1.78 and 20.6 percentage points higher than the percentage of violent felony arrest among juveniles who receive MST.

b) Since the entire interval is above 0, we can conclude that MST is successful in reducing the proportion of juvenile offenders who commit violent felonies. The population of interest is adolescents with mental health problems.

47. Infliximab.

H_0: The remission rates are the same for the three groups.

H_A: The remission rates are different for the three groups.

Counted data condition: The data are counts.
Randomization condition: Assume that these patients are representative of all patients.
Expected cell frequency condition: The expected counts are all greater than 5.

	Placebo (Obs / Exp)	5 mg (Obs / Exp)	10 mg (Obs / Exp)
Remission	23 / 38.418	44 / 39.466	50 / 39.116
No Remission	87 / 71.582	69 / 73.534	62 / 72.884

Under these conditions, the sampling distribution of the test statistic is χ^2 on 2 degrees of freedom. We will use a chi-square test for homogeneity.

$$\chi^2 = \sum_{all\,cells} \frac{(Obs - Exp)^2}{Exp} \approx 14.96,$$

and the *P*-value ≈ 0.0006.

Since the *P*-value ≈ 0.0006 is low, we reject the null hypothesis. There is strong evidence that the remission rates are different in the three groups. Patients receiving 10 mg of Infliximab have higher remission rates than the other groups. These data indicate that continued treatment with Infliximab is of value to Crohn's disease patients who exhibit a positive initial response to the drug.

49. Diet.

H₀: Cracker type and bloating are independent.

Hₐ: There is an association between cracker type and bloating.

Counted data condition: The data are counts.
Randomization condition: Assume that these women are representative of all women.
Expected cell frequency condition:
The expected counts are all (almost!) greater than 5.

Under these conditions, the sampling distribution of the test

	Bloat	
	Little/None (Obs / Exp)	Moderate/Severe (Obs / Exp)
Bran	11 / 7.6471	2 / 5.3529
Gum Fiber	4 / 7.6471	9 / 5.3529
Combination	7 / 7.6471	6 / 5.3529
Control	8 / 7.0588	4 / 4.9412

statistic is χ^2 on 3 degrees of freedom. We will use a chi-square test for independence.

$$\chi^2 = \sum_{all\,cells} \frac{(Obs - Exp)^2}{Exp} \approx 8.23,$$ and the *P*-value ≈ 0.0414.

Since the *P*-value is low, we reject the null hypothesis. There is evidence of an association between cracker type and bloating. The gum fiber crackers had a higher rate of moderate/severe bloating than expected. The company should head back to research and development and address the problem before attempting to market the crackers.

51. Hearing.

Paired data assumption: The data are paired by subject.
Randomization condition: The order of the tapes was randomized.
Normal population assumption: The histogram of differences between List A and List B is roughly unimodal and symmetric.

Since the conditions are satisfied, the sampling distribution of the difference can be modeled with a Student's *t*-model with 24 – 1 = 23 degrees of freedom. We will find a paired *t*-interval, with 95% confidence.

$$\bar{d} \pm t^*_{n-1} \left(\frac{s_d}{\sqrt{n}} \right) = -0.\overline{3} \pm t^*_{23} \left(\frac{8.12225}{\sqrt{24}} \right) \approx (-3.76, \, 3.10)$$

We are 95% confident that the mean difference in the number of words a person might misunderstand using these two lists is between –3.76 and 3.10 words. Since 0 is contained in the interval, there is no evidence to suggest that that the two lists are different for the purposes of the hearing test when there is background noise. It is reasonable to think that the two lists are still equivalent.

Chapter 26 – Analysis of Variance

Section 26.1

1. **Popcorn**

   a) H_0: The mean number of unpopped kernels is the same for all four brands of popcorn. $\left(\mu_1 = \mu_2 = \mu_3 = \mu_4\right)$

   H_A: The mean number of unpopped kernels is not the same for all four brands of popcorn.

   b) MS_T has $k - 1 = 4 - 1 = 3$ degrees of freedom.
   MS_E has $N - k = 16 - 4 = 12$ degrees of freedom.

   c) The F-statistic is 13.56 with 3 and 12 degrees of freedom, resulting in a P-value of 0.00037. We reject the null hypothesis and conclude that there is strong evidence that the mean number of unpopped kernels is not the same for all four brands of popcorn.

   d) To check the Similar Variance condition, look at side-by-side boxplots of the treatment groups to see whether they have similar spreads. To check the Nearly Normal condition, look to see if a normal probability plot of the residuals is straight, look to see that a histogram of the residuals is nearly normal, and look to see if a residuals plot shows no pattern, and no systematic change in spread.

3. **Gas mileage.**

   a) H_0: The mean gas mileage is the same for each muffler. $\left(\mu_1 = \mu_2 = \mu_3\right)$

   H_A: The mean gas mileages for each muffler are not all the same.

   b) MS_T has $k - 1 = 3 - 1 = 2$ degrees of freedom.
   MS_E has $N - k = 24 - 3 = 21$ degrees of freedom.

   c) The F-statistic is 2.35 with 2 and 21 degrees of freedom, resulting in a P-value of 0.1199. We fail to reject the null hypothesis and conclude that there is no evidence to suggest that gas mileage associated with any single muffler is different than the others.

   d) To check the Similar Variance condition, look at side-by-side boxplots of the treatment groups to see whether they have similar spreads. To check the Nearly Normal condition, look to see if a normal probability plot of the residuals is straight, look to see that a histogram of the residuals is nearly normal, and look to see if a residuals plot shows no pattern, and no systematic change in spread.

   e) By failing to notice that one of the mufflers resulted in significantly different gas mileage, you have committed a Type II error.

Section 26.2

5. Activating baking yeast.

a) H_0: The mean activation time is the same for all four recipes. $\left(\mu_1 = \mu_2 = \mu_3 = \mu_4\right)$

H_A: The mean activation times for each recipe are not all the same.

b) The *F*-statistic is 44.7392 with 3 and 12 degrees of freedom, resulting in a *P*-value less than 0.0001. We reject the null hypothesis and conclude that there is strong evidence that the mean activation times for each recipe are not all the same.

c) Yes, it would be appropriate to follow up with multiple comparisons, because we have rejected the null hypothesis.

Section 26.3

7. Fuel economy by cylinders, revisited.

a) H_0: The mean mileage is the same for each engine type (number of cylinders). $\left(\mu_4 = \mu_5 = \mu_6 = \mu_8\right)$

H_A: The mean mileages for each engine type are not all the same.

b) The Similar Variance condition is not met, because the boxplots show distributions with radically different spreads. A re-expression of the response variable may equalize the spreads, allowing us to proceed with an analysis of variance.

Section 26.4

9. Tellers.

a) H_0: The mean time to serve a customer is the same for each teller. $\left(\mu_1 = \mu_2 = \mu_3 = \mu_4 = \mu_5 = \mu_6\right)$

H_A: The mean times to serve a customer for each teller are not all the same.

b) The *F*-statistic is 1.508 with 5 and 134 degrees of freedom, resulting in a *P*-value equal to 0.1914. We fail to reject the null hypothesis and conclude that there is not enough evidence that the mean times to serve a customer for each teller are not all the same.

c) No, it would not be appropriate to follow up with multiple comparisons, because we have failed to reject the null hypothesis.

Section 26.5.

11. Eye and hair color.

An analysis of variance is not appropriate, because eye color is a categorical variable. The students could consider a chi-square test of independence.

Chapter Exercises.

13. Yogurt.

a) MS_T has $k - 1 = 3 - 1 = 2$ degrees of freedom.

$$MS_T = \frac{SS_T}{df_T} = \frac{17.300}{2} = 8.65$$

MS_E has $N - k = 9 - 3 = 6$ degrees of freedom.

$$MS_E = \frac{SS_E}{df_E} = \frac{0.4600}{6} \approx 0.0767$$

b) F-statistic $= \dfrac{MS_T}{MS_E} = \dfrac{8.65}{0.0767} \approx 112.78$

c) With a P-value equal to 0.000017, there is very strong evidence that the mean taste test scores for each method of preparation are not all the same.

d) We have assumed that the experimental runs were performed in random order, that the variances of the treatment groups are equal, and that the errors are Normal.

e) To check the Similar Variance condition, look at side-by-side boxplots of the treatment groups to see whether they have similar spreads. To check the Nearly Normal condition, look to see if a normal probability plot of the residuals is straight, look to see that a histogram of the residuals is nearly normal, and look to see if a residuals plot shows no pattern, and no systematic change in spread.

f) $s_p = \sqrt{MS_E} = \sqrt{0.0767} \approx 0.277$ points

15. Eggs.

a) H_0: The mean taste test scores are the same for both real and substitute eggs. $\left(\mu_R = \mu_S \right)$

H_A: The mean taste test scores are different. $\left(\mu_R \neq \mu_S \right)$

b) The F-statistic is 31.0712 with 1 and 6 degrees of freedom, resulting in a P-value equal to 0.0014. We reject the null hypothesis and conclude that there is strong evidence that the mean taste test scores for real and substitute eggs are different. The real eggs have a higher mean taste test score.

c) The Similar Variance assumption is a bit of a problem. The spread of the distribution of taste test scores for real eggs looks greater than the spread of the distribution of taste test scores for substitute eggs. Additionally, we cannot check the Nearly Normal condition because we don't have a residuals plot. Caution should be used in making any conclusions.

d) H₀: The mean taste test score of brownies made with real eggs is the same as the mean taste test score of brownies made with substitute eggs.
$$\left(\mu_R = \mu_S \ \text{ or } \ \mu_R - \mu_S = 0 \right)$$

H$_A$: The mean taste test score of brownies made with real eggs is different than the mean taste test score of brownies made with substitute eggs.
$$\left(\mu_R \neq \mu_S \ \text{ or } \ \mu_R - \mu_S \neq 0 \right)$$

Independent groups assumption: Taste test scores for each type of brownie should be independent.

Randomization condition: Brownies were tasted in random order.

Nearly Normal condition: We don't have the actual data, so we can't check the distribution of the sample. We will need to assume that the distribution of taste test scores is unimodal. The boxplots show distributions that are at least symmetric.

Since the conditions are satisfied, it is appropriate to model the sampling distribution of the difference in means with a Student's t-model, with 6 degrees of freedom ($n_R + n_S - 2 = 6$ for a pooled t-test). We will perform a two-sample t-test.

The pooled sample variance is $s_p^2 = \dfrac{(4-1)(0.651)^2 + (4-1)(0.395)^2}{(4-1)+(4-1)} \approx 0.2899$

The sampling distribution model has mean 0, with standard error:
$$SE_{pooled}(\bar{y}_R - \bar{y}_S) = \sqrt{\dfrac{0.2899}{4} + \dfrac{0.2899}{4}} \approx 0.3807 \ .$$

The observed difference between the mean scores is 6.78 – 4.66 = 2.12.

$$t = \dfrac{(\bar{y}_R - \bar{y}_T) - (0)}{SE_{pooled}(\bar{y}_R - \bar{y}_T)} \approx \dfrac{2.12 - 0}{0.3807} \approx 5.5687$$

With a $t = 5.5687$ and 6 degrees of freedom, the 2-sided P-value is 0.0014. Since the P-value is low, we reject the null hypothesis. There is strong evidence that the mean taste test score for brownies made with real eggs is different from the mean taste test score for brownies made with substitute eggs. In fact, there is evidence that the brownies made with real eggs taste better.

The P-value for the 2-sample pooled t-test, 0.0014, which is the same as the p-value for the analysis of variance test. The F-statistic, 31.0712, was approximately the same as $t^2 = (5.5727)^2 \approx 31.05499$. Also, the pooled estimate of the variance, 0.2899, is approximately equal to MS_E, the mean squared error. This is because an analysis of variance for two samples is equivalent to a 2-sample pooled t-test.

17. School system.

a) H_0: The mean math test score is the same at each of the 15 schools.
$$\left(\mu_A = \mu_B = \ldots = \mu_O\right)$$

H_A: The mean math test scores are not all the same at each of the 15 schools.

b) The *F*-statistic is 1.0735 with 14 and 105 degrees of freedom, resulting in a *P*-value equal to 0.3899. We fail to reject the null hypothesis and conclude that there is not enough evidence to suggest that the mean math scores are not all the same for the 15 schools.

c) This does not match our findings in part b. Because the intern performed so many *t*-tests, he may have committed several Type I errors. (Type I error is the probability of rejecting the null hypothesis when there is actually no difference between the schools.) Some tests would be expected to result in Type I error due to chance alone. The overall Type I error rate is higher for multiple *t*-tests than for an analysis of variance, or a multiple comparisons method. Because we failed to reject the null hypothesis in the analysis of variance, we should not do multiple comparison tests.

19. Cereals once more.

a) H_0: The mean sugar content is the same for each of the 3 shelves. $\left(\mu_1 = \mu_2 = \mu_3\right)$

H_A: The mean sugar contents are not all the same for each of the 3 shelves.

b) The *F*-statistic is 7.2721 with 2 and 74 degrees of freedom, resulting in a *P*-value equal to 0.0013. We reject the null hypothesis and conclude that there is strong evidence to suggest that the mean sugar content of cereal on at least one shelf differs from that on other shelves.

c) We cannot conclude that cereals on shelf two have a higher mean sugar content than cereals on shelf three, or that cereals on shelf two have a higher mean sugar content than cereals on shelf one. We can conclude only that the mean sugar contents are not all equal.

d) We can conclude that the mean sugar content on shelf two is significantly different from the mean sugar contents on shelves one and three. In fact, there is evidence that the mean sugar content on shelf two is greater than that of shelf one and shelf three.

21. Downloading.

a) H_0: The mean download time is the same for each of the three times of day.
$$\left(\mu_{Early} = \mu_{Evening} = \mu_{Late}\right)$$

H_A: The mean download times are not all the same for each of the three times of day.

b)

Analysis of Variance

Source	DF	Sum of Squares	Mean Square	F-ratio	P-Value
Time of Day	2	204641	102320	46.035	<0.0001
Error	45	100020	2222.67		
Total	47	304661			

The *F*-statistic is 46.035 with 2 and 45 degrees of freedom, resulting in a *P*-value less than 0.0001. We reject the null hypothesis and conclude that there is strong evidence to suggest that the mean download time is different in at least one of the three times of day.

c) **Randomization condition:** The runs were not randomized, but it is likely that they are representative of all download times at these times of day.
Similar Variance condition: The boxplots show similar spreads for the distributions of download times for the different times of day.
Nearly Normal condition: The normal probability plot of residuals is reasonably straight.

d) A Bonferroni test shows that all three pairs are different from each other at $\alpha = 0.05$.

Chapter 27 – Multifactor Analysis of Variance

Section 27.1

1. **Student TV time.**

   $$TV_{ijk} = \mu + Sex_j + Athlete_k + \varepsilon_{ijk}$$

Section 27.1

3. **Student TV assumptions.**

   The boxplots show different distributions for Women and Men, so the additivity assumption is probably not satisfied. The plots show that, at least for men, the equal variance assumption seems unlikely. And both plots show outliers, making the normality assumption questionable. If the survey was randomized, the responses may be mutually independent, but we don't have enough information to tell.

Section 27.3

5. **Student TV interactions.**

   The interaction term reflects the fact that there is very little difference in TV watching among women between athletes and non-athletes, but there is a difference for men with athletes watching more TV.

Chapter Exercises.

7. **Popcorn revisited.**

   a) H_0: The effect due to power level is the same for each level. $(\gamma_{Low} = \gamma_{Med} = \gamma_{High})$

      H_A: Not all of the power levels have the same effect on popcorn popping.

      H_0: The effect due to popping time is the same for each time. $(\tau_3 = \tau_4 = \tau_5)$

      H_A: Not all of the popping times have the same effect on popcorn popping.

   b) The *power* sum of squares has 3 – 1 = 2 degrees of freedom.
      The *popping time* sum of squares has 3 – 1 = 2 degrees of freedom.
      The error sum of squares has (9 – 1) – 2 – 2 = 4 degrees of freedom.

   c) Because the experiment did not include replication, there are no degrees of freedom left for the interaction term. The interaction term would require 2(2) = 4 degrees of freedom, leaving none for the error term, making any tests impossible.

9. **Popcorn again.**

   a) The *F*-statistic for *power* is 13.56 with 2 and 4 degrees of freedom, resulting in a *P*-value equal to 0.0165. The *F*-statistic for *time* is 9.36 with 2 and 4 degrees of freedom, resulting in a *P*-value equal to 0.0310.

b) With a *P*-value equal to 0.0165 we reject the null hypothesis that *power* has no effect and conclude that the mean number of uncooked kernels is not equal across all 3 *power* levels. With a *P*-value equal to 0.0310 we reject the null hypothesis that *time* has no effect and conclude that the mean number of uncooked kernels is not equal across all 3 *time* levels.

c) **Randomization condition:** The bags should be randomly assigned to treatments. **Similar Variance condition:** The side-by-side boxplots should have similar spreads. The residuals plot should show no pattern, and no change in spread. **Nearly Normal condition**: The Normal probability plot of the residuals should be straight, the histogram of the residuals should be unimodal and symmetric.

11. Crash analysis.

a) H_0: The effect on head injury severity is the same for both seats. $(\gamma_D = \gamma_P)$

H_A: The effects on head injury severity are different for the two seats.

H_0: The size of the vehicle has no effect on head injury severity.
$(\tau_1 = \tau_2 = \tau_3 = \tau_4 = \tau_5 = \tau_6)$

H_A: The size of the vehicle does have an effect on head injury severity.

b) **Randomization condition:** Assume that the cars are representative of all cars. **Additive Enough condition:** The interaction plot is reasonably parallel. **Similar Variance condition:** The side-by-side boxplots have similar spreads. The residuals plot shows no pattern, and no systematic change in spread. **Nearly Normal condition**: The Normal probability plot of the residuals should be straight, the histogram of the residuals should be unimodal and symmetric.

c) With a *P*-value equal to 0.838 the interaction term between *seat* and *vehicle size* is not significant. With a *P*-value less than 0.0001 for both *seat* and *car size* we reject both null hypotheses and conclude that both *seat* and *car size* affect the severity of head injury. By looking at one of the partial boxplots we see that the mean head injury severity is higher for the driver's side. The effect of driver's *seat* seems to be roughly the same for all six cars.

13. Baldness and Heart Disease.

a) A two-factor ANOVA must have a quantitative response variable. Here the response is whether they exhibited baldness or not, which is a categorical variable. A two-factor ANOVA is not appropriate.

b) We could use a chi-square analysis to test whether baldness and heart disease are independent.

15. Baldness and heart disease again.

a) A chi-square test of independence gives a chi-square statistic of 14.510 with a P-value equal to 0.0023. We reject the hypothesis that baldness and heart disease are independent.

b) No, the fact that these are not independent does not mean that one causes the other. There could be a lurking variable (such as age) that influences both.

17. Basketball shots.

a) H_0: The time of day has no effect on the number of shots made. $(\gamma_M = \gamma_N)$
H_A: The time of day does have an effect on the number of shots made.

H_0: The type of shoe has no effect on the number of shots made. $(\tau_F = \tau_O)$
H_A: The type of shoe does have an effect on the number of shots made.

b)

The partial boxplots show little effect of either *time of day* or *shoes* on the number of shots made.

Randomization condition: We assume that the number of shots made were independent from one treatment condition to the next.
Similar Variance condition: The side-by-side boxplots have similar spreads. The residuals plot shows no pattern, and no systematic change in spread.
Nearly Normal condition: The Normal probability plot of the residuals is straight.

The interaction plot shows a possible interaction effect. It looks as though the favorite shoes may make more of a difference at night. However, we fail to reject the null hypothesis that there is no interaction effect. In fact, none of the effects appears to be significant. It looks as though she cannot conclude that either *shoes* or *time of day* affect her mean free throw percentage.

19. Sprouts again.

a) H_0: The temperature level has no effect on the number of sprouts. ($\gamma_{32} = \gamma_{34} = \gamma_{36}$)

H_A: The temperature level does have an effect on the number of sprouts.

H_0: The salinity level has no effect on the number of sprouts. ($\tau_0 = \tau_4 = \tau_8 = \tau_{12}$)

H_A: The salinity level does have an effect on the number of sprouts.

b) **Randomization condition:** We assume that the sprouts were representative of all sprouts.
Similar Variance condition: There appears to be more spread in the number of sprouts for the lower salinity levels. This is cause for some concern, but most likely does not affect the conclusions.
Nearly Normal condition: The Normal probability plot of the residuals should be straight.

The partial boxplots show that *salinity* level appears to have an effect on the number of bean sprouts while *temperature* does not. The ANOVA supports this. With a p-value less than 0.0001 for *salinity*, there is strong evidence that the salinity level affects the number of bean sprouts. However, it appears that neither the interaction term nor *temperature* have a significant effect on the number of bean sprouts. The interaction term has a P-value equal to 0.7549 and *temperature* has a P-value equal to 0.3779.

21. Gas additives.

H_0: The car type has no effect on gas mileage. ($\gamma_H = \gamma_M = \gamma_S$)

H_A: The car type does have an effect on gas mileage.

H_0: The mean gas mileage is the same for both additives. ($\tau_G = \tau_R$)

H_A: The mean gas mileage is different for the additives.

The ANOVA table shows that both *car type* and *additive* affect gas mileage, with P-values less than 0.0001. There is a significant interaction effect as well that makes interpretation of the main effects problematic. However, the residual plot shows a strong increase in variance, which makes the whole analysis suspect. The Similar Variance condition appears to be violated.

23. Gas additives again.

After the re-expression of the response, gas mileage, the Normal probability plot of residuals looks straight and the residuals plot shows constant spread over the predicted values.

Analysis of Variance

Source	DF	Sum of Squares	Mean Square	F-ratio	P-Value
Type	2	10.1254	5.06268	5923.1	<0.0001
Additive	1	0.026092	0.0260915	30.526	<0.0001
Interaction	2	7.57E-05	3.78E-05	0.044265	0.9567
Error	54	0.046156	8.55E-04		
Total	59	10.1977			

After the re-expression, the ANOVA table only shows the main effects to be significant, while the interaction term is not. We can conclude that both the *car type* and *additive* have an effect on mileage and that the effects are constant (in log(*mpg*)) over the values of the various levels of the other factor.

25. Batteries again.

a) H_0: The mean times are the same under both environments. $(\gamma_C = \gamma_{RT})$

H_A: The mean times are different under the two environments.

H_0: The brand of battery has no effect on time. $(\tau_A = \tau_B = \tau_C = \tau_D)$

H_A: The brand of batter does have an effect on time.

b) From the partial boxplots it appears that *environment* does have an effect on time, but it is unclear whether *brand* has an effect on time.

c) Yes, the ANOVA does match our intuition based on the boxplots. The *brand* effect has a *P*-value equal to 0.099. While this is not significant at the $\alpha = 0.05$ level, it is significant at the $\alpha = 0.10$ level. As it appeared on the boxplot, the *environment* is clearly significant with a *P*-value less than 0.0001.

d) There is also an interaction, however, which makes the statement about *brands* problematic. Not all *brands* are affected by the environment in the same way. Brand C, which works best in the cold, performs worst at room temperature.

e) I would be uncomfortable recommending brand C because it performs the worst of the four at room temperature.

27. Batteries once more.

In this one-way ANOVA, we can see that the means vary across treatments. (However, boxplots with only 2 observations are not appropriate.) By looking closely, it seems obvious that the four flashlights at room temperature lasted much longer than the ones in the cold. It is much harder to see whether the means of the four brands are different, or whether they differ by the same amounts across both environmental conditions. The two-way ANOVA with interaction makes these distinctions clear.

Chapter 28 – Multiple Regression

Section 28.1

1. House prices.

a)

$$\widehat{Price} = 20{,}986.09 - 7483.10\,Bedrooms + 93.84\,LivingArea$$
$$= 20{,}986.09 - 7483.10(2) + 93.84(1000)$$
$$= 99{,}859.89$$

According to the multiple linear regression model, we would expect an Upstate New York home with 2 bedrooms with 1000 square feet of living space to have a price of approximately $99,859.89.

b) The residual is $135,000 – $99,859.89 = $35,140.11.

c) The house sold for about $35,100 more than our estimate.

Section 28.2

3. Movie profits.

a) $\widehat{USGross} = -22.9898 + 1.12442\,Budget + 24.9724\,Stars - 0.403296\,RunTime$

b) After allowing for the effects of *RunTime* and *Stars*, each million dollars spent making a film yields about 1.13 million dollars in gross revenue.

Section 28.3

5. More movies.

a) Linearity: The plot is reasonably linear with no bends.

b) Equal spread: The plot fails the Equal Spread condition. It is much more spread out to the right than on the left.

c) Normality: A scatterplot of two of the variables doesn't tell us anything about the distribution of the residuals.

Section 28.4

7. Movie tests.

a) $H_0 : \beta_{Stars} = 0$

b) $t = 4.24$

c) $P < 0.0001$

d) Since the *P*-value is low, reject the null hypothesis. There is strong evidence that the coefficient of *Stars* is not zero.

Section 28.5

9. Movie R-squared.

a) $R^2 = 0.474$ or 47.4%. About 47.4% of the variation *USGross* is accounted for by the least squares regression on *Budget*, *RunTime*, and *Stars*.

b) Adjusted R^2 accounts for the number of predictors, so it differs from R^2, which does not.

Chapter Exercises.

11. Interpretations.

a) There are two problems with this interpretation. First, the other predictors are not mentioned. Secondly, the prediction should be stated in terms of a mean, not a precise value.

b) This is a correct interpretation.

c) This interpretation attempts to predict in the wrong direction. This model cannot predict *lotsize* from *price*.

d) R^2 concerns the fraction of variability accounted for by the regression model, not the fraction of data values.

13. Predicting final exams.

a) $\widehat{Final} = -6.7210 + 0.2560(Test1) + 0.3912(Test2) + 0.9015(Test3)$

b) $R^2 = 77.7\%$, which means that 77.7% of the variation in *Final* grade is accounted for by the multiple regression model.

c) According to the multiple regression model, each additional point on *Test3* is associated with an average increase of 0.9015 points on the final, for students with given *Test1* and *Test2* scores.

d) Test scores are probably collinear. If we are only concerned about predicting the final exam score, *Test1* may not add much to the regression. However, we would expect it to be associated with the final exam score.

15. Home prices.

a) $\widehat{price} = -152037 + 9530(baths) + 139.87(sqft)$

b) $R^2 = 71.1\%$, which means that 71.1% of the variation in asking *price* is accounted for by the multiple regression model.

c) According to the multiple regression model, the asking *price* increases, on average, by about $139.87 for each additional square foot, for homes with the same number of bathrooms.

d) The number of bathrooms is probably correlated with the size of the house, even after considering the square footage of the bathroom itself. This correlation may account for the coefficient of *baths* not being discernibly different from 0. Moreover, the regression model does not predict what will happen when a house is modified, for example, by converting existing space into a bathroom.

17. Predicting finals II.

Straight enough condition: The plot of residuals versus predicted values looks curved, rising in the middle, and falling on both ends. This is a potential difficulty.
Randomization condition: It is reasonable to think of this class as a representative sample of all classes.
Nearly Normal condition: The Normal probability plot and the histogram of residuals suggest that the highest five residuals are extraordinarily high.
Does the plot thicken? condition: The spread is not consistent over the range of predicted values.

These data may benefit from a re-expression.

19. Admin performance.

a)
$$\widehat{salary} = 9.788 + 0.11(service) + 0.053(education)$$
$$+ 0.071(score) + 0.004(speed) + 0.065(dictation)$$

b) $\widehat{salary} = 9.788 + 0.11(120) + 0.053(9) + 0.071(50) + 0.004(60) + 0.065(30) \approx \$29,200$

c) H_0: When including the other potential predictors, typing speed does not increase our ability to predict salary. $(\beta_4 = 0)$

H_A: When including the other potential predictors, typing speed increases our ability to predict salary. $(\beta_4 \neq 0)$

With $t = 0.013$, and $30 - 6 = 24$ degrees of freedom, the *P*-value equals 0.9897. Since the *P*-value is so large, we fail to reject the null hypothesis. There is no evidence to suggest that the coefficient for typing *speed* is anything other than 0.

d) Omitting typing *speed* would simplify the model, and probably result in a model that was almost as good. Other predictors might also be omitted, but we cannot make that decision from the information given.

e) *Age* may be collinear with other predictors in the model. In particular, it is likely to be highly associated with months of *service*.

21. Body fat revisited.

a) H_0: There is no linear relationship between weight and percent body fat. $(\beta_W = 0)$

H_A: There is a linear relationship between weight and percent body fat. $(\beta_W \neq 0)$

With $t = 12.4$, and $250 - 2 = 248$ degrees of freedom, the P-value is less than 0.0001. Since the P-value is so small, we reject the null hypothesis. There is strong evidence of a linear relationship between *weight* and *%body fat*.

b) According to the linear model, each pound of *weight* is associated with a 0.189% increase in *%body fat*.

c) After removing the linear effects of *waist* and *height*, each pound of *weight* is associated, on average, with a decrease of 0.10% in *%body fat*. The change in coefficient and sign is a result of including the other predictors. We expect *weight* to be correlated with both *waist* and *height*. It may be collinear with them.

d) The P-value of 0.1567 says that if the coefficient of *height* in this model is truly 0, we could expect to observe a sample regression coefficient at least as far from 0 as the one we have here about 15.7% of the time.

23. Body fat one more time.

a) H_0: There is no linear relationship between chest size and percent body fat. $(\beta_C = 0)$

H_A: There is a linear relationship between chest size and percent body fat. $(\beta_C \neq 0)$

With $t = 15.5$, and $250 - 2 = 248$ degrees of freedom, the p-value is less than 0.0001. Since the p-value is so small, we reject the null hypothesis. There is strong evidence of a linear relationship between *chest* and *%body fat*.

b) According to the linear model, each additional inch in *chest* is associated with an average increase of 0.71272% in *%body fat*.

c) After allowing for the linear effects of *waist* and *height*, the multiple regression model predicts an average decrease of 0.233531% in *%body fat* for each additional inch *chest* size.

d) Each of the variables appears to contribute to the model. There does not appear to be an advantage to removing any of them.

25. Fifty states 2009.

a) The only model that seems to do poorly is the one that omits *murder*. The other

three are hard to choose among.

$$\widehat{Lifeexp} = 75.437 - 0.41458(Murder) - 0.00062(HSgrad) + 0.00010898(Income)$$
$$R^2 = 68.3\%$$

$$\widehat{Lifeexp} = 77.867 - 0.47180(Murder) + 0.00976(HSgrad) + 4.477(Illiteracy)$$
$$R^2 = 56.3\%$$

$$\widehat{Lifeexp} = 75.3472 - 0.42699(Murder) + 0.00010727(Income) + 1.335(Illiteracy)$$
$$R^2 = 68.5\%$$

b) Each of the models has at least one coefficient with a large *P*-value. This predictor variable could be omitted to simplify the model without degrading it too much.

c) No. Regression models cannot be interpreted that way. Association is not the same thing as causation.

d) Plots of the residuals highlight some states as possible outliers. You may want to consider setting them aside to see if the model changes.

27. Burger King.

a) With an $R^2 = 99.88\%$, the model should make excellent predictions.

b) The value of *s*, 8.178 calories, is very small compared to the initial standard deviation of *calories*. This means that the model fits the data quite well, leaving very little variation unaccounted for.

c) No, the residuals are not all 0. Indeed, we know that their standard deviation is *s* = 8.178 calories. They are very small compared with the original values.

Review of Part VII – Inference When Variables Are Related

1. Mutual funds 2013.

a) **Paired data assumption:** These data are paired by mutual fund.
Randomization condition: Assume that these funds are representative of all mutual funds.
Nearly Normal condition: The histogram of differences shows plausible normality, even if not symmetric.

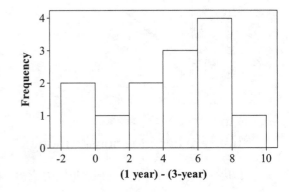

(1 year) - (3-year)

Since the conditions are satisfied, the sampling distribution of the difference can be modeled with a Student's t-model with $13 - 1 = 12$ degrees of freedom. We will find a paired t-interval, with 95% confidence.

$$\bar{d} \pm t^*_{n-1}\left(\frac{s_d}{\sqrt{n}}\right) = 4.3823 \pm t^*_{12}\left(\frac{3.14529}{\sqrt{13}}\right) \approx (2.48,\ 6.28)$$

Provided that these mutual funds are representative of all mutual funds, we are 95% confident that, on average, 1-year yields are between 2.48% and 6.28% higher than 3-year yields.

b) H_0: There is no linear relationship between 1-year and 3-year rates of return. $(\beta_1 = 0)$

H_A: There is a linear relationship between 1-year and 3-year rates of return. $(\beta_1 \neq 0)$

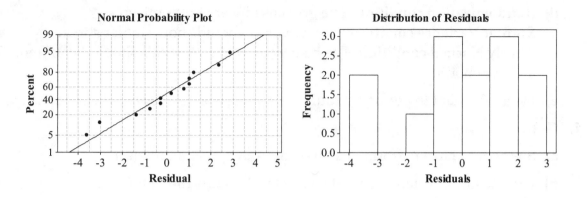

Straight enough condition: The scatterplot is straight enough to try linear regression, though one point seems to be influencing the association.
Independence assumption: The residuals plot shows no pattern.
Does the plot thicken? condition: The spread of the residuals is consistent.
Nearly Normal condition: The Normal probability plot of residuals isn't very straight and the histogram of residuals isn't symmetric. With a sample size of 13, it's difficult to determine the true nature of the distribution, so it is probably okay to proceed, but we won't have too much faith in our results.

Since the conditions for inference are satisfied, the sampling distribution of the regression slope can be modeled by a Student's t-model with $(13 - 2) = 11$ degrees of freedom. We will use a regression slope t-test.

The equation of the line of best fit for these data points is:

$$(\widehat{3year}) = 7.749 + 0.283(1year).$$

The value of $t \approx 1.74$. The P-value of 0.1095 means that the association we see in the data is likely to occur by chance. We fail to reject the null hypothesis, and conclude that there is no evidence of a linear relationship between the rates of return for 1-year

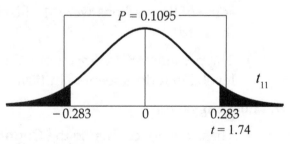

and 3-year periods. Furthermore, we don't know if this sample of mutual funds is a random sample (or even representative) of all mutual funds.

3. **Football.**

   a) He should have performed a one-way ANOVA and an F-test.

   b) H_0: The mean distance thrown is the same for each grip. $\left(\mu_1 = \mu_2 = \mu_3 = \mu_4\right)$

   H_A: The mean distances thrown are not all the same.

   c) With a P-value equal to 0.0032, we reject the null hypothesis and conclude that the distance thrown is not the same for all grips, on average.

d) Randomization condition: The grips used were randomized.
Similar Variance condition: The boxplots should show similar spread.
Nearly Normal condition: The histogram of the residuals should be unimodal and symmetric.

e) He might want to perform a multiple comparison test to see which grip is best.

4. **Golf.**

a) She should have performed a one-way ANOVA and an *F*-test.

b) H₀: The mean distance from the target is the same for all four clubs.
$$(\mu_1 = \mu_2 = \mu_3 = \mu_4)$$

Hₐ: The mean distances from the target are not all the same.

c) With a *P*-value equal to 0.0245, we reject the null hypothesis and conclude that the mean distances from the target are not the same for all four clubs.

d) Randomization condition: The clubs used were randomized.
Similar Variance condition: The boxplots should show similar spread.
Nearly Normal condition: The histogram of the residuals should be unimodal and symmetric.

e) She might want to perform a multiple comparison test to determine which club is best.

5. **Horses again, but less wild.**

a) *Sterilized* is an indicator variable.

b) According to the model, there were, on average, 6.4 fewer foals in herds in which some of the stallions were sterilized, after allowing for the number of adults in the herd.

c) The p-value for *sterilized* is equal to 0.096. That is not significant at the $\alpha = 0.05$ level, but it does seem to indicate some effect.

7. **Lost baggage.**

These are not counts, so the **Counted data condition** is not met. We cannot use a chi-square goodness-of-fit test.

9. **Video racing.**

a) H₀: The mean time is the same for all three types of mouse. $(\gamma_{Ergo} = \gamma_{Reg} = \gamma_{Cord})$

Hₐ: The mean times are not all the same for the three types of mouse.

H₀: The mean time is the same whether the lights are on or off. $(\tau_{On} = \tau_{Off})$

Hₐ: The mean times are not the same for lights on and lights off.

b) The *mouse* sum of squares has 3 – 1 = 2 degrees of freedom.
The *light* sum of squares has 2 – 1 = 1 degrees of freedom.
The error sum of squares has (6 – 1) – 2 – 1 = 2 degrees of freedom.

c) No, he should not fit an interaction term. The interaction term would use 2 degrees of freedom. This would not leave any degrees of freedom for the error sum of squares and thus would not allow any interpretation of the factors.

11. Paper airplanes.

a) It is reasonable to think that the flight distances are independent of one another. The histogram of flight distances (given) is unimodal and symmetric. Since the conditions are satisfied, the sampling distribution of the mean can be modeled by a Student's t model, with 11 – 1 = 10 degrees of freedom. We will use a one-sample t-interval with 95% confidence for the mean flight distance.

$$\bar{y} \pm t^*_{n-1}\left(\frac{s}{\sqrt{n}}\right) = 48.3636 \pm t^*_{10}\left(\frac{18.0846}{\sqrt{11}}\right) \approx (36.21, 60.51)$$

We are 95% confident that the mean distance the airplane may fly is between 36.21 and 60.51 feet.

b) Since 40 feet is contained within our 95% confidence interval, it is plausible that the mean distance is 40 feet.

c) A 99% confidence interval would be wider. Intervals with greater confidence are less precise.

d) In order to cut the margin of error in half, she would need a sample size four times as large, or 44 flights.

13. Barbershop music.

a) With an R^2 of 90.9%, your friend is right about being able to predict singing scores.

b) H_0: When including the other predictor, performance does not change our ability to predict singing scores. $\left(\beta_{Prs} = 0\right)$

H_A: When including the other predictor, performance changes our ability to predict singing scores. $\left(\beta_{Prs} \neq 0\right)$

With $t = 8.13$, and 31 degrees of freedom, the P-value is less than 0.0001, so we reject the null hypothesis and conclude that *performance* is useful in predicting singing scores.

c) Based on the multiple regression, both *performance* and *music* (even with a P-value equal to 0.0766) are useful in predicting singing scores. According to the residuals plot the spread is constant across the predicted values. The histogram of residuals is unimodal and symmetric. Based on the information provided we have met the Similar Variance and Nearly Normal conditions.

15. Study habits.

a) H_0: The mean hours studied is the same for both sexes. $(\gamma_F = \gamma_M)$

H_A: The mean hours studied not the same for both sexes.

H_0: The mean hours studied is the same for all classes. $(\tau_{Fr} = \tau_{So} = \tau_{Jr} = \tau_{Sr})$

H_A: The mean hours studied are not the same for all classes.

b) Based on the interaction plot, the lines are not parallel, so the Additive Enough condition is not met. The interaction term should be fit.

c) With all three P-values so large, none of the effects appear to be significant.

d) There are a few outliers. Based on the residuals plot, the Similar Variance condition appears to be met, but we do not have a Normal probability plot of residuals. The main concern is that we may not have enough power to detect differences between groups. We would need more data to increase the power.

17. Teen traffic deaths 2007.

a) **Straight enough condition:** The scatterplots of the response versus each predicted value should be reasonably straight.
Independence: These data are measured over time. We should plot the data and the residuals against time to look for failures of independence.
Nearly Normal condition: The Normal probability plot should be straight, and the histogram of residuals should be unimodal and symmetric.
Similar Variance condition: The spread of the residuals plot is constant.

b) $R^2 = 57.4\%$, so 57.4% of the variation in female teen traffic deaths is accounted for by the linear model.

c) $\widehat{deaths} = 45074.0 - 21.6006(year)$

d) The model predicts that female teen traffic deaths have been declining at a rate of approximately 21.6 per year.

19. Typing.

a) H₀: Typing speed is the same whether the gloves are on or off. ($\gamma_{On} = \gamma_{Off}$)

H_A: Typing speeds are not the same when the gloves are on and off.

H₀: Typing speed is the same at hot or cold temperatures. ($\tau_H = \tau_C$)

H_A: Typing speeds are not the same at hot and cold temperatures.

b) Based on the boxplots, it appears that both the effects of *temperature* and *gloves* affect typing speed.

c) The interaction plot is not parallel so the additive enough condition is not met. Therefore, an interaction term should be fit.

d) Yes, I think it will be significant because a difference in typing speed due to temperature seems to be significant only when he is not wearing the gloves.

e) The *P*-values for all three effects: *gloves, temperature* and the interaction term, are all very small, so all three effects are significant.

f) Gloves decreased typing speed at both temperatures. A hot temperature increases typing speed with the gloves off, but has little effect with the gloves on.

g) $s = \sqrt{\dfrac{58.75}{28}} = 1.45$ words per minute. Yes, this seems consistent with the size of the variation shown on the partial boxplots.

h) Tell him to type in a warm room without wearing his gloves.

i) The Normal probability plot of residuals should be straight. The residuals plot shows constant spread across the predicted values. The Similar Variance condition appears to be met. For the levels of the factors that he used, it seems clear that a warm room and not wearing gloves are beneficial for his typing.

21. New York Marathon.

a) The number of finishers per minute appears to increase by about 0.519037 finishers per minute.

b) There is a definite pattern in the residuals. This is a clear violation of the linearity assumption. While there does seem to be a strong association between time and number of finishers, it is not linear.

23. Pesticides.

H₀ : The percentage of males born to workers at the plant is 51.2%.($p = 0.512$)
H_A : The percentage of males is less than 51.2%. ($p < 0.512$)

Independence assumption: It is reasonable to think that the births are independent.

Success/Failure Condition: $np = (227)(0.512) = 116$ and $nq = (227)(0.488) = 111$ are both greater than 10, so the sample is large enough.

The conditions have been satisfied, so a Normal model can be used to model the sampling distribution of the proportion, with $\mu_{\hat{p}} = p = 0.512$ and

$$\sigma(\hat{p}) = \sqrt{\frac{pq}{n}} = \sqrt{\frac{(0.512)(0.488)}{227}} \approx 0.0332\,.$$ We can perform a one-proportion z-test.

The observed proportion of males is $\hat{p} = 0.40$.

The value of $z \approx -3.35$, meaning that the observed proportion of males is over 3 standard deviations below the expected proportion. The *P*-value associated with this z score is approximately 0.0004.

$$z = \frac{\hat{p} - p_0}{\sqrt{\frac{pq}{n}}}$$

$$z = \frac{0.4 - 0.512}{\sqrt{\frac{(0.512)(0.488)}{227}}}$$

$$z \approx -3.35$$

$P = 0.0004$

With a *P*-value this low, we reject the null hypothesis. There is strong evidence that the percentage of males born to workers is less than 51.2%. This provides evidence that human exposure to dioxin may result in the birth of more girls.

25. Javelin.

a) H_0: Type of javelin has no effect on the distance of her throw. $(\gamma_P = \gamma_S)$

H_A: Type of javelin does have an effect on the distance of her throw.

H_0: Preparation has no effect on the distance of her throw. $(\tau_W = \tau_C)$

H_A: Preparation does have an effect on the distance of her throw.

b) The partial boxplots show that both factors seem to have an effect on the distance of her throw.

The interaction plot is parallel. The additive enough condition is met. No interaction term is needed.

Randomization condition: The experiment was performed in random order.
Similar Variance condition: The side-by-side boxplots have similar spreads. The residuals plot shows no pattern, and no systematic change in spread.
Nearly Normal condition: The Normal probability plot of the residuals should be straight.

Analysis of Variance For   **Distance (m)**
No Selector

Source	df	Sums of Squares	Mean Square	F-ratio	Prob
Const	1	11552	11552	10502	≤ 0.0001
Prn	1	162	162	147.27	≤ 0.0001
Jvn	1	60.5000	60.5000	55.000	0.0007
Error	5	5.50000	1.10000		
Total	7	228			

With very small *P*-values, both effects are significant. It appears that depending on the cost, the premium javelin may be worth it – it increases the distance about 5.5 meters, on average. Warming up increases distance about 9 meters, on average. She should always warm up and consider using the premium javelin.

27. TV and athletics.

a) H_0: The mean hours of television watched is the same for all 3 groups.
$$\left(\mu_1 = \mu_2 = \mu_3 \right)$$

H_A: The mean hours of television watched are not the same for all 3 groups.

b) The variance for the *none* group appears to be slightly smaller, and there are outliers in all three groups. We do not have a Normal probability plot of the residuals, but we suspect that the data may not be Normal enough.

c) The ANOVA *F*-test indicates that athletic participation is significant with a p-value equal to 0.0167, we conclude that the number of television hours watched is not the same for all three types of athletic participation.

d) It seems that the differences are evident even when the outliers are removed. The conclusions seems valid.

29. Weight loss.

Randomization Condition: The respondents were randomly selected from among the clients of the weight loss clinic.

Nearly Normal Condition: The histogram of the number of pounds lost for each respondent is unimodal and symmetric, with no outliers.

The clients in the sample had a mean weight loss of 9.15 pounds, with a standard deviation of 1.94733 pounds. Since the conditions have been satisfied, construct a one-sample t-interval, with $20 - 1 = 19$ degrees of freedom, at 95% confidence.

$$\bar{y} \pm t^*_{n-1}\left(\frac{s}{\sqrt{n}}\right) = 9.15 \pm t^*_{19}\left(\frac{1.94733}{\sqrt{20}}\right) \approx (8.24,\ 10.06)$$

We are 95% confident that the mean weight loss experienced by clients of this clinic is between 8.24 and 10.06 pounds. Since 10 pounds is contained within the interval, the claim that the program will allow clients to lose 10 pounds in a month is plausible. Answers may vary, depending on the chosen level of confidence.

Chapter 29 – Multiple Regression Wisdom

Section 29.1

1. **Indicators.**

   a) Use and indicator. Code Male = 0, Female = 1 (or the reverse).

   b) Treat it as a quantitative predictor.

   c) Use and indicator. Code older than 65 = 1, Younger than 65 = 0.

Section 29.2

3. **Residual, leverage, influence.**

   a) Likely high leverage.

   b) Likely outlier.

Section 29.3

5. **Significant coefficient?**

   No. The predictors are almost certainly collinear, which would affect the coefficient of *Age*.

Chapter Exercises.

7. **Climate change 2013 again.**

   a) The distribution of Studentized residuals is unimodal and fairly symmetric. The association should be fine for regression.

   b) That would not be a correct interpretation. Since CO_2 levels have risen steadily over recent years, CO_2 and *Year* are likely to be collinear.

9. **Healthy breakfast, sick data.**

   a) The slope of a partial regression plot is the coefficient of the corresponding predictor, in this case, –1.020.

   b) Quaker oatmeal makes the slope more strongly negative. It appears to have substantial influence on this slope.

   c) Not surprisingly, omitting Quaker oatmeal changes the coefficient of *fiber*. It is now positive (although not significantly different from 0). This second regression model has a higher R^2, suggesting that it fits the data better. Without the influential point, the second regression is probably the better model.

   d) The coefficient of *fiber* is not discernibly different from 0. We have no evidence that it contributes significantly to calories.

11. Cereals, part 2.

a) After allowing for the effects of *sodium* and *sugars*, the model predicts a decrease of 0.019 calories for each additional gram of *potassium*.

b) Those points pull the slope of the relationship down. Omitting them should increase the value of the coefficient of *potassium*. It would likely become positive, since the remaining points show a positive slope in the partial regression plot.

c) These appear to be influential points. They have both high leverage and large residuals, and the partial regression plot shows their influence.

d) If our goal is to understand the relationships among these variables, then it might be best to omit these cereals because they seem to behave as if they are from a separate subgroup.

13. Traffic delays 2011.

a) This set of indicators uses *medium* size as its base. The coefficients of *small*, *large*, and *very large* estimate the average change in amount of *delay/person* relative to the amount of *delay/person* for *medium* size cities found for each of the other three sizes. If there were an indicator variable for *medium* as well, the four indicators would be collinear, so the coefficients could not be estimated.

b) The costs of congestion per commuter in *large* cities are, on average, about $106 greater than in *medium* size cities, after allowing for the effects of *congested%*.

15. More traffic 2011.

a) An assumption required for indicator variables to be useful is that the regression models fit for the different groups identified by the indicators are parallel, i.e. they have the same slope. These lines are not parallel.

b) The coefficient of $Sml \times C\%V$ adjusts the slope of the regression model fit for the small cities. We would say that the slope of *congestion cost per person* on *congested% of motor vehicle miles* for *small* cities (after allowing for the linear effects of the other variables in the model) is $27.815 – 4.42 = $23.395.

c) The regression model seems to do a good job. The R^2 shows that 65% of the variability in *congestion cost per person* is accounted for by the model. Most of the P-values for the coefficients are small.

17. Influential traffic?

Laredo has high leverage, but it does not have a particularly high Studentized residual. It appears that the point has influence but does not exert it. Removing this case from the regression may not result in a large change in the model, so the case is probably not influential.